MW00711029

STUDENT STUDY GUIDE

Richard N. Aufmann
Palomar College

Vernon C. Barker
Palomar College

Richard D. Nation
Palomar College

Christine S. Verity

COLLEGE TRIGONOMETRY

FIFTH EDITION

Aufmann/Barker/Nation

HOUGHTON MIFFLIN COMPANY   BOSTON NEW YORK

Senior Sponsoring Editor: Lynn Cox
Senior Development Editor: Dawn Nuttall
Assistant Editor: Melissa Parkin
Editorial Assistant: Noel Kamm
Manufacturing Manager: Florence Cadran
Senior Marketing Manager: Danielle Potvin

Copyright © 2005 by Houghton Mifflin Company. All rights reserved.

No part of this work may be reproduced or transmitted in any form or by any means, electronic or mechanical, including photocopying and recording, or by any information storage or retrieval system without the prior written permission of Houghton Mifflin Company unless such copying is expressly permitted by federal copyright law. Address inquiries to College Permissions, Houghton Mifflin Company, 222 Berkeley Street, Boston, MA 02116-3764.

Printed in the U.S.A.

ISBN: 0-618-38805-2

1 2 3 4 5 6 7 8 9 – MA – 08 07 06 05 04

Contents

Preface

The *Student Study Guide* for the Aufmann/Barker/Nation *College Trigonometry*, Fifth Edition text contains *Study Tips*, a *Solutions Manual*, and *Chapter Tests* with *Solutions to Chapter Tests*.

The *Study Tips* explain how to best utilize the text and your time in order to succeed in this course.

The *Solutions Manual* provides complete, worked-out solutions to the following text exercises:
- odd-numbered section exercises
- all *Prepare for Section* exercises (found within the section exercises)
- all *Chapter True/False Exercises*
- all *Chapter Review Exercises*
- all *Chapter Test* exercises
- all *Cumulative Review Exercises*

The *Chapter Tests* contain one practice test for each chapter in the text. They are modeled after the Chapter Tests found at the end of each chapter in the text and can be used to provide additional practice for an in-class chapter test. The *Chapter Tests* are immediately followed by *Solutions to Chapter Tests*.

Study Tips

The skills you will learn in any mathematics course will be important in your future career—no matter what career you choose. In your textbook, we have provided you with the tools to master these skills. There's no mystery to success in this course; a little hard work and attention to your instructor will pay off. Here are a few tips to help ensure your success in this class.

Know Your Instructor's Requirements

To do your best in this course, you must know exactly what your instructor requires. If you don't, you probably will not meet his or her expectations and are not likely to earn a good grade in the course.

Instructors ordinarily explain course requirements during the first few days of class. Course requirements may be stated in a *syllabus*, which is a printed outline of the main topics of the course, or they may be presented orally. When they are listed in a syllabus or on other printed pages, keep them in a safe place. When they are presented orally, make sure to take complete notes. In either case, understand them completely and follow them exactly.

Attend Every Class

Attending class is vital if you are to succeed in this course. Your instructor will provide not only information but also practice in the skills you are learning. Be sure to arrive on time. You are responsible for everything that happens in class, even if you are absent. If you *must* be absent from a class session:

1. Deliver due assignments to the instructor as soon as possible.
2. Contact a classmate to learn about assignments or tests announced in your absence.
3. Hand copy or photocopy notes taken by a classmate while you were absent.

Take Careful Notes in Class

You need a notebook in which to keep class notes and records about assignments and tests. Make sure to take complete and well-organized notes. Your instructor will explain text material that may be difficult for you to understand on your own and may supply important information that is not provided in the textbook. Be sure to include in your notes everything that is written on the chalkboard.

Information recorded in your notes about assignments should explain exactly what they are, how they are to be done, and when they are due. Information about tests should include exactly what text material and topics will be covered on each test and the dates on which the tests will be given.

Survey the Chapter

Before you begin reading a chapter, take a few minutes to survey it. Glancing through the chapter will give you an overview of its contents and help you see how the pieces fit together as you read.

Begin by reading the chapter title. The title summarizes what the chapter is about. Next, read the section headings. The section headings summarize the major topics presented in the chapter. Then read the topic headings that appear in green within each section. The topic headings describe the concepts for that section. Keep these headings in mind as you work through the material. They provide direction as you study.

Use the Textbook to Learn the Material

For each concept studied, read very carefully all of the material from the topic heading to the examples provided for that concept. As you read, note carefully the formulas and words printed in **boldface** type. It is important for you to know these formulas and the definitions of these words.

Copyright © Houghton Mifflin Company. All rights reserved.

You will note that each example references an exercise. The example is worked out for you; the exercise, which is highlighted in red in the section's exercise set, is left for you to do. After studying the example, do the exercise. Immediately look up the answer to this exercise in the Solutions section at the back of the text. If your answer is correct, continue. If your answer is incorrect, check your solution against the one given in the Solutions section. It may be helpful to review the worked-out example also. Determine where you made your mistakes.

Next, do the other problems in the exercise set that correspond to the concept just studied. The answers to all the odd-numbered exercises appear in the answer section in the back of the text, and the solutions appear in this Study Guide. Check your answers to the exercises against these.

If you have difficulty solving problems in the exercise set, review the material in the text. Many examples are solved within the text material. Review the solutions to these problems. Reread the examples provided for the concept. If, after checking these sources and trying to find your mistakes, you are still unable to solve a problem correctly, make a note of the exercise number so that you can ask someone for help with that problem.

Review Material

Reviewing material is the repetition that is essential for learning. Much of what we learn is soon forgotten unless we review it. If you find that you do not remember information that you studied previously, you probably have not reviewed it sufficiently. *You will remember best what you review most.*

One method of reviewing material is to begin a study session by reviewing a concept you have studied previously. For example, before trying to solve a new type of problem, spend a few minutes solving a kind of problem you already know how to solve. Not only will you provide yourself with the review practice you need, but you are also likely to put yourself in the right frame of mind for learning how to solve the new type of problem.

Use the End-of-Chapter Material

To help you review the material presented within a chapter, a Chapter Review appears at the end of each chapter. In the Chapter Review, the main concepts of each section are summarized. Included are important definitions and formulas. After completing a chapter, be sure to read the Chapter Review. Use it to check your understanding of the material presented and to determine what concepts you need to review. Return to any section that contains a concept you need to study again.

Each chapter ends with Chapter Review Exercises and a Chapter Test. The problems these contain summarize what you should have learned when you have finished the chapter. Do these exercises as you prepare for an examination. Check your answers against those in the back of the text. Answers to the odd-numbered Chapter Review Exercises and all the Chapter Test exercises are provided there. For any problem you answer incorrectly, review the material corresponding to that concept in the textbook. Determine *why* your answer was wrong.

Finding Good Study Areas

Find a place to study where you are comfortable and can concentrate well. Many students find the campus library to be a good place. You might select two or three places at the college library where you like to study. Or there may be a small, quiet lounge on the third floor of a building where you find you can study well. Take the time to find places that promote good study habits.

Determining When to Study

Spaced practice is generally superior to massed practice. For example, four half-hour study periods will produce more learning than one two-hour study session. The following suggestions may help you decide when to study.

1. A free period immediately before class is the best time to study about the lecture topic for the class.
2. A free period immediately after class is the best time to review notes taken during the class.
3. A brief period of time is good for reciting or reviewing information.

Copyright © Houghton Mifflin Company. All rights reserved.

4. A long period of an hour or more is good for doing challenging activities such as learning to solve a new type of problem.

5. Free periods just before you go to sleep are good times for learning information. (There is evidence that information learned just before sleep is remembered longer than information learned at other times.)

Determining How Much to Study

Instructors often advise students to spend twice as much time outside of class studying as they spend in the classroom. For example, if a course meets for three hours each week, instructors customarily advise students to study for six hours each week outside of class.

Any mathematics course requires the learning of skills, which are abilities acquired through practice. It is often necessary to practice a skill more than a teacher requires. For example, this textbook may provide 50 practice problems on a specific concept, and the instructor may assign only 25 of them. However, some students may need to do 30, 40, or all 50 problems.

If you are an accomplished athlete, musician, or dancer, you know that long hours of practice are necessary to acquire a skill. Do not cheat yourself of the practice you need to develop the abilities taught in this course.

Study followed by reward is usually productive. Schedule something enjoyable to do following study sessions. If you know that you only have two hours to study because you have scheduled an enjoyable activity for yourself, you may be inspired to make the best use of the two hours that you have set aside for studying.

Keep Up to Date with Course Work

College terms start out slowly. Then they gradually get busier and busier, reaching a peak of activity at final examination time. If you fall behind in the work for a course, you will find yourself trying to catch up at a time when you are very busy with all of your other courses. Don't fall behind—keep up to date with course work.

Keeping up with course work is doubly important for a course in which information and skills learned early in the course are needed to learn information and skills later in the course. Any mathematics course falls into this category. Skills must be learned immediately and reviewed often.

Your instructor gives assignments to help you acquire a skill or understand a concept. Do each assignment as it is assigned, or you may well fall behind and have great difficulty catching up. Keeping up with course work also makes it easier to prepare for each exam.

Be Prepared for Tests

The Chapter Test at the end of a chapter should be used to prepare for an examination. Additional Chapter Tests are also provided in this Study Guide. We suggest that you try a Chapter Test a few days before your actual exam. Do these exercises in a quiet place, and try to complete the exercises in the same amount of time as you will be allowed for your exam. When completing the exercises, practice the strategies of successful test takers: 1) look over the entire test before you begin to solve any problem; 2) write down any rules or formulas you may need so they are readily available; 3) read the directions carefully; 4) work the problems that are easiest for you first; 5) check your work, looking particularly for careless errors.

When you have completed the exercises in the Chapter Test, check your answers. If you missed a question, review the material in the appropriate section and then rework some of the exercises from that concept. This will strengthen your ability to perform the skills in that concept.

Get Help for Academic Difficulties

If you do have trouble in this course, teachers, counselors, and advisers can help. They usually know of study groups, tutors, or other sources of help that are available. They may suggest visiting an office of academic skills, a learning center, a tutorial service or some other department or service on campus.

Students who have already taken the course and who have done well in it may be a source of assistance. If they have a good understanding of the material, they may be able to help by explaining it to you.

Copyright © Houghton Mifflin Company. All rights reserved.

Solutions Manual

Chapter 1
Functions and Graphs

1.
$$2x + 10 = 40$$
$$2x + 10 = 40$$
$$2x = 40 - 10$$
$$2x = 30$$
$$x = 15$$

3.
$$5x + 2 = 2x - 10$$
$$5x - 2x = -10 - 2$$
$$3x = -12$$
$$x = -4$$

5.
$$2(x - 3) - 5 = 4(x - 5)$$
$$2x - 6 - 5 = 4x - 20$$
$$2x - 11 = 4x - 20$$
$$2x - 4x = -20 + 11$$
$$-2x = -9$$
$$x = \frac{9}{2}$$

7.
$$\frac{3}{4}x + \frac{1}{2} = \frac{2}{3}$$
$$12\left(\frac{3}{4}x + \frac{1}{2}\right) = 12\left(\frac{2}{3}\right)$$
$$9x + 6 = 8$$
$$9x = 8 - 6$$
$$9x = 2$$
$$x = \frac{2}{9}$$

9.
$$\frac{2}{3}x - 5 = \frac{1}{2}x - 3$$
$$6\left(\frac{2}{3}x - 5\right) = 6\left(\frac{1}{2}x - 3\right)$$
$$4x - 30 = 3x - 18$$
$$4x - 3x = -18 + 30$$
$$x = 12$$

11.
$$0.2x + 0.4 = 3.6$$
$$0.2x = 3.2$$
$$x = 16$$

13.
$$\frac{3}{5}(n + 5) - \frac{3}{4}(n - 11) = 0$$
$$20\left[\frac{3}{5}(n + 5) - \frac{3}{4}(n - 11)\right] = 20 - 0$$
$$12(n + 5) - 15(n - 11) = 0$$
$$12n + 60 - 15n + 165 = 0$$
$$-3n = -225$$
$$n = 75$$

15.
$$3(x + 5)(x - 1) = (3x + 4)(x - 2)$$
$$3x^2 + 12x - 15 = 3x^2 - 2x - 8$$
$$14x = 7$$
$$x = \frac{1}{2}$$

17.
$$0.08x + 0.12(4000 - x) = 432$$
$$0.08x + 480 - 0.12x = 432$$
$$-0.04x = -48$$
$$x = 1200$$

19.
$$x^2 - 2x - 15 = 0$$
$$a = 1 \quad b = -2 \quad c = -15$$
$$x = \frac{-(-2) \pm \sqrt{(-2)^2 - 4(1)(-15)}}{2(1)}$$
$$x = \frac{2 \pm \sqrt{4 + 60}}{2}$$
$$x = \frac{2 \pm \sqrt{64}}{2} = \frac{2 \pm 8}{2}$$
$$x = \frac{2 + 8}{2} = \frac{10}{2} = 5 \quad \text{or}$$
$$x = \frac{2 - 8}{2} = \frac{-6}{2} = -3$$

21.
$$x^2 + x - 1 = 0$$
$$a = 1 \quad b = 1 \quad c = -1$$
$$x = \frac{-1 \pm \sqrt{1^2 - 4(1)(-1)}}{2}$$
$$x = \frac{-1 \pm \sqrt{1 + 4}}{2}$$
$$x = \frac{-1 \pm \sqrt{5}}{2}$$

Copyright © Houghton Mifflin Company. All rights reserved.

23. $2x^2 + 4x + 1 = 0$

$a = 2 \quad b = 4 \quad c = 1$

$x = \dfrac{-4 \pm \sqrt{4^2 - 4(2)(1)}}{2(2)}$

$x = \dfrac{-4 \pm \sqrt{16 - 8}}{4}$

$x = \dfrac{-4 \pm \sqrt{8}}{4} = \dfrac{-4 \pm 2\sqrt{2}}{4}$

$x = \dfrac{-2 \pm \sqrt{2}}{2}$

25. $3x^2 - 5x - 3 = 0$

$a = 3 \quad b = -5 \quad c = -3$

$x = \dfrac{-(-5) \pm \sqrt{(-5)^2 - 4(3)(-3)}}{2(3)}$

$x = \dfrac{5 \pm \sqrt{25 + 36}}{6}$

$x = \dfrac{5 \pm \sqrt{61}}{6}$

27. $\dfrac{1}{2}x^2 + \dfrac{3}{4}x - 1 = 0$

$a = \dfrac{1}{2} \quad b = \dfrac{3}{4} \quad c = -1$

$x = \dfrac{-\dfrac{3}{4} \pm \sqrt{\left(\dfrac{3}{4}\right)^2 - 4\left(\dfrac{1}{2}\right)(-1)}}{2\left(\dfrac{1}{2}\right)}$

$x = \dfrac{-\dfrac{3}{4} \pm \sqrt{\dfrac{9}{16} + 2}}{1}$

$x = -\dfrac{3}{4} \pm \sqrt{\dfrac{41}{16}}$

$x = -\dfrac{3}{4} \pm \dfrac{\sqrt{41}}{4}$

$x = \dfrac{-3 \pm \sqrt{41}}{4}$

29. $\sqrt{2}x^2 + 3x + \sqrt{2} = 0$

$a = \sqrt{2} \quad b = 3 \quad c = \sqrt{2}$

$x = \dfrac{-3 \pm \sqrt{3^2 - 4 \cdot \sqrt{2} \cdot \sqrt{2}}}{2\sqrt{2}}$

$x = \dfrac{-3 \pm \sqrt{9 - 8}}{2\sqrt{2}}$

$x = \dfrac{-3 + 1}{2\sqrt{2}} = \dfrac{-2}{2\sqrt{2}} = -\dfrac{\sqrt{2}}{2}$ or

$x = \dfrac{-3 - 1}{2\sqrt{2}} = \dfrac{-4}{2\sqrt{2}} = -\sqrt{2}$

31. $x^2 = 3x + 5$

$x^2 - 3x - 5 = 0$

$a = 1 \quad b = -3 \quad c = -5$

$x = \dfrac{-(-3) \pm \sqrt{(-3)^2 - 4(1)(5)}}{2(1)}$

$x = \dfrac{3 \pm \sqrt{9 + 20}}{2}$

$x = \dfrac{3 \pm \sqrt{29}}{2}$

33. $x^2 - 2x - 15 = 0$

$(x + 3)(x - 5) = 0$

$x + 3 = 0 \quad$ or $\quad x - 5 = 0$

$\quad x = -3 \qquad\qquad x = 5$

35. $8y^2 + 189y - 72 = 0$

$(8y - 3)(y + 24) = 0$

$8y - 3 = 0 \quad$ or $\quad y + 24 = 0$

$\quad y = \dfrac{3}{8} \qquad\qquad y = -24$

37. $3x^2 - 7x = 0$

$x(3x - 7) = 0$

$x = 0 \quad$ or $\quad 3x - 7 = 0$

$\qquad\qquad\qquad x = \dfrac{7}{3}$

39. $(x - 5)^2 - 9 = 0$

$[(x - 5) - 3][(x - 5) + 3] = 0$

$x - 8 = 0 \quad$ or $\quad x - 2 = 0$

$\quad x = 8 \qquad\qquad x = 2$

41. $2x + 3 < 11$

$\quad 2x < 8$

$\quad\, x < 4$

43. $x + 4 > 3x + 16$

$\quad -2x > 12$

$\quad\,\, x < -6$

45. $-6x + 1 \ge 19$

$\quad -6x \ge 18$

$\quad\,\, x \le -3$

47. $-3(x + 2) \le 5x + 7$

$\quad -3x - 6 \le 5x + 7$

$\quad\quad -8 \le 13$

$\quad\quad\quad x \ge -\dfrac{13}{8}$

49. $-4(3x - 5) > 2(x - 4)$

$\quad -12x + 20 > 2x - 8$

$\quad\quad -14x > -28$

$\quad\quad\quad x < 2$

Copyright © Houghton Mifflin Company. All rights reserved.

51.
$$x^2 + 7x > 0$$
$$x(x + 7) > 0$$

The product is positive.
The critical values are 0 and –7.

$x(x - 7)$ $+ + + + | - - - - - - - | + + + +$

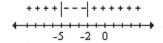
$$-7 \qquad 0$$

$(-\infty, -7) \cup (0, \infty)$

53.
$$x^2 + 7x + 10 < 0$$
$$(x + 5)(x + 2) < 0$$

The product is negative.
The critical values are –5 and –2.

$(x + 5)(x + 2)$ $+ + + + | - - - | + + + + + +$

$$-5 \quad -2 \quad 0$$

$(-5, -2)$

55.
$$x^2 - 3x \geq 28$$
$$x^2 - 3x - 28 \geq 0$$
$$(x + 4)(x - 7) \geq 0$$

The product is positive or zero.
The critical values are –4 and 7.

$(x + 4)(x - 7)$ $+ + | - - - - - - - - - - - | + +$

$$-4 \qquad 0 \qquad 7$$

$(-\infty, -4] \cup [7, \infty)$

57.
$$6x^2 - 4 \leq 5x$$
$$6x^2 - 5x - 4 \leq 0$$
$$(3x - 4)(2x + 1) \leq 0$$

The product is negative or zero.
The critical values are $\frac{4}{3}$ and $-\frac{1}{2}$.

$(3x - 4)(2x + 1)$ $+ + + + | - - - - | + + + + +$

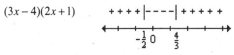
$$-\frac{1}{2} \, 0 \qquad \frac{4}{3}$$

$$\left[-\frac{1}{2}, \frac{4}{3} \right]$$

59.
$$|x| < 4$$

$$-4 < x < 4$$

$(-4, 4)$

61.
$$|x - 1| < 9$$
$$-9 < x - 1 < 9$$
$$-8 < x < 10$$

$(-8, 10)$

63.
$$|x + 3| > 30$$

$$x + 3 < -30 \quad \text{or} \quad x + 3 > 30$$
$$x < -33 \qquad \qquad x > 27$$

$(-\infty, -33) \cup (27, \infty)$

65.
$$|2x - 1| > 4$$

$$2x - 1 < -4 \quad \text{or} \quad 2x - 1 > 4$$
$$2x < -3 \qquad \qquad 2x > 5$$
$$x < -\frac{3}{2} \qquad \qquad x > \frac{5}{2}$$

$$\left(-\infty, -\frac{3}{2} \right) \cup \left(\frac{5}{2}, \infty \right)$$

67.
$$|x + 3| \geq 5$$

$$x + 3 \leq -5 \quad \text{or} \quad x + 3 \geq 5$$
$$x \leq -8 \qquad \qquad x \geq 2$$

$(-\infty, -8] \cup [2, \infty)$

69.
$$|3x - 10| \leq 14$$

$$-14 \leq 3x - 10 \leq 14$$
$$-4 \leq 3x \leq 24$$
$$-\frac{4}{3} \leq x \leq 8$$

$$\left[-\frac{4}{3}, 8 \right]$$

71.
$$|4 - 5x| \geq 24$$

$$4 - 5x \leq 24 \qquad \text{or} \qquad 4 - 5x \geq 24$$
$$-5x \leq -28 \qquad \qquad -5x \geq 20$$
$$x \geq \frac{28}{5} \qquad \qquad x \leq -4$$

$$(-\infty, -4] \cup \left[\frac{28}{5}, \infty \right)$$

73.
$$|x - 5| \geq 0$$

Because an absolute value is always nonnegative, the inequality is always true. The solution set consists of all real numbers.

$(-\infty, \infty)$

75.
$$|x - 4| \leq 0$$

Because an absolute value is always nonnegative, the inequality $|x - 4| < 0$ has no solution. Thus the only solution of the inequality $|x - 4| \leq 0$ is the solution of the equation $x - 4 = 0$.

$$x = 4$$

Copyright © Houghton Mifflin Company. All rights reserved.

77.
$$A = 35$$
$$A = LW$$
$$LW = 35$$
$$L = \frac{35}{W}$$

$$P = 27$$
$$P = 2L + 2W$$
$$2L + 2W = 27$$
$$2\left(\frac{35}{W}\right) + 2W = 27$$
$$70 + 2W^2 = 27W$$
$$2W^2 - 27W + 70 = 0$$
$$(2W - 7)(W - 10) = 0$$

$$W = \frac{7}{2} \quad \text{or} \quad W = 10$$

$$35 = \frac{7}{2}L \qquad 35 = LW$$
$$70 = 7L \qquad 35 = 10L$$
$$10 = L \qquad 3.5 = L$$

The rectangle measures 3.5 cm by 10 cm.

79.
$$A = 1500 = lw$$
$$P = 600 - 2l + 3w$$
$$l = \frac{15000}{w}$$

$$2l + 3w = 600$$
$$2\left(\frac{15000}{w}\right) + 3w = 600$$
$$30,000 + 3w^2 = 600w$$
$$3w^2 - 600w + 30,000 = 0$$
$$3(w^2 - 200w + 10,000) = 0$$
$$3(w - 100)(w - 100) = 0$$

$$w = 100 \text{ ft}$$
$$l = \frac{15000}{100} = 150 \text{ ft}$$

The dimensions are 100 feet by 150 feet.

81. Plan A: $5 + 0.01x$

Plan B: $1 + 0.08x$

$$5 + 0.01x < 1 + 0.08x$$
$$4 < .07x$$
$$57.1 < x$$

Plan A is less expensive if you use at least 58 checks.

83. Plan A: $100 + 8x$

Plan B: $250 + 3.5x$

$$100 + 8x > 250 + 3.5x$$
$$4.5x > 150$$
$$x > 33.3$$

Plan A pays better if at least 34 sales are made.

85. $68 \leq F \leq 104$

$$68 \leq \frac{9}{5}C + 32 \leq 104$$
$$36 \leq \frac{9}{5}C \leq 72$$
$$20 \leq C \leq 40$$

●●●●●●●●●●●●●●●●●●●●●●●●●●●●●●●●●●●

Connecting Concepts

87.
$$253 = \frac{1}{2}n(n + 1)$$
$$n^2 + n - 506 = 0$$
$$(n + 23)(n - 22) = 0$$
$$n = 22$$

So $1 + 2 + 3 + \cdots + 21 + 22 = 253$.

89. $R = 420x - 2x^2$

$$420x - 2x^2 > 0$$
$$2x(210 - x) > 0$$

The product is positive.
The critical values are 0 and 210.

$(0, 210)$

Copyright © Houghton Mifflin Company. All rights reserved.

91. $s = -16t^2 + v_0t + s_0 \quad s > 48, \quad v_0 = 64, \quad s_0 = 0$

$$-16t + 64t > 48$$

$$-16t^2 + 64t - 48 > 0$$

$$-16(t^2 - 4t + 3) > 0$$

$$-16(t - 3)(t - 1) > 0$$

The product is positive.
The critical values are $t = 3$ and $t = 1$.

$-16(t - 3)(t - 1)$ $|-|+ +|- - - -$

                       0 1 3

1 second $< t <$ 3 seconds

The ball is higher than 48 ft between 1 and 3 seconds.

93. **a.** $|s - 4.25| \le 0.01$

b. $s - 4.25 = 0.01, \quad$ or $\quad s - 4.25 = -0.01$
            $s = 4.26 \qquad\qquad\qquad s = 4.24$ critical values

               $4.24 \le s \le 4.26$

. . . .

Prepare for **S**ection **1.2**

94. $\dfrac{4 + (-7)}{2} = \dfrac{-3}{2}$

95. $\sqrt{50} = \sqrt{25}\sqrt{2} = 5\sqrt{2}$

96. $y = 3x - 2$

     $\overset{?}{5 = 3(-1) - 2}$

     $5 \ne -5 \qquad$ No, the equation is not true.

97. $y = (-3)^2 - 3(-3) - 2$
     $y = 9 + 9 - 2$
     $y = 16$

98. $|-3 - (-1)| = |-3 + 1| = |-2| = 2$

99. $\sqrt{(-3)^2 - 4(-2)(2)} = \sqrt{9 + 16} = \sqrt{25} = 5$

Section 1.2

1.

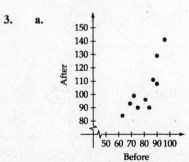

3. **a.**

b. $\text{average} = \dfrac{(84 - 63) + (99 - 72) + (111 - 87) + (129 - 90) + (108 - 90) + (141 - 96) + (93 - 69) + (96 - 81) + (90 - 75) + (90 - 84)}{10}$

$= \dfrac{21 + 27 + 24 + 39 + 18 + 45 + 24 + 15 + 15 + 6}{10} = \dfrac{234}{10} = 23.4$

The average increase in heart rate is 23.4 beats per minute.

Copyright © Houghton Mifflin Company. All rights reserved.

5.
$$d = \sqrt{(-8-6)^2 + (11-4)^2}$$
$$= \sqrt{(-14)^2 + (7)^2}$$
$$= \sqrt{196+49}$$
$$= \sqrt{245}$$
$$= 7\sqrt{5}$$

7.
$$d = \sqrt{(-10-(-4))^2 + (15-(-20))^2}$$
$$= \sqrt{(-6)^2 + (35)^2}$$
$$= \sqrt{36+1225}$$
$$= \sqrt{1261}$$

9.
$$d = \sqrt{(0-5)^2 + (0-(-8))^2}$$
$$= \sqrt{(-5)^2 + (8)^2}$$
$$= \sqrt{25+64}$$
$$= \sqrt{89}$$

11.
$$d = \sqrt{(\sqrt{12}-\sqrt{3})^2 + (\sqrt{27}-\sqrt{8})^2}$$
$$= \sqrt{(2\sqrt{3}-\sqrt{3})^2 + (3\sqrt{3}-2\sqrt{2})^2}$$
$$= \sqrt{(\sqrt{3})^2 + (3\sqrt{3}-2\sqrt{2})^2}$$
$$= \sqrt{3+(27-12\sqrt{6}+8)}$$
$$= \sqrt{3+27-12\sqrt{6}+8}$$
$$= \sqrt{38-12\sqrt{6}}$$

13.
$$d = \sqrt{(-a-a)^2 + (-b-b)^2}$$
$$= \sqrt{(-2a)^2 + (-2b)^2}$$
$$= \sqrt{4a^2 + 4b^2}$$
$$= \sqrt{4(a^2+b^2)}$$
$$= 2\sqrt{a^2+b^2}$$

15.
$$d = \sqrt{(-2x-x)^2 + (3x-4x)^2} \text{ with } x < 0$$
$$= \sqrt{(-3x)^2 + (-x)^2}$$
$$= \sqrt{9x^2 + x^2}$$
$$= \sqrt{10x^2}$$
$$= -x\sqrt{10} \quad \text{(Note: since } x < 0, \sqrt{x^2} = -x\text{)}$$

17.
$$\sqrt{(4-x)^2 + (6-0)^2} = 10$$
$$\left(\sqrt{(4-x)^2 + (6-0)^2}\right)^2 = 10^2$$
$$16 - 8x + x^2 + 36 = 100$$
$$x^2 - 8x - 48 = 0$$
$$(x-12)(x+4) = 0$$
$$x = 12 \quad \text{or} \quad x = -4$$
The points are $(12, 0)$, $(-4, 0)$.

19.
$$M = \left(\frac{x_1 + x_2}{2}, \frac{y_1 + y_2}{2}\right)$$
$$= \left(\frac{1+5}{2}, \frac{-1+5}{2}\right)$$
$$= \left(\frac{6}{2}, \frac{4}{2}\right)$$
$$= (3, 2)$$

21.
$$M = \left(\frac{6+6}{2}, \frac{-3+11}{2}\right)$$
$$= \left(\frac{12}{2}, \frac{8}{2}\right)$$
$$= (6, 4)$$

23.
$$M = \left(\frac{1.75+(-3.5)}{2}, \frac{2.25+5.57}{2}\right)$$
$$= \left(-\frac{1.75}{2}, \frac{7.82}{2}\right)$$
$$= (-0.875, 3.91)$$

25.
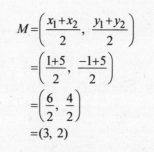

27.
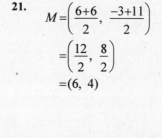

29.
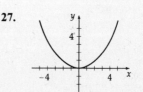

31.

33.
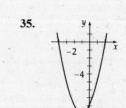

35.

Copyright © Houghton Mifflin Company. All rights reserved.

37.

39. Intercepts: $\left(0, \dfrac{12}{5}\right), (6, 0)$

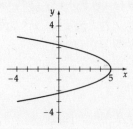

41. $\left(0, \sqrt{5}\right), \left(0, -\sqrt{5}\right), (5, 0)$

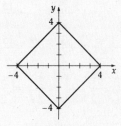

$x = -y^2 + 5$

43. $(0, 4), (0, -4), (-4, 0)$

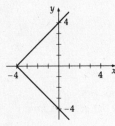

$x = |y| - 4$

45. $(0, \pm 2), (\pm 2, 0)$

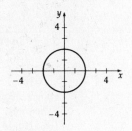

$x^2 + y^2 = 4$

47. $(0, \pm 4), (\pm 4, 0)$

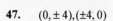

$|x| + |y| = 4$

49. center $(0, 0)$, radius 6

51. center $(1, 3)$, radius 7

53. center $(-2, -5)$, radius 5

55. center $(8, 0)$, radius $\dfrac{1}{2}$

57. $(x - 4)^2 + (y - 1)^2 = 2^2$

59. $\left(x - \dfrac{1}{2}\right)^2 + \left(y - \dfrac{1}{4}\right)^2 = \left(\sqrt{5}\right)^2$

61.
$$(x - 0)^2 + (y - 0)^2 = r^2$$
$$(-3 - 0)^2 + (4 - 0)^2 = r^2$$
$$(-3)^2 + 4^2 = r^2$$
$$9 + 16 = r^2$$
$$25 = 5^2 = r^2$$
$$(x - 0)^2 + (y - 0)^2 = 5^2$$

63.
$$(x + 2)^2 + (y - 5)^2 = r^2$$
$$(x - 1)^2 + (y - 3)^2 = r^2$$
$$(4 - 1)^2 + (-1 - 3)^2 = r^2$$
$$3^2 + (-4)^2 = r^2$$
$$9 + 16 = r^2$$
$$25 = 5^2 = r^2$$
$$(x - 1)^2 + (y - 3)^2 = 5^2$$

65.
$$x^2 - 6x \quad + y^2 = -5$$
$$x^2 - 6x + 9 + y^2 = -5 + 9$$
$$(x - 3)^2 + y^2 = 2^2$$
center $(3, 0)$, radius 2

Copyright © Houghton Mifflin Company. All rights reserved.

67.
$$x^2 - 14x \qquad + y^2 + 8y \qquad = -56$$
$$x^2 - 14x + 49 + y^2 + 8y + 16 = -56 + 49 + 16$$
$$(x-7)^2 + (y+4)^2 = 3^2$$
center $(7, -4)$, radius 3

69.
$$4x^2 + 4x \qquad + 4y^2 = 63$$
$$x^2 + x \qquad + y^2 = \frac{63}{4}$$
$$x^2 + x + \frac{1}{4} \qquad + y^2 = \frac{63}{4} + \frac{1}{4}$$
$$\left(x + \frac{1}{2}\right)^2 + y^2 = 16$$
$$\left(x + \frac{1}{2}\right)^2 + (y - 0)^2 = 4^2$$
center $\left(-\frac{1}{2}, 0\right)$, radius 4

71.
$$x^2 - x \qquad + y^2 + \frac{3}{2}y \qquad = \frac{15}{4}$$
$$x^2 - x + \frac{1}{4} + y^2 + \frac{3}{2}y + \frac{9}{4} = \frac{15}{4} + \frac{1}{4} + \frac{9}{4}$$
$$\left(x - \frac{1}{2}\right)^2 + \left(y + \frac{3}{2}\right)^2 = \left(\frac{5}{2}\right)^2$$
center $\left(\frac{1}{2}, -\frac{3}{2}\right)$, radius $\frac{5}{2}$

73.
$$d = \sqrt{(-4-2)^2 + (11-3)^2}$$
$$= \sqrt{36 + 64}$$
$$= \sqrt{100}$$
$$= 10$$
Since the diameter is 10, the radius is 5.
The center is the midpoint of the line segment from $(2,3)$ to $(-4,11)$.
$$\left(\frac{2 + (-4)}{2}, \frac{3 + 11}{2}\right) = (-1, 7) \text{ center}$$
$$(x + 1)^2 + (y - 7)^2 = 5^2$$

75.   Since it is tangent to the x-axis, its radius is 11.
$$(x - 7)^2 + (y - 11)^2 = 11^2$$

· ·

Connecting Concepts

77.

79.

81.

83.

85.

87.
$$\left(\frac{x+5}{2}, \frac{y+1}{2}\right) = (9, 3)$$

therefore $\dfrac{x+5}{2} = 9$ and $\dfrac{y+1}{2} = 3$

$\qquad\qquad x + 5 = 18 \qquad y + 1 = 6$

$\qquad\qquad\quad x = 13 \qquad\quad y = 5$

Thus $(13, 5)$ is the other endpoint.

Copyright © Houghton Mifflin Company. All rights reserved.

89. $\left(\dfrac{x+(-3)}{2}, \dfrac{y+(-8)}{2}\right) = (2, -7)$

therefore $\dfrac{x-3}{2} = 2$ and $\dfrac{y-8}{2} = -7$

$\qquad x-3 = 4 \qquad y-8 = -14$

$\qquad\qquad x = 7 \qquad\qquad y = -6$

Thus $(7, -6)$ is the other endpoint.

91.

$\sqrt{(3-x)^2 + (4-y)^2} = 5$

$(3-x)^2 + (4-y)^2 = 5^2$

$9 - 6x + x^2 + 16 - 18y + y^2 = 25$

$x^2 - 6x + y^2 - 8y = 0$

93. $\sqrt{(4-x)^2 + (0-y)^2} + \sqrt{(-4-x)^2 + (0-y)^2} = 10$

$(4-x)^2 + (0-y)^2 = 100 - 20\sqrt{(-4-x)^2 + (0-y)^2} + (-4-x)^2 + (-y)^2$

$16 - 8x + x^2 + y^2 = 100 - 20\sqrt{(-4-x)^2 + (-y)^2} + 16 + 8x + x^2 + y^2$

$-16x - 100 = -20\sqrt{(-4-x)^2 + (-y)^2}$

$4x + 25 = 5\sqrt{(-4-x)^2 + (-y)^2}$

$16x^2 + 200x + 625 = 25\left[(-4-x)^2 + (-y)^2\right]$

$16x^2 + 200x + 625 = 25\left[16 + 8x + x^2 + y^2\right]$

$16x^2 + 200x + 625 = 400 + 200x + 25x^2 + 25y^2$

Simplifying yields $9x^2 + 25y^2 = 225$.

95. The center is $(-3, 3)$. The radius is 3.

$(x+3)^2 + (y-3)^2 = 3^2$

••

Prepare for Section 1.3

97. $x^2 + 3x - 4$

$(-3)^2 + 3(-3) - 4 = 9 - 9 - 4 = -4$

98. $D = \{-3, -2, -1, 0, 2\}$

$R = \{1, 2, 4, 5\}$

99. $d = \sqrt{(3-(-4))^2 + (-2-1)^2} = \sqrt{49+9} = \sqrt{58}$

100. $2x - 6 \geq 0$

$\qquad 2x \geq 6$

$\qquad\ x \geq 3$

101. $x^2 - x - 6 = 0$

$\quad (x+2)(x-3) = 0$

$x+2 = 0 \quad x-3 = 0$

$\quad x = -2 \quad\ x = 3$

$-2, 3$

102. $a = 3x+4, \quad a = 6x-5$

$3x+4 = 6x-5$

$\qquad 9 = 3x$

$\qquad 3 = x$

$a = 3(3)+4 = 13$

Copyright © Houghton Mifflin Company. All rights reserved.

Section 1.3

1. Given $f(x) = 3x - 1$,

   a. $f(2) = 3(2) - 1$
$$= 6 - 1$$
$$= 5$$

   b. $f(-1) = 3(-1) - 1$
$$= -3 - 1$$
$$= -4$$

   c. $f(0) = 3(0) - 1$
$$= 0 - 1$$
$$= -1$$

   d. $f\left(\dfrac{2}{3}\right) = 3\left(\dfrac{2}{3}\right) - 1$
$$= 2 - 1$$
$$= 1$$

   e. $f(k) = 3(k) - 1$
$$= 3k - 1$$

   f. $f(k+2) = 3(k+2) - 1$
$$= 3k + 6 - 1$$
$$= 3k + 5$$

3. Given $A(w) = \sqrt{w^2 + 5}$,

   a. $A(0) = \sqrt{(0)^2 + 5}$
$$= \sqrt{5}$$

   b. $A(2) = \sqrt{(2)^2 + 5}$
$$= \sqrt{9}$$
$$= 3$$

   c. $A(-2) = \sqrt{(-2)^2 + 5}$
$$= \sqrt{9}$$
$$= 3$$

   d. $A(4) = \sqrt{4^2 + 5}$
$$= \sqrt{21}$$

   e. $A(r+1) = \sqrt{(r+1)^2 + 5}$
$$= \sqrt{r^2 + 2r + 1 + 5}$$
$$= \sqrt{r^2 + 2r + 6}$$

   f. $A(-c) = \sqrt{(-c)^2 + 5}$
$$= \sqrt{c^2 + 5}$$

5. Given $f(x) = \dfrac{1}{|x|}$,

   a. $f(2) = \dfrac{1}{|2|} = \dfrac{1}{2}$

   b. $f(-2) = \dfrac{1}{|-2|} = \dfrac{1}{2}$

   c. $f\left(-\dfrac{3}{5}\right) = \dfrac{1}{\left|-\dfrac{3}{5}\right|}$
$$= \dfrac{1}{\frac{3}{5}}$$
$$= 1 \div \dfrac{3}{5} = 1 \cdot \dfrac{5}{3} = \dfrac{5}{3}$$

   d. $f(2) + f(-2) = \dfrac{1}{2} + \dfrac{1}{2} = 1$

   e. $f(c^2 + 4) = \dfrac{1}{|c^2 + 4|} = \dfrac{1}{c^2 + 4}$

   f. $f(2 + h) = \dfrac{1}{|2 + h|}$

7. Given $s(x) = \dfrac{x}{|x|}$,

   a. $s(4) = \dfrac{4}{|4|} = \dfrac{4}{4} = 1$

   b. $s(5) = \dfrac{5}{|5|} = \dfrac{5}{5} = 1$

   c. $s(-2) = \dfrac{-2}{|-2|} = \dfrac{-2}{2} = -1$

   d. $s(-3) = \dfrac{-3}{|-3|} = \dfrac{-3}{3} = -1$

   e. Since $t > 0, |t| = t$.
$$s(t) = \dfrac{t}{|t|} = \dfrac{t}{t} = 1$$

   f. Since $t < 0, |t| = -t$.
$$s(t) = \dfrac{t}{|t|} = \dfrac{t}{-t} = -1$$

9.   **a.** Since $x = -4 < 2$, use $P(x) = 3x + 1$.
$$P(-4) = 3(-4) + 1 = -12 + 1 = -11$$

   b. Since $x = \sqrt{5} \geq 2$, use $P(x) = -x^2 + 11$.
$$P\left(\sqrt{5}\right) = -\left(\sqrt{5}\right)^2 + 11 = -5 + 11 = 6$$

   c. Since $x = c < 2$, use $P(x) = 3x + 1$.
$$P(c) = 3c + 1$$

   d. Since $k \geq 1$, then $x = k + 1 \geq 2$,
so use $P(x) = -x^2 + 11$.
$$P(k + 1) = -(k + 1)^2 + 11 = -(k^2 + 2k + 1) + 11$$
$$= -k^2 - 2k - 1 + 11$$
$$= -k^2 - 2k + 10$$

11. $2x + 3y = 7$
$$3y = -2x + 7$$
$$y = -\dfrac{2}{3}x + \dfrac{7}{3}, \quad y \text{ is a function of } x.$$

13. $-x + y^2 = 2$
$$y^2 = x + 2$$
$$y = \pm\sqrt{x + 2}, \quad y \text{ is a not function of } x.$$

Copyright © Houghton Mifflin Company. All rights reserved.

15. $y = 4 \pm \sqrt{x}$, y is not a function of x since for each $x > 0$ there are two values of x.

17. $y = \sqrt[3]{x}$, y is a function of x.

19. $y^2 = x^2$

$y = \pm\sqrt{x^2}$, y is a not function of x.

21. Function; each x is paired with exactly one y.

23. Function; each x is paired with exactly one y.

25. Function; each x is paired with exactly one y.

27. $f(x) = 3x - 4$ Domain is the set of all real numbers.

29. $f(x) = x^2 + 2$ Domain is the set of all real numbers.

31. $f(x) = \dfrac{4}{x+2}$ Domain is $\{x \mid x \neq -2\}$

33. $f(x) = \sqrt{7 + x}$ Domain is $\{x \mid x \geq -7\}$

35. $f(x) = \sqrt{4 - x^2}$ Domain is $\{x \mid -2 \leq x \leq 2\}$

37. $f(x) = \dfrac{1}{\sqrt{x+4}}$ Domain is $\{x \mid x > -4\}$

39.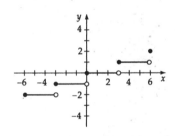

Domain: the set of all real numbers

41.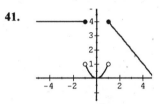

Domain: the set of all real numbers

43.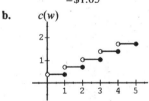

Domain: $\{x \mid -6 \leq x \leq 6\}$

45.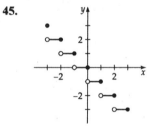

Domain: $\{x \mid -3 \leq x \leq 3\}$

47. **a.**
$$C(2.8) = 0.37 - 0.34\text{int}(1 - 2.8)$$
$$= 0.37 - 0.34\text{int}(-1.8)$$
$$= 0.37 - 0.34(-2)$$
$$= 0.37 + 0.68$$
$$= \$1.05$$

b.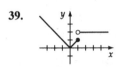

49. **a.** Yes; every vertical line intersects the graph in one point.
b. Yes; every vertical line intersects the graph in one point.
c. No; some vertical lines intersect the graph at more than one point.
d. Yes; every vertical line intersects the graph in at most one point.

51. Decreasing on $(-\infty, 0]$; increasing on $[0, \infty)$

53. Increasing on $(-\infty, \infty)$

55. Decreasing on $(-\infty, -3]$; increasing on $[-3, 0]$; decreasing on $[0, 3]$; increasing on $[3, \infty)$

Copyright © Houghton Mifflin Company. All rights reserved.

57. Constant on $(-\infty, 0]$; increasing on $[0, \infty)$

59. Decreasing on $(-\infty, 0]$; constant on $[0, 1]$; increasing on $[1, \infty)$

61. g and F are one-to-one since every horizontal line intersects the graph at one point.
f, V, and p are not one-to-one since some horizontal lines intersect the graph at more than one point.

63. **a.** $2l + 2w = 50$
$2w = 50 - 2l$
$w = 25 - l$

   b. $A = lw$
$A = l(25 - l)$
$A = 25l - l^2$

65. $v(t) = 80,000 - 6500t, \quad 0 \le t \le 10$

67. **a.** $C(x) = 5(400) + 22.80x$
$= 2000 + 22.80x$

   b. $R(x) = 37.00x$

   c. $P(x) = 37.00x - C(x)$
$= 37.00 - [2000 + 22.80x]$
$= 37.00x - 2000 - 22.80x$
$= 14.20x - 2000$

   Note x is a natural number.

69. $\dfrac{15}{3} = \dfrac{15 - h}{r}$

$5 = \dfrac{15 - h}{r}$

$5r = 15 - h$

$h = 15 - 5r$

$h(r) = 15 - 5r$

71. $d = \sqrt{(3t)^2 + (50)^2}$

$d = \sqrt{9t^2 + 2500}$ meters, $0 \le t \le 60$

73. $d = \sqrt{(45 - 8t)^2 + (6t)^2}$ miles

where t is the number of hours after 12:00 noon

75. **a.**

Circle	Square
$C = 2\pi r$	$C = 4s$
$x = 2\pi r$	$20 - x = 4s$
$r = \dfrac{x}{2\pi}$	$s = 5 - \dfrac{x}{4}$
$\text{Area} = \pi r^2 = \pi\left(\dfrac{x}{2\pi}\right)^2$	$\text{Area} = s^2 = \left(5 - \dfrac{x}{4}\right)^2$
$= \dfrac{x^2}{4\pi}$	$= 25 - \dfrac{5}{2}x + \dfrac{x^2}{16}$

$\text{Total Area} = \dfrac{x^2}{4\pi} + 25 - \dfrac{5}{2}x + \dfrac{x^2}{16}$

$= \left(\dfrac{1}{4\pi} + \dfrac{1}{16}\right)x^2 - \dfrac{5}{2}x + 25$

   b.

x	0	4	8	12	16	20
Total Area	25	17.27	14.09	15.46	21.37	31.83

   c. Domain: $[0, 20]$.

77. **a.**

Left side triangle	Right side triangle
$c^2 = 20^2 + (40 - x)^2$	$c^2 = 30^2 + x^2$
$c = \sqrt{400 + (40 - x)^2}$	$c = \sqrt{900 + x^2}$

   Total length $= \sqrt{900 + x^2} + \sqrt{400 + (40 - x)^2}$

   b.

x	0	10	20	30	40
Total Length	74.72	67.68	64.34	64.79	70

   c. Domain: $(0, 40)$.

Copyright © Houghton Mifflin Company. All rights reserved.

79.

x	5	10	12.5	15	20
$Y(x)$	275	375	385	390	394

answers accurate to nearest apple

81.
$$f(c)=c^2-c-5=1$$
$$c^2-c-6=0$$
$$(c-3)(c+2)=0$$
$$c-3=0 \quad \text{or} \quad c+2=0$$
$$c=3 \qquad\qquad c=-2$$

83. 1 is not in the range of $f(x)$, since

$1 = \dfrac{x-1}{x+1}$ only if $x+1=x-1$ or $1=-1$.

85. Set the graphing utility to "dot" mode.

```
Y1◼int X/abs X
Y2 =
Y3 =
Y4 =
Y5 =
Y6 =
Y7 =
Y8 =
```

```
WINDOW FORMAT
  Xmin=-4.7
  Xmax=4.7
  Xscl=1
  Ymin=-5
  Ymax=2
  Yscl=1
```

87.
```
Y1◼X²-2abs X-3
Y2 =
Y3 =
Y4 =
Y5 =
Y6 =
Y7 =
Y8 =
```

```
WINDOW FORMAT
  Xmin=-4.7
  Xmax=4.7
  Xscl=1
  Ymin=-5
  Ymax=1
  Yscl=1
```

89.
```
Y1◼abs (X²-1)-
    abs (X-2)
Y2 =
Y3 =
Y4 =
Y5 =
Y6 =
Y7 =
```

```
WINDOW FORMAT
  Xmin=-4.7
  Xmax=4.7
  Xscl=1
  Ymin=-4.7
  Ymax=4.7
  Yscl=1
```

Connecting Concepts

91. $f(x)\Big|_2^3 = (9-3)-(4-2)=6-2=4$

93. $f(x)\Big|_0^2 = (16-12-2)-0=2$

95.
a. $f(1,7)=3(1)+5(7)-2=3+35-2=36$
b. $f(0,3)=3(0)+5(3)-2=13$
c. $f(-2,4)=3(-2)+5(4)-2=12$
d. $f(4,4)=3(4)+5(4)-2=30$
e. $f(k,2k)=3(k)+5(2k)-2=13k-2$
f. $f(k+2,k-3)=3(k+2)+5(k-3)-2=3k+6+5k-15-2=8k-11$

97. $s = \dfrac{5+8+11}{2}=12$

$A(5,8,11) = \sqrt{12(12-5)(12-8)(12-11)}$

$\qquad\qquad = \sqrt{12(7)(4)(1)} = \sqrt{336}=4\sqrt{21}$

99.
$$a^2+3a-3=a$$
$$a^2+2a-3=0$$
$$(a-1)(a+3)=0$$
$$a=1 \quad \text{or} \quad a=-3$$

101.

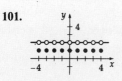

Copyright © Houghton Mifflin Company. All rights reserved.

103.
$$f(x)=x^2+4x-6$$
$$-\frac{b}{2a}=-\frac{4}{2(1)}=-2$$
$$x=-2$$

104.
$$f(3)=\frac{3(3)^4}{(3)^2+1}=\frac{243}{10}=24.3$$
$$f(-3)=\frac{3(-3)^4}{(-3)^2+1}=\frac{243}{10}=24.3$$
$$f(3)=f(-3)$$

105.
$$f(-2)=2(-2)^3-5(-2)=-16+10=-6$$
$$-f(2)=-[2(2)^3-5(2)]=-[16-10]=-6$$
$$f(-2)=-f(2)$$

106.
$$f(-2)-g(-2)=(-2)^2-[-2+3]=4-1=3$$
$$f(-1)-g(-1)=(-1)^2-[-1+3]=1-2=-1$$
$$f(0)-g(0)=(0)^2-[0+3]=0-3=-3$$
$$f(1)-g(1)=(1)^2-[1+3]=1-4=-3$$
$$f(2)-g(2)=(2)^2-[2+3]=4-5=-1$$

107.
$$\frac{-a+a}{2}=0,\quad \frac{b+b}{2}=b$$
midpoint is $(0, b)$

108.
$$\frac{-a+a}{2}=0,\quad \frac{-b+b}{2}=0$$
midpoint is $(0, 0)$

Section 1.4

1.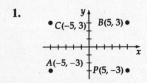

3.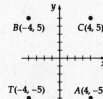

5.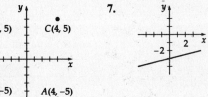

7.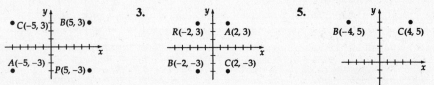

9.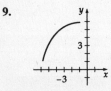

11.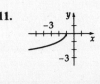

13. Replacing x by $-x$ leaves the equation unaltered. Thus the graph is symmetric with respect to the y-axis.

15. Not symmetric with respect to either axis. (neither)

17. Symmetric with respect to both the x- and the y-axes.

19. Symmetric with respect to both the x- and the y-axes.

21. Symmetric with respect to both the x- and the y-axes.

23. No, since $(-y)=3(-x)-2$ simplifies to $(-y)=-3x-2$, which is not equivalent to the original equation $y=3x-2$.

25. Yes, since $(-y)=-(-x)^3$ implies $-y=x^3$ or $y=-x^3$, which is the original equation.

27. Yes, since $(-x)^2+(-y)^2=10$ simplifies to the original equation.

29. Yes, since $-y=\dfrac{-x}{|-x|}$ simplifies to the original equation.

Copyright © Houghton Mifflin Company. All rights reserved.

31.

symmetric with
respect to the y-axis

33.

symmetric with
respect to the origin

35.

symmetric with
respect to the origin

37.

symmetric with
respect to the line $x = 4$

39.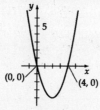

symmetric with
respect to the line $x = 2$

41.

no symmetry

43. Even since $g(-x) = (-x)^2 - 7 = x^2 - 7 = g(x)$.

45. Odd, since $F(-x) = (-x)^5 + (-x)^3$
$$= -x^5 - x^3$$
$$= -F(x).$$

47. Even   **49.** Even   **51.** Even   **53.** Even   **55.** Neither

57.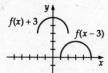

59. **a.** $f(x+2)$

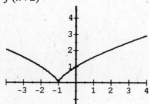

b. $f(x)+2$

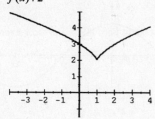

61. **a.** $f(x+3)$
$(-2-3,5) = (-5,5)$
$(0-3,-2) = (-3,-2)$
$(1-3,0) = (-2,0)$

b. $f(x)+1$
$(-2,5+1) = (-2,6)$
$(0,-2+1) = (0,-1)$
$(1,0+1) = (1,1)$

Copyright © Houghton Mifflin Company. All rights reserved.

24

63. **a.** $f(-x)$

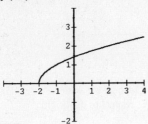

b. $-f(x)$

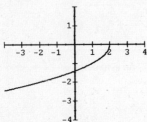

65. **a.** $f(-x)$

$(--1,3)=(1,3)$

$(-2,-4)$

b. $-f(x)$

$(-1,-3)$

$(2,--4)=(2,4)$

67.

69.

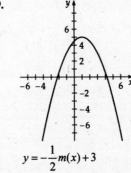

$$y=-\frac{1}{2}m(x)+3$$

71. **a.**

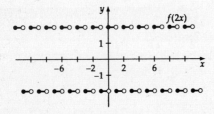

b.

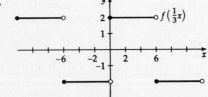

73. **a.**

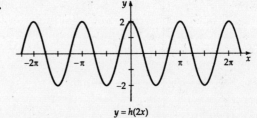

$y=h(2x)$

b.

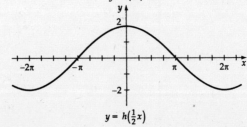

$$y=h\left(\frac{1}{2}x\right)$$

75.

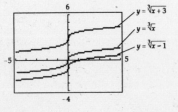

77.

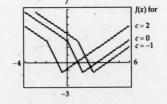

79.

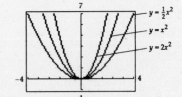

Copyright © Houghton Mifflin Company. All rights reserved.

81.

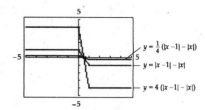

83. a.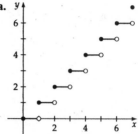

b.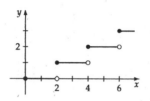

c.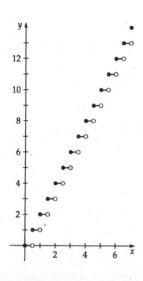

85. a.
$$f(x) = \frac{2}{(x+1)^2 + 1} + 1$$

Connecting Concepts

b.
$$f(x) = -\frac{2}{(x-2)^2 + 1}$$

Prepare for Section 1.5

87.
$$(2x^2 + 3x - 4) - (x^2 + 3x - 5) = x^2 + 1$$

88.
$$(3x^2 - x + 2)(2x - 3) = 6x^3 - 2x^2 + 4x - 9x^2 + 3x - 6$$
$$= 6x^3 - 11x^2 + 7x - 6$$

89.
$$f(3a) = 2(3a)^2 - 5(3a) + 2$$
$$= 18a^2 - 15a + 2$$

90.
$$f(2+h) = 2(2+h)^2 - 5(2+h) + 2$$
$$= 2h^2 + 8h + 8 - 5h - 10 + 2$$
$$= 2h^2 + 3h$$

91. Domain: all real numbers except $x = 1$

92.
$$2x - 8 = 0$$
$$x = 4$$
Domain: $x \geq 4$

Copyright © Houghton Mifflin Company. All rights reserved.

Section 1.5

1.
$$f(x)+g(x)=(x^2-2x-15)+(x+3)$$
$$=x^2-x-12 \text{ Domain all real numbers}$$
$$f(x)-g(x)=(x^2-2x-15)-(x+3)$$
$$=x^2-3x-18 \text{ Domain all real numbers}$$
$$f(x)g(x)=(x^2-2x-15)(x+3)$$
$$=x^3+x^2-21x-45 \text{ Domain all real numbers}$$
$$f(x)/g(x)=(x^2-2x-15)/(x+3)$$
$$=x-5 \text{ Domain } \{x|x\neq-3\}$$

3.
$$f(x)+g(x)=(2x^2+8)+(x+4)$$
$$=3x+12 \text{ Domain all real numbers}$$
$$f(x)-g(x)=(2x+8)-(x+4)$$
$$=x+4 \text{ Domain all real numbers}$$
$$f(x)g(x)=(2x+8)(x+4)$$
$$=2x^2+16x+32 \text{ Domain all real numbers}$$
$$f(x)/g(x)=(2x+8)/(x+4)$$
$$=[2(x+4)]/(x+4)$$
$$=2 \text{ Domain } \{x|x\neq-4\}$$

5.
$$f(x)+g(x)=(x^3-2x^2+7x)+x$$
$$=x^3-2x^2+8x \text{ Domain all real numbers}$$
$$f(x)-g(x)=(x^3-2x^2+7x)-x$$
$$=x^3-2x^2+6x \text{ Domain all real numbers}$$
$$f(x)g(x)=(x^3-2x^2+7x)x$$
$$=x^4-2x^3+7x^2 \text{ Domain all real numbers}$$
$$f(x)/g(x)=(x^3-2x^2+7x)/x$$
$$=x^2-2x+7 \text{ Domain } \{x|x\neq0\}$$

7.
$$f(x)+g(x)=(2x^2+4x-7)+(2x^2+3x-5)$$
$$=4x^2+7x-12 \text{ Domain all real numbers}$$
$$f(x)-g(x)=(2x^2+4x-7)-(2x^2+3x-5)$$
$$=x-2 \text{ Domain all real numbers}$$
$$f(x)g(x)=(2x^2+4x-7)(2x^2+3x-5)$$
$$=4x^4+6x^3-10x^2+8x^3+12x^2-20x-14x^2-21x+35$$
$$=4x^4+14x^3-12x^2-41x+35 \text{ Domain all real numbers}$$
$$f(x)/g(x)=(2x^2+4x-7)/(2x^2+3x-5)$$
$$=1+\frac{x-2}{2x^2+3x-5} \text{ Domain } \left\{x|x\neq1, x\neq-\frac{5}{2}\right\}$$

9.
$$f(x)+g(x)=\sqrt{x-3}+x \quad \text{Domain } \{x\,|\,x\geq3\}$$
$$f(x)-g(x)=\sqrt{x-3}-x \quad \text{Domain } \{x\,|\,x\geq3\}$$
$$f(x)g(x)=x\sqrt{x-3} \quad \text{Domain } \{x\,|\,x\geq3\}$$
$$f(x)/g(x)=\frac{\sqrt{x-3}}{x}+x \quad \text{Domain } \{x\,|\,x\geq3\}$$

11.
$$f(x)+g(x)=\sqrt{4-x^2}+2+x \quad \text{Domain } \{x\,|-2\leq x\leq2\}$$
$$f(x)-g(x)=\sqrt{4-x^2}-2-x \quad \text{Domain } \{x\,|-2\leq x\leq2\}$$
$$f(x)g(x)=\left(\sqrt{4-x^2}\right)(2+x) \quad \text{Domain } \{x\,|-2\leq x\leq2\}$$
$$f(x)/g(x)=\frac{\sqrt{4-x^2}}{2+x} \quad \text{Domain } \{x\,|-2\leq x\leq2\}$$

13.
$$(f+g)(x)=x^2-x-2$$
$$(f+g)(5)=(5)^2-(5)-2$$
$$=25-5-2$$
$$=18$$

15.
$$(f+g)(x)=x^2-x-2$$
$$(f+g)\left(\frac{1}{2}\right)=\left(\frac{1}{2}\right)^2-\left(\frac{1}{2}\right)-2$$
$$=\frac{1}{4}-\frac{1}{2}-2$$
$$=-\frac{9}{4}$$

17.
$$(f-g)(x)=x^2-5x+6$$
$$(f-g)(-3)=(-3)^2-5(-3)+6$$
$$=9+15+6$$
$$=30$$

Copyright © Houghton Mifflin Company. All rights reserved.

19.
$$(f-g)(x) = x^2 - 5x + 6$$
$$(f-g)(-1) = (-1)^2 - 5(-1) + 6$$
$$= 1 + 5 + 6$$
$$= 12$$

21.
$$(fg)(x) = \left(x^2 - 3x + 2\right)(2x - 4)$$
$$= 2x^3 - 6x^2 + 4x - 4x^2 + 12x - 8$$
$$= 2x^3 - 10x^2 + 16x - 8$$
$$(fg)(7) = 2(7)^3 - 10(7)^2 + 16(7) - 8$$
$$= 686 - 490 + 112 - 8$$
$$= 300$$

23.
$$(fg)(x) = 2x^3 - 10x^2 + 16x - 8$$
$$(fg)\left(\frac{2}{5}\right) = 2\left(\frac{2}{5}\right)^3 - 10\left(\frac{2}{5}\right)^2 + 16\left(\frac{2}{5}\right) - 8$$
$$= \frac{16}{125} - \frac{40}{25} + \frac{32}{5} - 8$$
$$= \frac{-384}{125} = -3.072$$

25.
$$\left(\frac{f}{g}\right)(x) = \frac{x^2 - 3x + 2}{2x - 4}$$
$$\left(\frac{f}{g}\right)(x) = \frac{1}{2}x - \frac{1}{2}$$
$$\left(\frac{f}{g}\right)(-4) = \frac{1}{2}(-4) - \frac{1}{2}$$
$$= -2 - \frac{1}{2}$$
$$= -2\frac{1}{2} \text{ or } -\frac{5}{2}$$

27.
$$\left(\frac{f}{g}\right)(x) = \frac{1}{2}x - \frac{1}{2}$$
$$\left(\frac{f}{g}\right)\left(\frac{1}{2}\right) = \frac{1}{2}\left(\frac{1}{2}\right) - \frac{1}{2}$$
$$= \frac{1}{4} - \frac{1}{2}$$
$$= -\frac{1}{4}$$

29.
$$\frac{f(x+h) - f(x)}{h} = \frac{[2(x+h)+4] - (2x+4)}{h}$$
$$= \frac{2x + 2(h) + 4 - 2x - 4}{h}$$
$$= \frac{2h}{h}$$
$$= 2$$

31.
$$\frac{f(x+h) - f(x)}{h} = \frac{[(x+h)-6] - (x^2-6)}{h}$$
$$= \frac{x^2 + 2x(h) + (h)^2 - 6 - x^2 + 6}{h}$$
$$= \frac{2x(h) + h^2}{h}$$
$$= 2x + h$$

33.
$$\frac{f(x+h) - f(x)}{h} = \frac{2(x+h)^2 + 4(x+h) - 3 - (2x^2 + 4x - 3)}{h}$$
$$= \frac{2x^2 + 4xh + 2h^2 + 4x + 4h - 3 - 2x^2 - 4x + 3}{h}$$
$$= \frac{4xh + 2h^2 + 4h}{h}$$
$$= 4x + 2h + 4$$

35.
$$\frac{f(x+h) - f(x)}{h} = \frac{-4(x+h)^2 + 6 - (-4x^2 + 6)}{h}$$
$$= \frac{-4x^2 - 8xh - 4h^2 + 6 + 4x^2 - 6}{h}$$
$$= \frac{-8xh - 4h^2}{h}$$
$$= -8x - 4h$$

37.
$$(g \circ f)(x) = g[f(x)] \qquad (f \circ g)(x) = f[g(x)]$$
$$= g[3x+5] \qquad\qquad = f[2x-7]$$
$$= 2[3x+5] \qquad\qquad = 3[2x-7]+5$$
$$= 6x + 10 - 7 \qquad\quad = 6x - 21 + 5$$
$$= 6x + 3 \qquad\qquad\quad = 6x - 16$$

39.
$$(g \circ f)(x) = g\left[x^2 + 4x - 1\right] \qquad (f \circ g)(x) = f[x+2]$$
$$= \left[x^2 + 4x - 1\right] + 2 \qquad\qquad = [x+2]^2 + 4[x+2] - 1$$
$$= x^2 + 4x + 1 \qquad\qquad\qquad = x^2 + 4x + 4 + 4x + 8 - 1$$
$$\qquad\qquad\qquad\qquad\qquad\qquad = x^2 + 8x + 11$$

Copyright © Houghton Mifflin Company. All rights reserved.

41.

$$(g \circ f)(x) = g[f(x)]$$
$$= g\left[x^3 + 2x\right]$$
$$= -5\left[x^3 + 2x\right]$$
$$= -5x^3 - 10x$$

$$(f \circ g)(x) = f[g(x)]$$
$$= f[-5x]$$
$$= [-5x]^3 + 2[-5x]$$
$$= -125x^3 - 10x$$

43.

$$(g \circ f)(x) = g[f(x)]$$
$$= g\left[\frac{2}{x+1}\right]$$
$$= 3\left[\frac{2}{x+1}\right] - 5$$
$$= \frac{6}{x+1} - \frac{5(x+1)}{x+1}$$
$$= \frac{6 - 5x - 5}{x+1}$$
$$= \frac{1 - 5x}{x+1}$$

$$(f \circ g)(x) = f[g(x)]$$
$$= f[3x - 5]$$
$$= \frac{2}{[3x-5]+1}$$
$$= \frac{2}{3x-4}$$

45.

$$(g \circ f)(x) = g[f(x)]$$
$$= g\left[\frac{1}{x^2}\right]$$
$$= \sqrt{\left[\frac{1}{x^2}\right] - 1}$$
$$= \sqrt{\frac{1 - x^2}{x^2}}$$
$$= \frac{\sqrt{1 - x^2}}{|x|}$$

$$(f \circ g)(x) = f[g(x)]$$
$$= f\left[\sqrt{x-1}\right]$$
$$= \frac{1}{\left[\sqrt{x-1}\right]^2}$$
$$= \frac{1}{x-1}$$

47.

$$(g \circ f)(x) = g\left[\frac{3}{|5-x|}\right]$$
$$= -\frac{2}{\left[\dfrac{3}{|5-x|}\right]}$$
$$= \frac{-2|5-x|}{3}$$

$$(f \circ g)(x) = f\left[-\frac{2}{x}\right]$$
$$= \frac{3}{\left|5 - \left[-\dfrac{2}{x}\right]\right|}$$
$$= \frac{3}{\left|5 + \dfrac{2}{x}\right|}$$
$$= \frac{3}{\dfrac{|5x+2|}{|x|}}$$
$$= \frac{3|x|}{|5x+2|}$$

Use the results to work Exercises 49 to 64.

49.
$$(g \circ f)(x) = 4x^2 + 2x - 6$$
$$(g \circ f)(4) = 4(4)^2 + 2(4) - 6$$
$$= 64 + 8 - 6$$
$$= 66$$

51.
$$(f \circ g)(x) = 2x^2 - 10x + 3$$
$$(f \circ g)(-3) = 2(-3)^2 - 10(-3) + 3$$
$$= 18 + 30 + 3$$
$$= 51$$

Copyright © Houghton Mifflin Company. All rights reserved.

53.
$$(g \circ h)(x) = 9x^4 - 9x^2 - 4$$
$$(g \circ h)(0) = 9(0)^4 - 9(0)^2 - 4$$
$$= -4$$

55.
$$(f \circ f)(x) = 4x + 9$$
$$(f \circ f)(8) = 4(8) + 9$$
$$= 41$$

57.
$$(h \circ g)(x) = -3x^4 + 30x^3 - 75x^2 + 4$$
$$(h \circ g)\left(\frac{2}{5}\right) = -3\left(\frac{2}{5}\right)^4 + 30\left(\frac{2}{5}\right)^3 - 75\left(\frac{2}{5}\right)^2 + 4$$
$$= -\frac{48}{625} + \frac{240}{125} - \frac{300}{25} + 4$$
$$= \frac{-48 + 1200 - 7500 + 2500}{625}$$
$$= -\frac{3848}{625}$$

59.
$$(g \circ f)(x) = 4x^2 + 2x - 6$$
$$(g \circ f)(\sqrt{3}) = 4(\sqrt{3})^2 + 2(\sqrt{3}) - 6$$
$$= 12 + 2\sqrt{3} - 6$$
$$= 6 + 2\sqrt{3}$$

61.
$$(g \circ f)(x) = 4x^2 + 2x - 6$$
$$(g \circ f)(2c) = 4(2c)^2 + 2(2c) - 6$$
$$= 16c^2 + 4c - 6$$

63.
$$(g \circ h)(x) = 9x^4 - 9x^2 - 4$$
$$(g \circ h)(k+1) = 9(k+1)^4 - 9(k+1)^2 - 4$$
$$= 9(k^4 + 4k^3 + 6k^2 + 4k + 1) - 9k^2 - 18k - 9 - 4$$
$$= 9k^4 + 36k^3 + 54k^2 + 36k + 9 - 9k^2 - 18k - 13$$
$$= 9k^4 + 36k^3 + 45k^2 + 18k - 4$$

65.
a.
$$r = 1.5t \text{ and } A = \pi r^2$$
$$\text{so } A(t) = \pi\left[r(t)\right]^2$$
$$= \pi(1.5t)^2$$
$$= 2.25\pi(2)^2$$
$$= 9\pi \text{ square feet}$$
$$\approx 28.27 \text{ square feet}$$

b.
$$r = 1.5t$$
$$h = 2r = 2(1.5t) = 3t \quad \text{and}$$
$$V = \frac{1}{3}\pi r^2 h \quad \text{so}$$
$$V(t) = \frac{1}{3}\pi(1.5t)^2[3t]$$
$$= 2.25\pi t^3$$

Note: $V = \frac{1}{3}\pi r^2 h = \frac{1}{3}(\pi r^2) = \frac{1}{3}hA$
$$= \frac{1}{3}(3t)(2.25\pi t^2) = 2.25\pi t^3$$
$$V(3) = 2.25\pi(3)^3$$
$$= 60.75\pi \text{ cubic feet}$$
$$\approx 190.85 \text{ cubic feet}$$

67.
a. Since $d^2 + 4^2 = s^2$,
$$d^2 = s^2 - 16$$
$$d = \sqrt{s^2 - 16}$$
$$d = \sqrt{(48-t)^2 - 16} \quad \cdot \quad s = 48 - t$$
$$= \sqrt{2304 - 96t + t^2 - 16}$$
$$= \sqrt{t^2 - 96t + 2288}$$

b.
$$s(35) = 48 - 35 + 13$$
$$d(35) = \sqrt{35^2 - 96(35) + 2288}$$
$$= \sqrt{153} \approx 12.37 \text{ ft}$$

69. $(Y \circ F)(x) = Y(F(x))$ converts x inches to yards.

F takes x inches to feet, and then Y takes feet to yards.

Copyright © Houghton Mifflin Company. All rights reserved.

71. **a.** On $[0, 1]$, $a = 0$

$\Delta t = 1 - 0 = 1$

$C(a + \Delta t) = C(1) = 99.8$

$C(a) = C(0) = 0$

Average rate of change $= \dfrac{C(1) - C(0)}{1} = 99.8 - 0 = 99.8$

This is identical to the slope of the line through

$(0, C(0))$ and $(1, C(1))$ since $m = \dfrac{C(1) - C(0)}{1 - 0} = C(1) - C(0)$

b. On $[0, 0.5]$, $a = 0$

$\Delta t = 0.5$

Average rate of change $= \dfrac{C(0.5) - C(0)}{0.5} = \dfrac{78.1 - 0}{0.5} = 156.2$

c. On $[1, 2]$, $a = 1$

$\Delta t = 2 - 1 = 1$

Average rate of change $= \dfrac{C(2) - C(1)}{1} = \dfrac{50.1 - 99.8}{1} = -49.7$

d. On $[1, 1.5]$, $a = 1$

$\Delta t = 1.5 - 1 = 0.5$

Average rate of change $= \dfrac{C(1.5) - C(1)}{0.5} = \dfrac{84.4 - 99.8}{0.5} = \dfrac{-15.4}{0.5} = -30.8$

e. On $[1, 1.25]$, $a = 1$

$\Delta t = 1.25 - 1 = 0.25$

Average rate of change $= \dfrac{C(1.25) - C(1)}{0.25} = \dfrac{95.7 - 99.8}{0.25} = \dfrac{-4.1}{0.25} = -16.4$

f. On $[1, 1 + \Delta t]$, $\;Con(1 + \Delta t) = 25(1 + \Delta t)^3 - 150(1 + \Delta t)^2 + 225(1 + \Delta t)$

$= 25(1 + 3(\Delta t) + 3(\Delta t)^3) - 150(1 + 2(\Delta t) + (\Delta t)^2) + 225(1 + \Delta t)$

$= 25 + 75(\Delta t) + 75(\Delta t)^2) + 25(\Delta t)^3 - 150 - 300(\Delta t) - 150(\Delta t)^2 + 225 + 225(\Delta t)$

$= 100 - 75(\Delta t)^2 + 25(\Delta t)^3$

$Con(1) = 100$

Average rate of change $= \dfrac{Con(1 + \Delta t) - Con(1)}{\Delta t} = \dfrac{100 - 75(\Delta t)^2 + 25(\Delta t)^3 - 100}{\Delta t}$

$= \dfrac{-75(\Delta t)^2 + 25(\Delta t)^3}{\Delta t}$

$= -75(\Delta t) + 25(\Delta t)^2$

As Δt approaches 0, the average rate of change over $[1, 1 + \Delta t]$ seems to approach 0.

Connecting Concepts

73. $(g \circ f)(x) = g[f(x)]$

$= g[2x + 3]$

$= 5(2x + 3) + 12$

$= 10x + 15 + 12$

$= 10x + 27$

$(g \circ f)(x) = (f \circ g)(x)$

$(f \circ g)(x) = f[g(x)]$

$= f[5x + 12]$

$= 2(5x + 12) + 3$

$= 10x + 24 + 3$

$= 10x + 27$

Copyright © Houghton Mifflin Company. All rights reserved.

75.
$$(g \circ f)(x) = g[f(x)]$$
$$= g\left[\frac{6x}{x-1}\right]$$
$$= \frac{5\left(\frac{6x}{x-1}\right)}{\frac{6x}{x-1} - 2}$$
$$= \frac{\frac{30x}{x-1}}{\frac{6x-2x+2}{x-1}} = \frac{\frac{30x}{x-1}}{\frac{4x+2}{x-1}}$$
$$= \frac{30x}{x-1} \cdot \frac{x-1}{2(2x+1)}$$
$$= \frac{15x}{2x+1}$$

$$(f \circ g)(x) = f[g(x)]$$
$$= f\left[\frac{5x}{x-2}\right]$$
$$= \frac{6\left(\frac{5x}{x-2}\right)}{\frac{5x}{x-2} - 1}$$
$$= \frac{\frac{30x}{x-2}}{\frac{5x-x+2}{x-2}} = \frac{\frac{30x}{x-2}}{\frac{4x+2}{x-2}}$$
$$= \frac{30x}{x-2} \cdot \frac{x-2}{2(2x+1)}$$
$$= \frac{15x}{2x+1}$$

$$(g \circ f)(x) = (f \circ g)(x)$$

77.
$$(g \circ f)(x) = g[f(x)]$$
$$= g[2x+3]$$
$$= \frac{[2x+3]-3}{2}$$
$$= \frac{2x}{2}$$
$$= x$$

$$(f \circ g)(x) = f[g(x)]$$
$$= f\left[\frac{x-3}{2}\right]$$
$$= 2\left[\frac{x-3}{2}\right] + 3$$
$$= x-3+3$$
$$= x$$

79.
$$(g \circ f)(x) = g[f(x)]$$
$$= g\left[\frac{4}{x+1}\right]$$
$$= \frac{4-\left[\frac{4}{x+1}\right]}{\left[\frac{4}{x+1}\right]}$$
$$= \frac{\frac{4x+4-4}{x+1}}{\frac{4}{x+1}}$$
$$= \frac{4x}{x+1} \cdot \frac{x+1}{4}$$
$$= x$$

$$(f \circ g)(x) = f[g(x)]$$
$$= f\left[\frac{4-x}{x}\right]$$
$$= \frac{4}{\left[\frac{4-x}{x}\right]+1}$$
$$= \frac{4}{\frac{4-x+x}{x}}$$
$$= \frac{4}{\frac{4}{x}}$$
$$= 4 \cdot \frac{x}{4}$$
$$= x$$

81.
$$(g \circ f)(x) = g[f(x)]$$
$$= g\left[x^3-1\right]$$
$$= \sqrt[3]{\left[x^3-1\right]+1}$$
$$= \sqrt[3]{x^3}$$
$$= x$$

$$(f \circ g)(x) = f[g(x)]$$
$$= f\left[\sqrt[3]{x+1}\right]$$
$$= \left[\sqrt[3]{x+1}\right]^3 - 1$$
$$= x+1-1$$
$$= x$$

Copyright © Houghton Mifflin Company. All rights reserved.

83.
$$2x + 5y = 15$$
$$5y = -2x + 15$$
$$y = -\frac{2}{5}x + 3$$

84.
$$x = \frac{y+1}{y}$$
$$xy = y + 1$$
$$xy - y = 1$$
$$y(x - 1) = 1$$
$$y = \frac{1}{x-1}$$

85.
$$f(-1) = \frac{2(-1)^2}{(-1)-1} = \frac{2}{-2} = -1$$

86. $5 - 4a < 5 - 4b$

87. For $a^2 > b^2$, $a < b$ and $b \le 0$.

88.
$$x + 2 \ge 0$$
$$x \ge -2$$
$$\{x \mid x \ge -2\}$$

Section 1.6

1. If $f(3) = 7$, then $f^{-1}(7) = 3$.

3. If $h^{-1}(-3) = -4$, then $h(-4) = -3$.

5. If 3 is in the domain of f^{-1}, then $f[f^{-1}(3)] = 3$.

7. The domain of the inverse function f^{-1} is the <u>range</u> of f.

9.

Yes, the inverse is a function.

11.

Yes, the inverse is a function.

13.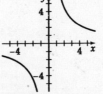

Yes, the inverse is a function.

15.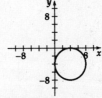

No, the inverse relation is not a function.

17.
$$f(x) = 4x; \ g(x) = \frac{x}{4}$$
$$f[g(x)] = f\left(\frac{x}{4}\right) = 4\left(\frac{x}{4}\right) = x$$
$$g[f(x)] = g(4x) = \frac{4x}{4} = x$$

Yes, f and g are inverses of each other.

19.
$$f(x) = 4x - 1; \ g(x) = \frac{1}{4}x + \frac{1}{4}$$
$$f[g(x)] = f\left(\frac{1}{4}x + \frac{1}{4}\right)$$
$$= 4\left(\frac{1}{4}x + \frac{1}{4}\right) - 1 = x + 1 - 1$$
$$= x$$
$$g[f(x)] = g(4x - 1)$$
$$= \frac{1}{4}(4x - 1) + \frac{1}{4} = x - \frac{1}{4} + \frac{1}{4}$$
$$= x$$

Yes, f and g are inverses of each other.

21.
$$f(x) = -\frac{1}{2}x - \frac{1}{2}; \ g(x) = -2x + 1$$
$$f[g(x)] = f(-2x + 1)$$
$$= -\frac{1}{2}(-2x + 1) - \frac{1}{2} = x - \frac{1}{2} - \frac{1}{2}$$
$$= x - 1$$
$$\neq x$$

No, f and g are not inverses of each other.

23. The inverse of $\{(-3, 1), (-2, 2), (1, 5), (4, -7)\}$ is $\{(1, -3), (2, -2), (5, 1), (-7, 4)\}$.

Copyright © Houghton Mifflin Company. All rights reserved.

25. The inverse of $\{(0, 1), (1, 2), (2, 4), (3, 8), (4, 16)\}$ is $\{(1, 0), (2, 1), (4, 2), (8, 3), (16, 4)\}$.

27.
$$f(x) = 2x + 4$$
$$x = 2y + 4$$
$$x - 4 = 2y$$
$$\tfrac{1}{2}x - 2 = y$$
$$f^{-1}(x) = \tfrac{1}{2}x - 2$$

29.
$$f(x) = 3x - 7$$
$$x = 3y - 7$$
$$x + 7 = 3y$$
$$\tfrac{1}{3}x + \tfrac{7}{3} = y$$
$$f^{-1}(x) = \tfrac{1}{3}x + \tfrac{7}{3}$$

31.
$$f(x) = -2x + 5$$
$$x = -2y + 5$$
$$x - 5 = -2y$$
$$-\tfrac{1}{2}x + \tfrac{5}{2} = y$$
$$f^{-1}(x) = -\tfrac{1}{2}x + \tfrac{5}{2}$$

33.
$$f(x) = \frac{2x}{x-1}, \; x \neq 1$$
$$x = \frac{2y}{y-1}$$
$$x(y-1) = xy - x = 2y$$
$$xy - 2y = y(x-2) = x$$
$$y = \frac{x}{x-2}$$
$$f^{-1}(x) = \frac{x}{x-2}, \; x \neq 2$$

35.
$$f(x) = \frac{x-1}{x+1}, \; x \neq -1$$
$$x = \frac{y-1}{y+1}$$
$$x(y+1) = xy + x = y - 1$$
$$xy - y = -x - 1$$
$$y - xy = y(1-x) = x + 1$$
$$y = \frac{x+1}{1-x}$$
$$f^{-1}(x) = \frac{x+1}{1-x}, \; x \neq 1$$

37.
$$f(x) = x^2 + 1, \; x \geq 0$$
$$x = y^2 + 1$$
$$x - 1 = y^2$$
$$\sqrt{x-1} = y$$
$$f^{-1}(x) = \sqrt{x-1}, \; x \geq 1$$
Note: Do not use $\pm$ with the radical because the domain of f, and thus the range of f^{-1}, is nonnegative.

39.
$$f(x) = \sqrt{x-2}, \; x \geq 2$$
$$x = \sqrt{y-2}$$
$$x^2 = y - 2$$
$$x^2 + 2 = y$$
$$f^{-1}(x) = x^2 + 2, \; x \geq 0$$
Note: The range of f is nonnegative, therefore the domain of f^{-1} is also nonnegative.

41.
$$f(x) = x^2 + 4x, \; x \geq -2$$
$$x = y^2 + 4y$$
$$x + 4 = y^2 + 4y + 4$$
$$x + 4 = (y+2)^2$$
$$\sqrt{x+4} = y + 2$$
$$y = \sqrt{x+4} - 2$$
$$f^{-1}(x) = \sqrt{x+4} - 2, \; x \geq -4$$
Note: The range of f is non-negative, therefore the domain of f^{-1} is also non-negative.

43.
$$f(x) = x^2 + 4x - 1, \; x \leq -2$$
$$x = y^2 + 4y - 1$$
$$x + 1 = y^2 + 4y$$
$$x + 1 + 4 = y^2 + 4y + 4$$
$$x + 5 = (y+2)^2$$
$$-\sqrt{x+5} = y + 2$$
$$-\sqrt{x+5} - 2 = y$$
$$f^{-1}(x) = -\sqrt{x+5} - 2, \; x \geq -5$$
Note: Because the range of f is non-positive, the range of f^{-1} must also be non-positive.

45.
$$V(x) = x^3$$
$$x = y^3$$
$$\sqrt[3]{x} = y$$
$$V^{-1}(x) = \sqrt[3]{x}$$
$V^{-1}(x)$ finds the length of a side of a cube given the volume.

Copyright © Houghton Mifflin Company. All rights reserved.

47. Yes, a conversion function is always a one-to-one function. Yes, a conversion function always has an inverse function. A conversion function is a nonconstant linear function. All nonconstant linear functions have inverses which are also functions.

49.
$$s(x) = 2x + 24$$
$$x = 2y + 24$$
$$x - 24 = 2y$$
$$\tfrac{1}{2}x - 12 = y$$
$$s^{-1}(x) = \tfrac{1}{2}x - 12$$

51.
$$E(s) = 0.05s + 2500$$
$$s = 0.05y + 2500$$
$$s - 2500 = 0.05y$$
$$\frac{1}{0.05}s - \frac{2500}{0.05} = y$$
$$20s - 50{,}000 = y$$
$$E^{-1}(s) = 20s - 50{,}000$$

From the monthly earnings, the executive can find the value of the software sold.

53.
$$f(x) = 2x - 1$$
$$f(22102917) = 2(22102917) - 1$$
$$= 44205833$$

Find the inverse:
$$f(x) = 2x - 1$$
$$x = 2y - 1$$
$$x + 1 = 2y$$
$$\frac{x+1}{2} = y$$
$$f^{-1}(x) = \frac{x+1}{2}$$

Apply the inverse to the code:
$$f^{-1}(44205833) = \frac{44205833 + 1}{2}$$
$$= \frac{44205834}{2}$$
$$= 22102917$$

55. $f(2) = 7,\ f(5) = 12,\ \text{and}\ f(4) = c.$ Because f is an increasing linear function, and 4 is between 2 and 5, then $f(4)$ is between $f(2)$ and $f(5)$. Thus, c is between 7 and 12.

57. f is a linear function, therefore f^{-1} is a linear function.
$$f(2) = 3 \Rightarrow f^{-1}(3) = 2$$
$$f(5) = 9 \Rightarrow f^{-1}(9) = 5$$
Since 6 is between 3 and 9, $f^{-1}(6)$ is between 2 and 5.

59. g is a linear function, therefore g^{-1} is a linear function.
$$g^{-1}(3) = 4 \Rightarrow g(4) = 3$$
$$g^{-1}(7) = 8 \Rightarrow g(8) = 7$$
Since 5 is between 4 and 8, $g(5)$ is between 3 and 7.

•••

Connecting Concepts

61.
$$f(x) = ax + b,\ a \neq 0$$
$$y = ax + b$$
$$x = ay + b$$
$$x - b = ay$$
$$\frac{x-b}{a} = y$$
Thus $f^{-1}(x) = \frac{x-b}{a},\ a \neq 0$

63. The reflection of f across the line given by $y=x$ yields f. Thus f is its own inverse.

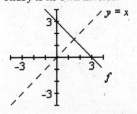

65. There is at most one point where each horizontal line intersects the graph of the function. The function is a one-to-one function.

67. A horizontal line intersects the graph of the function at more than one point. Thus, the function is not a one-to-one function.

Copyright © Houghton Mifflin Company. All rights reserved.

69. Slope: $-\dfrac{1}{3}$; y-intercept: $(0, 4)$

70. $3x - 4y = 12$

$y = \dfrac{3}{4}x - 3$

Slope: $\dfrac{3}{4}$; y-intercept: $(0, -3)$

71. $y = -0.45x + 2.3$

72. $y + 4 = -\dfrac{2}{3}(x - 3)$

$y = -\dfrac{2}{3}x - 2$

73. $f(2) = 3(2)^2 + 4(2) - 1 = 12 + 8 - 1 = 19$

74. $|f(x_1) - y_1| + |f(x_2) - y_2| = |(2)^2 - 3 - (-1)| + |(4)^2 - 3 - 14|$

$= |4 - 3 + 1| + |16 - 3 - 14|$

$= 2 + 1$

$= 3$

Section 1.7

1. The scatter diagram suggests no relationship between x and y.

3. The scatter diagram suggests a linear relationship between x and y.

5. Figure A more approximates a graph that can be modeled by an equation than does Figure B. Thus Figure A has a coefficient of determination closer to 1.

7. Enter the data on your calculator. The technique for a TI-83 calculator is illustrated here.

 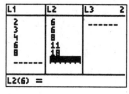 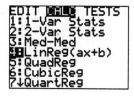 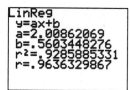

$y = 2.00862069x + 0.5603448276$

9. Enter the data on your calculator. The technique for a TI-83 calculator is illustrated here.

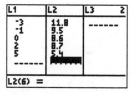 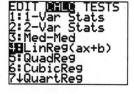 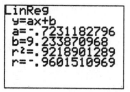

$y = -0.7231182796x + 9.233870968$

11. Enter the data on your calculator. The technique for a TI-83 calculator is illustrated here.

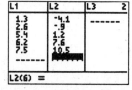 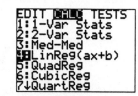

$y = 2.222641509x - 7.364150943$

Copyright © Houghton Mifflin Company. All rights reserved.

13. Enter the data on your calculator. The technique for a TI-83 calculator is illustrated here.

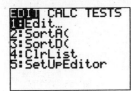 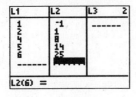 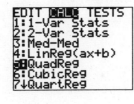 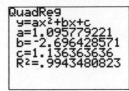

$y = 1.095779221x^2 - 2.69642857x + 1.136363636$

15. Enter the data on your calculator. The technique for a TI-83 calculator is illustrated here.

 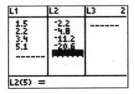 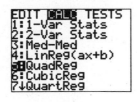 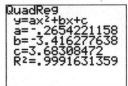

$y = -0.2987274717x^2 - 3.20998141x + 3.416463667$

17. Enter the data on your calculator. The technique for a TI-83 calculator is illustrated here.

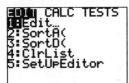 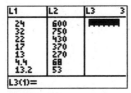 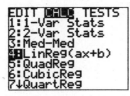 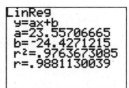

   a. $y = 23.55706665x - 24.4271215$

   b. $y = 23.55706665(54) - 24.4271215 \approx 1247.7$ cm

19. Enter the data on your calculator. The technique for a TI-83 calculator is illustrated here.

 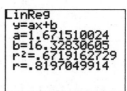

   a. $y = 1.671510024x + 16.32830605$

   b. $y = 1.671510024(18) + 16.32830605 \approx 46.4$ cm

21. Enter the data on your calculator. The technique for a TI-83 calculator is illustrated here.

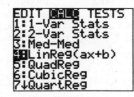 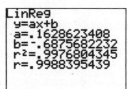

   a. $y = 0.1628623408x - 0.6875682232$

   b. $y = 0.1628623408(158) - 0.6875682232 \approx 25$

Copyright © Houghton Mifflin Company. All rights reserved.

23. Enter the data on your calculator. The technique for a TI-83 calculator is illustrated here.

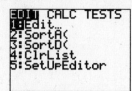 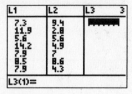 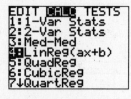 

The value of r is close to 0. Therefore, no, there is not a strong linear relationship between the current and the torque.

25. Enter the data on your calculator. The technique for a TI-83 calculator is illustrated here.

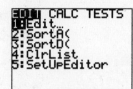 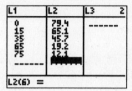 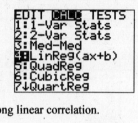 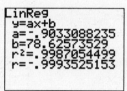

a. The value of r is close to –1, so, yes, there is a strong linear correlation.

b. $y = -0.9033088235x + 78.62573529$

c. $y = -0.9033088235(25) + 78.62573529 \approx 56$ years

27. Enter the data on your calculator. The technique for a TI-83 calculator is illustrated here.

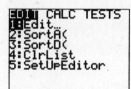 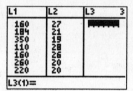 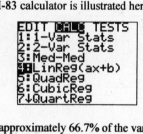 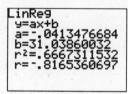

$r^2 \approx 0.667$ The coefficient of determination means that approximately 66.7% of the variation in EPA mileage estimates can be attributed to the horsepower of a car.

29. Enter the data on your calculator. The technique for a TI-83 calculator is illustrated here.

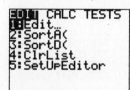 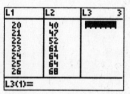 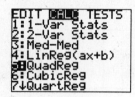 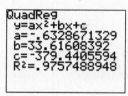

$y = -0.6328671329x^2 + 33.6160839x - 379.4405594$

31. Enter the data on your calculator. The technique for a TI-83 calculator is illustrated here.

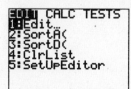 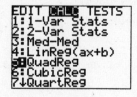

a. $y = -0.0165034965x^2 + 1.366713287x + 5.685314685$

b. $y = -0.0165034965(50)^2 + 1.366713287(50) + 5.685314685 \approx 32.8$ mpg

Copyright © Houghton Mifflin Company. All rights reserved.

33. **a.** Enter the data on your calculator. The technique for a TI-83 calculator is illustrated here.

5-lb ball

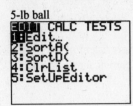 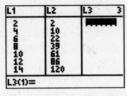 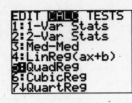 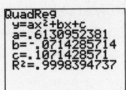

$y = 0.6130952381t^2 - 0.0714285714t + 0.1071428571$

10-lb ball

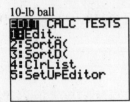 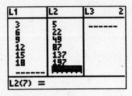  

$y = 0.6091269841t^2 - 0.0011904762t - 0.3$

15-lb ball

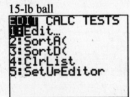 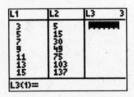 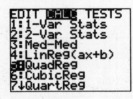 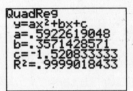

$y = 0.5922619048t^2 + 0.3571428571t - 1.520833333$

b. All the regression equations are approximately the same. Therefore, there is one equation of motion.

35. Enter the data on your calculator. The technique for a TI-83 calculator is illustrated here.
Compute r for the linear regression model and R^2 for the quadratic model.

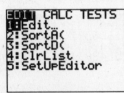 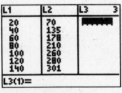 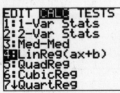 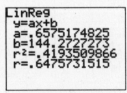

The quadratic regression model is the better fit for this data. R^2 is closer to 1 for the quadratic model than r^2 is for the linear model.

Copyright © Houghton Mifflin Company. All rights reserved.

1. False. Let $f(x) = x^2$. Then $f(3) = f(-3) = 9$, but $3 \neq -3$.

2. False. $f(x)^2 = x^2$ does not have an inverse function.

3. False. Let $f(x) = 2x$, $g(x) = 3x$. Then $f(g(0)) = 0$ and $g(f(0)) = 0$, but f and g are not inverse functions.

4. True

5. False. Let $f(x) = 3x$. $[f(x)]^2 = [3x]^2 = 9x^2$, whereas $f[f(x)] = f(3x) = 3(3x) = 9x$.

6. False. Let $f(x) = x^2$. Then $\dfrac{f(2)}{f(1)} = \dfrac{2^2}{1^2} = \dfrac{4}{1} = 4 \neq \dfrac{2}{1}$.

7. True

8. False. Let $f(x) = |x|$. Then $f(-1+3) = f(2) = 2$, whereas $f(-1) + f(3) = 1 + 3 = 4$.

9. True **10.** True **11.** True **12.** True **13.** True

1. $3 - 4z = 12$
$-4z = 9$
$z = -\dfrac{9}{4}$

2. $4y - 3 = 6y + 5$
$-2y = 8$
$y = -4$

3. $2x - 3(2 - 3x) = 14x$
$2x - 6 + 9x = 14x$
$-6 = 3x$
$-2 = x$

4. $5 - 2(3m + 2) = 3(1 - m)$
$5 - 6m - 4 = 3 - 3m$
$1 - 6m = 3 - 3m$
$-3m = 2$
$m = -\dfrac{2}{3}$

5. $y^2 - 3y - 18 = 0$
$(y - 6)(y + 3) = 0$
$y - 6 = 0$ or $y + 3 = 0$
$y = 6$ or $y = -3$

6. $2z^2 - 9z + 4 = 0$
$(2z - 1)(z - 4) = 0$
$2z - 1 = 0$ or $z - 4 = 0$
$2z = 1$ $\quad z = 4$
$z = \dfrac{1}{2}$

7. $3v^2 + v = 1$
$3v^2 + v - 1 = 0$
$v = \dfrac{-1 \pm \sqrt{1^2 - 4(3)(-1)}}{2(3)}$
$v = \dfrac{-1 \pm \sqrt{13}}{6}$

8. $3s = 4 - 2s^2$
$2s^2 + 3s - 4 = 0$
$s = \dfrac{-3 \pm \sqrt{3^2 - 4(2)(-4)}}{2(2)}$
$s = \dfrac{-3 \pm \sqrt{9 + 32}}{4}$
$s = \dfrac{-3 \pm \sqrt{41}}{4}$

9. $3c - 5 \leq 5c + 7$
$-2c \leq 12$
$c \geq -6$

10. $7a > 5 - 2(3a - 4)$
$7a > 5 - 6a + 8$
$13a > 13$
$a > 1$

11. $x^2 - x - 12 \geq 0$
$(x - 4)(x + 3) \geq 0$
Critical values are 4 and -3.
$(-\infty, -3] \cup [4, \infty)$

12. $2x^2 - x < 1$
$2x^2 - x - 1 < 0$
$(2x + 1)(x - 1) < 0$
Critical values are $-\dfrac{1}{2}$ and 1.
$-\dfrac{1}{2} < x < 1$

Copyright © Houghton Mifflin Company. All rights reserved.

13. $|2x-5| > 3$

$2x-5 > 3$ or $2x-5 < -3$

$2x > 8$ $\qquad$ $2x < 2$

$x > 4$ $\qquad$ $x < 1$

$(-\infty, 1) \cup (4, \infty)$

14. $|1-3x| \le 4$

$-4 \le 1-3x \le 4$

$-5 \le -3x \le 3$

$\dfrac{5}{3} \ge x \ge -1$

15. $c^2 = a^2 + b^2$

$c^2 = 6^2 + 8^2$

$c = \sqrt{36+64}$

$c = \sqrt{100} = 10$

16. $c^2 = a^2 + b^2$

$20^2 = 11^2 + b^2$

$b = \sqrt{400-121}$

$b = \sqrt{279}$

$b \approx 16.7$

17. $c^2 = a^2 + b^2$

$13^2 = a^2 + 12^2$

$a = \sqrt{169-144}$

$a = \sqrt{25}$

$a = 5$

18. $c^2 = a^2 + b^2$

$c^2 = 7^2 + 14^2$

$c = \sqrt{49+196}$

$c = \sqrt{245}$

$c \approx 15.7$

19. $d = \sqrt{(7-(-3))^2 + (11-2)^2}$

$= \sqrt{100+81}$

$= \sqrt{181}$

20. $d = \sqrt{(-3-5)^2 + (-8-(-4))^2}$

$= \sqrt{64+16}$

$= \sqrt{80} = 4\sqrt{5}$

21. $\left(\dfrac{2+(-3)}{2}, \dfrac{8+12}{2}\right) = \left(-\dfrac{1}{2}, 10\right)$

22. $\left(\dfrac{-4+8}{2}, \dfrac{7+(-11)}{2}\right) = (2, -2)$

23.

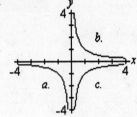

24.

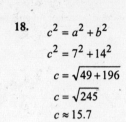

25. $y = x^2 - 7$

Replace y with $-y$.

$-y = x^2 - 7$

$y = -x^2 + 7$

Thus, y is not symmetric with respect to the x-axis.

Replace x with $-x$.

$y = (-x)^2 - 7$

$y = x^2 - 7$

Thus, y is symmetric with respect to the y-axis.

Replace x with $-x$ and replace y with $-y$.

$-y = (-x)^2 - 7$

$y = -x^2 + 7$

Thus, y is not symmetric with respect to the origin.

Therefore, the graph of $y = x^2 - 7$ is symmetric with respect to the y-axis.

26. $x = y^2 + 3$

$y^2 = x - 3$

$y = \pm\sqrt{x-3}$

Replace y with $-y$.

$x = (-y)^2 + 3$

$x = y^2 + 3$

$y^2 = x - 3$

$y = \pm\sqrt{x-3}$

Thus, y is symmetric with respect to the x-axis.

Replace x with $-x$.

$-x = y^2 + 3$

$y^2 = -x - 3$

$y = \pm\sqrt{-x-3}$

Thus, y is not symmetric with respect to the y-axis.

Replace x with $-x$ and replace y with $-y$.

$-x = (-y)^2 + 3$

$-x = y^2 + 3$

$y^2 = -x - 3$

$y = \pm\sqrt{-x-3}$

Thus, y is not symmetric with respect to the origin.

Therefore, the graph of $x = y^2 + 3$ is symmetric with respect to the x-axis.

Copyright © Houghton Mifflin Company. All rights reserved.

27.

$$y = x^3 - 4x$$

Replace y with $-y$.

$$-y = x^3 - 4x$$

$$y = -x^3 + 4x$$

Thus, y is not symmetric with respect to the x-axis.

Replace x with $-x$.

$$y = (-x)^3 - 4(-x)$$

$$y = -x^3 + 4x$$

Thus, y is not symmetric with respect to the y-axis.

Replace x with $-x$ and replace y with $-y$.

$$-y = (-x)^3 - 4(-x)$$

$$-y = -x^3 + 4x$$

$$y = x^3 - 4x$$

Thus, y is symmetric with respect to the origin.

Therefore, the graph of $y = x^3 - 4x$ is symmetric with respect to the origin.

28.

$$y^2 = x^2 + 4$$

$$y = \pm\sqrt{x^2 + 4}$$

Replace y with $-y$.

$$(-y)^2 = x^2 + 4$$

$$y^2 = x^2 + 4$$

$$y = \pm\sqrt{x^2 + 4}$$

y is symmetric with respect to the x-axis.

Replace x with $-x$.

$$y^2 = (-x)^2 + 4$$

$$y^2 = x^2 + 4$$

$$y = \pm\sqrt{x^2 + 4}$$

Thus, y is symmetric with respect to the y-axis.

Replace x with $-x$ and replace y with $-y$.

$$(-y)^2 = (-x)^2 + 4$$

$$y^2 = x^2 + 4$$

$$y = \pm\sqrt{x^2 + 4}$$

Thus, y is symmetric with respect to the origin.

Therefore, the graph of $y^2 = x^2 + 4$ is symmetric with respect to the x-axis, the y-axis, and the origin.

29.

$$\frac{x^2}{3^2} + \frac{y^2}{4^2} = 1$$

Replace y with $-y$.

$$\frac{x^2}{3^2} + \frac{(-y)^2}{4^2} = 1 \Rightarrow \frac{x^2}{3^2} + \frac{y^2}{4^2} = 1$$

Thus, y is symmetric with respect to the x-axis.

Replace x with $-x$.

$$\frac{(-x)^2}{3^2} + \frac{y^2}{4^2} = 1 \Rightarrow \frac{x^2}{3^2} + \frac{y^2}{4^2} = 1$$

Thus, y is symmetric with respect to the y-axis.

Replace x with $-x$ and replace y with $-y$.

$$\frac{(-x)^2}{3^2} + \frac{(-y)^2}{4^2} = 1 \Rightarrow \frac{x^2}{3^2} + \frac{y^2}{4^2} = 1$$

Thus, y is symmetric with respect to the origin.

Therefore, the graph of $\dfrac{x^2}{3^2} + \dfrac{y^2}{4^2} = 1$ is symmetric with respect to the x-axis, the y-axis, and the origin.

30.

$$xy = 8$$

$$y = \frac{8}{x}$$

Replace y with $-y$.

$$x(-y) = 8$$

$$y = -\frac{8}{x}$$

Thus, y is not symmetric with respect to the x-axis.

Replace x with $-x$.

$$-xy = 8$$

$$y = -\frac{8}{x}$$

Thus, y is not symmetric with respect to the y-axis.

Replace x with $-x$ and replace y with $-y$.

$$-x(-y) = 8$$

$$y = \frac{8}{x}$$

Thus, y is symmetric with respect to the origin.

Therefore, the graph of $xy = 8$ is symmetric with respect to the origin.

Copyright © Houghton Mifflin Company. All rights reserved.

31. $|y| = |x|$

Replace y with $-y$.

$$|-y| = |x| \Rightarrow |y| = |x|$$

Thus, y is symmetric with respect to the x-axis.

Replace x with $-x$.

$$|y| = |-x| \Rightarrow |y| = |x|$$

Thus, y is symmetric with respect to the y-axis.

Replace x with $-x$ and replace y with $-y$.

$$|-y| = |-x| \Rightarrow |y| = |x|$$

Thus, y is symmetric with respect to the origin.

Therefore, the graph of $|y| = |x|$ is symmetric with respect to the x-axis, the y-axis, and the origin.

33. center $(3, -4)$, radius 9

35. $(x-2)^2 + (y+3)^2 = 5^2$

37.

a. $f(1) = 3(1)^2 + 4(1) - 5$

$\qquad = 2$

b. $f(-3) = 27 - 12 - 5$

$\qquad = 10$

c. $f(t) = 3t^2 + 4t - 5$

d. $f(x+h) = 3(x+h)^2 + 4(x+h) - 5$

$\qquad = 3x^2 + 6xh + 3h^2 + 4x + 4h - 5$

e. $3f(t) = 9t^2 + 12t - 15$

f. $f(3t) = 3(3t)^2 + 4(3t) - 5$

$\qquad = 27t^2 + 12t - 5$

32. $|x + y| = 4$

Replace y with $-y$.

$$|x + (-y)| = 4 \Rightarrow |x - y| = 4$$

Thus, y is not symmetric with respect to the x-axis.

Replace x with $-x$.

$$|(-x) + y| = 4 \Rightarrow |y - x| = 4$$

Thus, y is not symmetric with respect to the y-axis.

Replace x with $-x$ and replace y with $-y$.

$$|(-x) + (-y)| = 4 \Rightarrow |-(x+y)| = 4 \Rightarrow |x + y| = 4$$

Thus, y is symmetric with respect to the origin.

Therefore, the graph of $|x + y| = 4$ is symmetric with re-spect to the origin.

34.

$$x^2 + 10x \qquad + y^2 + 4y \qquad = -20$$

$$x^2 + 10x + 25 + y^2 + 4y + 4 = 25 + 4 - 20$$

$$(x+5)^2 \qquad + (y+2)^2 \qquad = 9$$

center $(-5, -2)$, radius 3

36. $(x+5)^2 + (y-1)^2 = 8^2$, radius $= |-5 - (3)| = 8$

38.

a. $g(3) = \sqrt{64 - 3^2}$

$\qquad = \sqrt{55}$

b. $g(-5) = \sqrt{64 - (-5)^2}$

$\qquad = \sqrt{64 - 25}$

$\qquad = \sqrt{39}$

c. $g(8) = \sqrt{64 - 8^2}$

$\qquad = 0$

d. $g(-x) = \sqrt{64 - (-x)^2}$

$\qquad = \sqrt{64 - x^2}$

e. $2g(t) = 2\sqrt{64 - t^2}$

f. $g(2t) = \sqrt{64 - (2t)^2}$

$\qquad = \sqrt{64 - 4t^2}$

$\qquad = 2\sqrt{16 - t^2}$

Copyright © Houghton Mifflin Company. All rights reserved.

39.   **a.** $(f \circ g)(3) = f[g(3)]$

$$= f[3-8]$$
$$= f[-5]$$
$$= (-5)^2 + 4(-5)$$
$$= 5$$

b. $(g \circ f)(-3) = g[f(-3)]$

$$= g[(-3)^2 + 4(-3)]$$
$$= g[-3]$$
$$= [-3-8]$$
$$= -11$$

c. $(f \circ g)(x) = f[g(x)]$

$$= f[x-8]$$
$$= (x-8)^2 + 4(x-8)$$
$$= x^2 - 16x + 64 + 4x - 32$$
$$= x^2 - 12x + 32$$

d. $(g \circ f)(x) = g[f(x)]$

$$= g[x^2 + 4x]$$
$$= [x^2 + 4x] - 8$$
$$= x^2 + 4x - 8$$

40.   **a.** $(f \circ g)(-5) = f[g(-5)]$

$$= f\left[\, |(-5)-1| \, \right]$$
$$= f[6]$$
$$= 2(6)^2 + 7$$
$$= 72 + 7$$
$$= 79$$

b. $(g \circ f)(-5) = g[f(-5)]$

$$= g[2(-5)^2 + 7]$$
$$= g[57]$$
$$= |57 - 1|$$
$$= 56$$

c. $(f \circ g)(x) = f[g(x)]$

$$= f\left[\, |x-1| \, \right]$$
$$= 2\left(|x-1| \right)^2 + 7$$
$$= 2(x-1)^2 + 7$$
$$= 2(x^2 - 2x + 1) + 7$$
$$= 2x^2 - 4x + 2 + 7$$
$$= 2x^2 - 4x + 9$$

d. $(g \circ f)(x) = g[f(x)]$

$$= g[2x^2 + 7]$$
$$= |2x^2 + 7 - 1|$$
$$= |2x^2 + 6|$$
$$= 2x^2 + 6$$

41. $\dfrac{f(x+h) - f(x)}{h} = \dfrac{4(x+h)^2 - 3(x+h) - 1 - (4x^2 - 3x - 1)}{h}$

$$= \frac{4x^2 + 8xh + 4h^2 - 3x - 3h - 1 - 4x^2 + 3x + 1}{h}$$
$$= \frac{8xh + 4h^2 - 3h}{h}$$
$$= 8x + 4h - 3$$

42. $\dfrac{g(x+h) - g(x)}{h} = \dfrac{(x+h)^3 - (x+h) - (x^3 - x)}{h}$

$$= \frac{x^3 + 3x^2 h + 3xh^2 + h^3 - x - h - x^3 + x}{h}$$
$$= \frac{3x^2 h + 3xh^2 + h^3 - h}{h}$$
$$= 3x^2 + 3xh + h^2 - 1$$

43. $f(x) = -2x^2 + 3$

Domain: All real numbers

Copyright © Houghton Mifflin Company. All rights reserved.

44. $f(x) = \sqrt{6-x}$

$6 - x \geq 0$

$-x \geq -6$

$x \leq 6$

Domain: $\{x \mid x \leq 6\}$

45. $f(x) = \sqrt{25 - x^2}$

$25 - x^2 \geq 0$

$(5 - x)(5 + x) \geq 0$

Critical values -5 and 5.

Domain: $\{x \mid -5 \leq x \leq 5\}$

46. $f(x) = \dfrac{3}{x^2 - 2x - 15}$

$x^2 - 2x - 15 = 0$

$(x + 3)(x - 5) = 0$

$x = -3 \qquad x = 5$

Domain: $\{x \mid x \neq -3 \text{ and } x \neq 5\}$

47.

f is increasing on $[3, \infty)$

f is decreasing on $(-\infty, 3]$

48.

f is increasing on $[0, \infty)$

f is decreasing on $(-\infty, 0]$

49.

f is increasing on $[-2, 2]$

f is constant on $(-\infty, -2] \cup [2, \infty)$

50.

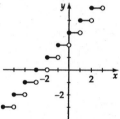

f is constant on $...,[-6, -5), [-5, -4), [-4, -3),$
$[-3, -2), [-2, -1), [-1, 0), [0, 1),...$

51.

f is increasing on $(-\infty, \infty)$

52.

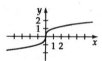

f is increasing on $(-\infty, \infty)$

53.

a.   Domain $\{x \mid x \text{ is a real number}\}$

Range $\{y \mid y \leq 4\}$

b. g is an even function

54.

a.   Domain all real numbers
Range all real numbers

b. g is neither even nor odd

55.

a.   Domain all real numbers
Range $\{y \mid y \geq 4\}$

b. g is an even function

56.

a.   Domain $\{x \mid -4 \leq x \leq 4\}$

Range $\{y \mid 0 \leq y \leq 4\}$

b. g is an even function

Copyright © Houghton Mifflin Company. All rights reserved.

57.

a. Domain $\{x | x \text{ is a real number}\}$

Range $\{y | y \text{ is a real number}\}$

b. g is an odd function

58.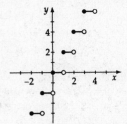

a. Domain $\{x | x \text{ is a real number}\}$

Range $\{y | y \text{ is an even integer}\}$

b. g is neither even or odd.

59. $(f+g)(x) = x^2 - 9 + x + 3$
$$= x^2 + x - 6$$

Domain of $(f+g)(x)$ is $\{x | x \text{ is a real number}\}$.

$(f-g)(x) = x^2 - 9 - (x+3)$
$$= x^2 - x - 12$$

Domain of $(f-g)(x)$ is $\{x | x \text{ is a real number}\}$.

$(fg)(x) = (x^2 - 9)(x+3)$
$$= x^3 + 3x^2 - 9x - 27$$

Domain of $(fg)(x)$ is $\{x | x \text{ is a real number}\}$.

$\left(\dfrac{f}{g}\right)(x) = \dfrac{x^2 - 9}{x + 3}$
$$= x - 3$$

Domain of $\left(\dfrac{f}{g}\right)(x)$ is $\{x | x \neq -3\}$.

60. $(f+g)(x) = x^3 + 8 + x^2 - 2x + 4$
$$= x^3 + x^2 - 2x + 12$$

Domain of $(f+g)(x)$ is the set of all real numbers.

$(f-g)(x) = x^3 + 8 - (x^2 - 2x + 4)$
$$= x^3 + 8 - x^2 + 2x - 4$$
$$= x^3 - x^2 + 2x + 4$$

Domain of $(f-g)(x)$ is the set of all real numbers.

$(fg)(x) = (x^3 + 8)(x^2 - 2x + 4)$
$$= x^5 - 2x^4 + 4x^3 + 8x^2 - 16x + 32$$

Domain of $(fg)(x)$ is the set of all real numbers.

$\left(\dfrac{f}{g}\right)(x) = \dfrac{x^3 + 8}{x^2 - 2x + 4}$
$$= \dfrac{(x+2)(x^2 - 2x + 4)}{x^2 - 2x + 4}$$
$$= x + 2$$

Domain of $\left(\dfrac{f}{g}\right)(x)$ is the set of all real numbers.

61. $F[G(x)] = 2\left(\dfrac{x+5}{2}\right) - 5$ and $G[F(x)] = \dfrac{(2x-5)+5}{2}$
$\qquad = x + 5 - 5 \qquad\qquad\qquad\qquad = \dfrac{2x}{2}$
$\qquad = x \qquad\qquad\qquad\qquad\qquad\quad = x$

Because $F[G(x)] = x$ and $G[F(x)] = x$ for all real numbers x, F and G are inverses.

Copyright © Houghton Mifflin Company. All rights reserved.

62.
$$h[k(x)] = \sqrt{x^2} = x \quad \text{Since } x \geq 0$$
$$k[h(x)] = \left(\sqrt{x}\right)^2 = x$$

Because $h[k(x)] = x$ for all x in the domain of k and $k[h(x)] = x$ for all x in the domain of h, we have shown that h and k are inverses.

63.
$$l[m(x)] = \frac{\frac{3}{x-1}+3}{\frac{3}{x-1}} \qquad m[l(x)] = m\left[\frac{x+3}{x}\right]$$
$$= \frac{3+3x-3}{x-1} \cdot \frac{x-1}{3} \qquad = \frac{3}{\frac{x+3}{x}-1}$$
$$= \frac{3x}{3} \qquad\qquad = \frac{3}{\frac{x+3-x}{x}}$$
$$= x \qquad\qquad = 3 \cdot \frac{x}{3} = x$$

Thus, l and m are inverse functions.

64.
$$p[q(x)] = \frac{\frac{2x}{x-5}-5}{2\left(\frac{2x}{x-5}\right)}$$
$$= \frac{\frac{2x-5x+25}{x-5}}{\frac{4x}{x-5}}$$
$$= \frac{-3x+25}{x-5} \cdot \frac{x-5}{4x}$$
$$= \frac{-3x+25}{4x}$$

Thus, p and q are not inverse functions.

65.
$$f(x) = 3x - 4$$
$$y = 3x - 4$$
$$x = 3y - 4$$
$$x + 4 = 3y$$
$$\frac{x+4}{3} = y$$

Thus $f^{-1}(x) = \frac{x+4}{3} = \frac{1}{3}x + \frac{4}{3}$.

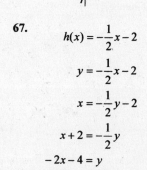

66.
$$g(x) = -2x + 3$$
$$y = -2x + 3$$
$$x = -2y + 3$$
$$x - 3 = -2y$$
$$\frac{x-3}{-2} = y$$
$$\frac{3-x}{2} = y$$

Thus, $g^{-1}(x) = \frac{3-x}{2} = -\frac{1}{2}x + \frac{3}{2}$.

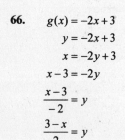

67.
$$h(x) = -\frac{1}{2}x - 2$$
$$y = -\frac{1}{2}x - 2$$
$$x = -\frac{1}{2}y - 2$$
$$x + 2 = -\frac{1}{2}y$$
$$-2x - 4 = y$$

Thus, $h^{-1}(x) = -2x - 4$.

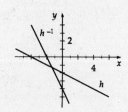

Copyright © Houghton Mifflin Company. All rights reserved.

68.

$$k(x) = \frac{1}{x}$$

$$y = \frac{1}{x}$$

$$x = \frac{1}{y}$$

$$y = \frac{1}{x}$$

Thus $k^{-1}(x) = k(x) = \frac{1}{x}$.

both k
and k^{-1}

69. Let $x =$ one of the numbers and $50 - x =$ the other number. Their product y is given by

$$y = x(50 - x) = 50x - x^2 = x^2 + 50x$$

Now y takes on its maximum value when

$$x = \frac{-b}{2a} = \frac{-50}{2(-1)} = 25$$

Thus, the two numbers are 25 and $(50 - 25) = 25$. That is, both numbers are 25.

70. Let $x =$ the smaller number. Let $x + 10 =$ the larger number. The sum of their squares y is given by

$$y = x^2 + (x + 10)^2$$
$$= x^2 + x^2 + 20x + 100$$
$$= 2x^2 + 20x + 100$$

Now y takes on its minimum value when

$$x = \frac{-b}{2a} = \frac{-20}{2(2)} = -5$$

Thus, the numbers are -5 and $(-5 + 10) = 5$.

71. $h(t) = -16t^2 + 220$ and $h(t) = 0$ when

$$-16t^2 + 220 = 0$$
$$220 = 16t^2$$
$$\frac{220}{16} = t^2$$
$$\sqrt{\frac{220}{16}} = t$$
$$\frac{2\sqrt{55}}{4} = t$$

Thus, $t \approx 3.7$ seconds.

72. **a.**

$$h(1050) = \frac{1}{8820}(1050)^2 + 25$$
$$= \frac{1,102,500}{8,820} + 25$$
$$= 125 + 25$$
$$= 150 \text{ feet}$$

b.

$$h(2100) = \frac{1}{8820}(2100)^2 + 25$$
$$= \frac{4,410,000}{8,820} + 25$$
$$= 500 + 25$$
$$= 525 \text{ feet}$$

73. **a.** Enter the data on your calculator. The technique for a TI-83 is illustrated here.

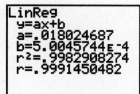

$$y = 0.018024687x + 0.00050045744$$

b. Yes. $r \approx 0.999$, which is very close to 1.

c.

$$y = 0.018024687(100) + 0.00050045744$$
$$= 1.8024687 + 0.00050045744$$
$$\approx 1.8 \text{ seconds}$$

Copyright © Houghton Mifflin Company. All rights reserved.

74.   **a.**   Enter the data on your calculator. The technique for a TI-83 is illustrated here.

L1	L2	L3	3
0	180	▬▬▬	
10	163		
20	147		
30	133		
40	118		
50	105		
60	93		

L3(1)=

```
EDIT CALC TESTS
1:1-Var Stats
2:2-Var Stats
3:Med-Med
4:LinReg(ax+b)
5:QuadReg
6:CubicReg
7↓QuartReg
```

```
QuadReg
y=ax²+bx+c
a=.0047952048
b=-1.756843157
c=180.4065934
R²=.999144351
```

$$h = 0.0047952048t^2 - 1.756843157t + 180.4065934$$

b.   If the can is empty, then height = 0.

$$0 = 0.0047952048t^2 - 1.756843157t + 180.4065934$$

$$t = \frac{-(-1.756843157) \pm \sqrt{(-1.756843157)^2 - 4(0.0047952048)(180.4065934)}}{2(0.0047952048)}$$

$$= \frac{1.756843157 \pm \sqrt{3.086497878 - 3.46034625}}{0.0095904096}$$

$$= \frac{1.756843157 \pm \sqrt{-0.373848372}}{0.0095904096}$$

This does not represent a real number. Thus, no, according to the model, the can will never empty.

c.   The regression line is a model of the data and is not based on physical principles.

●●●   **C**hapter **T**est

1. $4x - 2(2 - x) = 5 - 3(2x + 1)$
$4x - 4 + 2x = 5 - 6x - 3$
$6x - 4 = 2 - 6x$
$12x = 6$
$x = \frac{1}{2}$

2. $6 - 3x \geq 3 - 4(2 - 2x)$
$6 - 3x \geq 3 - 8 + 8x$
$-11x \geq -11$
$x \leq 1$

3. $2x^2 - 3x = 2$
$2x^2 - 3x - 2 = 0$
$(2x + 1)(x - 2) = 0$
$x = -\frac{1}{2}$ or $x = 2$

4. $3x^2 - x = 2$
$3x^2 - x - 2 = 0$
$(3x + 2)(x - 1) = 0$
$x = -\frac{2}{3}$ or $x = 1$

5. $|4 - 5x| > 6$
$4 - 5x < -6$ or $4 - 5x > 6$
$-5x < -10$ $-5x > 2$
$x > 2$ $x < -\frac{2}{5}$

$\left(-\infty, -\frac{2}{5}\right) \cup (2, \infty)$

6. $d = \sqrt{(x_2 - x_1)^2 + (y_2 - y_1)^2}$
$d = \sqrt{[4 - (-2)]^2 + (-2 - 5)^2}$
$d = \sqrt{36 + 49}$
$d = \sqrt{85}$

7. $x_m = \frac{4 - 2}{2} = 1$ length $= \sqrt{(x_2 - x_1)^2 + (y_2 - y_1)^2}$
$y_m = \frac{-1 + 3}{2} = 1$ $= \sqrt{[4 - (-2)]^2 + (-1 - 3)^2}$
                               $= \sqrt{6^2 + (-4)^2}$
midpoint = (1, 1) $= \sqrt{36 + 16} = \sqrt{52}$
                               $= 2\sqrt{13}$

Copyright © Houghton Mifflin Company. All rights reserved.

8. $x = 2y^2 - 4$

$x = 0 \qquad 2y^2 - 4 = 0$

$\qquad\qquad\qquad 2y^2 = 4$

$\qquad\qquad\qquad y^2 = 2$

$\qquad\qquad\qquad y = \pm\sqrt{2}$

If $y = 0, \qquad x = -4$

intercepts $(0, -\sqrt{2}), (0, \sqrt{2}), (-4, 0)$

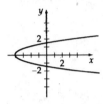

9.

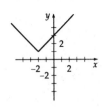

10. $x^2 - 4x + y^2 + 2y - 4 = 0$

$(x^2 - 4x + 4) + (y^2 + 2y + 1) = 4 + 4 + 1$

$\qquad\qquad (x - 2)^2 + (y + 1)^2 = 9$

center: $(2, -1)$ \qquad radius: 3

11. $f(x) = -\sqrt{25 - x^2}$

$f(-3) = -\sqrt{25 - (-3)^2}$

$f(-3) = -\sqrt{16}$

$f(-3) = -4$

12. $x^2 - 16 \geq 0$

$(x - 4)(x + 4) \geq 0$

Critical values 4 and -4.

Domain $\{x \mid x \geq 4 \text{ or } x \leq -4\}$

13.

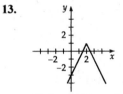

f is increasing on $(-\infty, 2]$

f is decreasing on $[2, \infty)$

14.   First shift the graph of $f(x)$ horizontally 2 units to the left. Next, reflect the graph across the x-axis. Finally, shift the graph vertically down 1 unit.

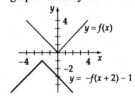

15.

a. $f(-x) = (-x)^4 - (-x)^2$

$\qquad\qquad = x^4 - x^2$

$\qquad\qquad = f(x)$

The function of $f(x) = x^4 - x^2$ is an even function.

b. $f(-x) = (-x)^3 - (-x)$

$\qquad\qquad = -x^3 + x$

$\qquad\qquad = -(x^3 - x)$

$\qquad\qquad = -f(x)$

The function $f(x) = x^3 - x$ is an odd function.

c. $f(-x) = -x - 1$

The function $f(x) = x - 1$ is neither an even nor an odd function.

Thus, only **b** defines an odd function.

Copyright © Houghton Mifflin Company. All rights reserved.

16.
$$(f+g)(x) = (x^2-1)+(x-2)$$
$$= x^2+x-3$$
$$\left(\frac{f}{g}\right)(x) = \frac{x^2-1}{x-2}, \quad x \neq 2$$

17.
$$\frac{f(x+h)-f(x)}{h} = \frac{\left[(x+h)^2+1\right]-(x^2+1)}{h}$$
$$= \frac{x^2+2xh+h^2+1-x^2-1}{h}$$
$$= \frac{2xh+h^2}{h}$$
$$= 2x+h$$

18.
$$(f \circ g)(x) = (\sqrt{x-2})^2 - 2\sqrt{x-2}+1$$
$$= x-2-2\sqrt{x-2}+1$$
$$= x-2\sqrt{x-2}-1$$

19.
$$y = \frac{x}{x+1}$$

Interchange x and y. Then solve for y.
$$x = \frac{y}{y+1}$$
$$x(y+1) = y$$
$$xy+x = y$$
$$xy-y = -x$$
$$y(x-1) = -x$$
$$y = \frac{-x}{x-1} = \frac{x}{1-x}$$
$$f^{-1}(x) = \frac{x}{1-x}$$

20. a. Enter the data on your calculator. The technique for a TI-83 is illustrated here.

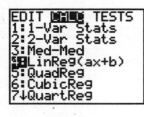 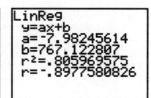

$$y = -7.98245614x + 767.122807$$

b.
$$y = -7.98245614(89) + 767.122807$$
$$\approx 57 \text{ Calories}$$

Copyright © Houghton Mifflin Company. All rights reserved.

Chapter 2
Trigonometric Functions

Section 2.1

1.
$$90° - 15° = 75°$$
$$180° - 15° = 165°$$

3.
$$\frac{\begin{array}{r}90°\\-70°15'\end{array}}{} = \frac{\begin{array}{r}89°60'\\-70°15'\\\hline 19°45'\end{array}}{} \qquad \frac{\begin{array}{r}180°\\-70°15'\end{array}}{} = \frac{\begin{array}{r}179°60'\\-70°15'\\\hline 109°45'\end{array}}{}$$

5.
$$\frac{\begin{array}{r}90°\\-56°33'15''\end{array}}{} = \frac{\begin{array}{r}89°59'60''\\-56°33'15''\\\hline 33°26'45''\end{array}}{} \qquad \frac{\begin{array}{r}180°\\-56°33'15''\end{array}}{} = \frac{\begin{array}{r}179°59'60''\\-56°33'15''\\\hline 123°26'45''\end{array}}{}$$

7.
$$\frac{\pi}{2} - 1$$
$$\pi - 1$$

9.
$$\frac{\pi}{2} - \frac{\pi}{4} = \frac{\pi}{4}$$
$$\pi - \frac{\pi}{4} = \frac{3\pi}{4}$$

11.
$$\frac{\pi}{2} - \frac{2\pi}{5} = \frac{\pi}{10}$$
$$\pi - \frac{2\pi}{5} = \frac{3\pi}{5}$$

13.
$$610° = 250° + 360°$$
α is a quadrant III angle coterminal with an angle of measure 250°.

15.
$$-975° = 105° - 3 \cdot 360°$$
α is a quadrant II angle coterminal with an angle of measure 105°.

17.
$$2456° = 296° + 6 \cdot 360°$$
α is a quadrant IV angle coterminal with an angle of measure 296°.

19. On a TI-83 graphing calculator, the degree symbol, °, and the DMS function are located in the ANGLE menu.

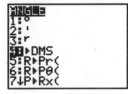 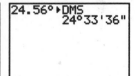

$$24.56° = 24°33'36''$$

21. On a TI-83 graphing calculator, the degree symbol, °, and the DMS function are located in the ANGLE menu.

$$64.158° = 64°9'28.8''$$

23. On a TI-83 graphing calculator, the degree symbol, °, and the DMS function are located in the ANGLE menu.

$$3.402° = 3°24'7.2''$$

25. A TI-83 calculator needs to be in degree mode to convert a DMS measure to its equivalent degree measure. On a TI-83 both the degree symbol, °, and the minute symbol, ', are located in the ANGLE menu. The second symbol, ", is entered by pressing ALPHA followed by ["] which is located on the plus sign, [+], key.

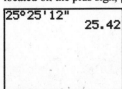

$$25°25'12'' = 25.42°$$

27. A TI-83 calculator needs to be in degree mode to convert a DMS measure to its equivalent degree measure. On a TI-83 both the degree symbol, °, and the minute symbol, ', are located in the ANGLE menu. The second symbol, ", is entered by pressing ALPHA followed by ["] which is located on the plus sign, [+], key.

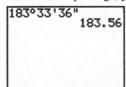

$$183°33'36'' = 183.56°$$

Copyright © Houghton Mifflin Company. All rights reserved.

29. A TI-83 calculator needs to be in degree mode to convert a DMS measure to its equivalent degree measure. On a TI-83 both the degree symbol, °, and the minute symbol, ', are located in the ANGLE menu. The second symbol, ", is entered by pressing ALPHA followed by ["] which is located on the plus sign, [+], key.

```
211°46'48"
          211.78
```

$211°46'48'' = 211.78°$

31.
$$30° = 30°\left(\frac{\pi}{180°}\right) = \frac{\pi}{6}$$

33.
$$90° = 90°\left(\frac{\pi}{180°}\right) = \frac{\pi}{2}$$

35.
$$165° = 165°\left(\frac{\pi}{180°}\right) = \frac{11\pi}{12}$$

37.
$$420° = 420°\left(\frac{\pi}{180°}\right) = \frac{7\pi}{3}$$

39.
$$585° = 585°\left(\frac{\pi}{180°}\right) = \frac{13\pi}{4}$$

41.
$$\frac{\pi}{5} = \frac{\pi}{5}\left(\frac{180°}{\pi}\right) = 36°$$

43.
$$\frac{\pi}{6} = \frac{\pi}{6}\left(\frac{180°}{\pi}\right) = 30°$$

45.
$$\frac{3\pi}{8} = \frac{3\pi}{8}\left(\frac{180°}{\pi}\right) = 67.5°$$

47.
$$\frac{11\pi}{3} = \frac{11\pi}{3}\left(\frac{180°}{\pi}\right) = 660°$$

49.
$$1.5 = 1.5\left(\frac{180°}{\pi}\right) \approx 85.94°$$

51.
$$133° = 133°\left(\frac{\pi}{180°}\right) \approx 2.32$$

53.
$$8.25 = 8.25\left(\frac{180°}{\pi}\right) \approx 472.69°$$

55.
$$\theta = \frac{s}{r}$$
$$= \frac{8}{2} = 4$$
$$= 4\left(\frac{180°}{\pi}\right) \approx 229.18°$$

57.
$$\theta = \frac{s}{r}$$
$$= \frac{12.4}{5.2} \approx 2.38$$
$$= 2.38\left(\frac{180°}{\pi}\right) \approx 136.63°$$

59.
$$s = r\theta$$
$$= (8)\frac{\pi}{4}$$
$$\approx 6.28 \text{ in.}$$

61.
$$s = r\theta$$
$$= 25\,(42°)\left(\frac{\pi}{180°}\right)$$
$$\approx 18.33 \text{ cm}$$

63.
$$\theta = \frac{3}{2}(2\pi)$$
$$= 3\pi$$

65.
$$\theta_2 = \frac{r_1}{r_2}\theta_1$$
$$= \frac{14}{28}(150°)\left(\frac{\pi}{180°}\right)$$
$$= \frac{5\pi}{12} \text{ radians or } 75°$$

67.
$$\omega = \frac{\theta}{t}$$
$$= \frac{2\pi}{60}$$
$$= \frac{\pi}{30} \text{ radian/sec}$$

69.
$$\omega = \frac{\theta}{t}$$
$$= \frac{50(2\pi)}{60}$$
$$= \frac{5\pi}{3} \text{ radians/sec}$$

71.
$$\omega = \frac{\theta}{t}$$
$$= \frac{2\pi\left(33\frac{1}{3}\right)}{60}$$
$$= \frac{10\pi}{9} \text{ radians/sec}$$
$$\approx 3.49 \text{ radians per second}$$

73.
$$v = \omega r$$
$$= \frac{450 \cdot 2\pi \cdot 60 \cdot 15}{12 \cdot 5280}$$
$$\approx 40 \text{ mph}$$

Copyright © Houghton Mifflin Company. All rights reserved.

75.
$$r_1\theta_1 = r_2\theta_2$$
$$(3.5)(150 \cdot 2\pi) = (1.75)\theta_2$$
$$600\pi = \theta_2$$
$$300(2\pi) = \theta_2$$

The rear gear is making 300 revolutions.
The rear gear and tire are making same number of revolutions
Tire is 12 inches = 1 ft
$$s = 1\text{ ft}(300)(2\pi) = 1885\text{ ft}$$

79. $s = r\theta$

$$= 93,000,000(31')\left(\frac{1°}{60'}\right)\left(\frac{\pi}{180°}\right)$$

$$\approx 840,000\text{ mi}$$

77.

a. $v = \dfrac{s}{t}$

$$= \frac{3460}{2\frac{59}{60}}$$

$$= 1160\text{ mph}$$

b. $s = r\theta$

$$\theta = \frac{s}{r} = \frac{3460}{(3960+10)} = 0.8715$$

$$\omega = \frac{\theta}{t} = \frac{0.8715}{2\frac{59}{60}} \approx 0.29\text{ radians per hour}$$

c. 1 PM + 2hr 59min = 3:59 PM London time zone
3:59 − 5hr = 10:59 AM NY time zone

81.

a. $\omega = \dfrac{\theta}{t}$

$$= \frac{2\pi}{1.61\text{ hours}}$$

$$\approx 3.9\text{ radians per hour}$$

b. $v = \dfrac{s}{t}$

$$= \frac{2\pi r}{1.61\text{ hours}}$$

$$= \frac{2\pi(625\text{ km} + 6370\text{ km})}{1.61\text{ hours}}$$

$$= \frac{2\pi(6995\text{ km})}{1.61\text{ hours}}$$

$$\approx 27,300\text{ km per hour}$$

83.

a. When the rear tire makes one revolution, the bicycle
travels $s = r\theta = 2\pi r = 2\pi(30\text{ inches}) = 60\pi$ inches.

The angular velocity of point A is
$$\omega = \frac{\theta}{t} = \frac{2\pi}{t} = 2\left(\frac{\pi}{t}\right).$$

When the bicycle travels 60π inches, point B on the front
tire travels though an angle of
$$\theta = \frac{s}{r} = \frac{60\pi\text{ inches}}{20\text{ inches}} = 3\pi.$$

The angular velocity of point B is
$$\omega = \frac{\theta}{t} = \frac{3\pi}{t} = 3\left(\frac{\pi}{t}\right).$$

Thus, point B has the greater angular velocity.

b. Point A and point B travel a linear distance of
60π inches in the same amount of time. Therefore,
both points have the same linear velocity.

Copyright © Houghton Mifflin Company. All rights reserved.

85. **a.**

$$1' = 1' \cdot \left(\frac{1°}{60'}\right) \cdot \left(\frac{\pi}{180°}\right) = \frac{\pi}{10,800} \text{ radians}$$

1 nautical mile $= s = r\theta = (3960 \text{ statute miles})\left(\frac{\pi}{10,800}\right) \approx 1.15 \text{ statute miles}$

b. Earth's circumference $= 2\pi r \approx 2\pi(3960 \text{ statute miles})\left(\frac{1 \text{ nautical mile}}{1.15 \text{ statute miles}}\right) \approx \frac{7920\pi}{1.15} \text{ nautical miles}$

The question, then, is what percent of $\frac{7920\pi}{1.15}$ is 2217.

$$\frac{2217}{7920\pi/1.15} \approx 0.10 = 10\%$$

●   **Connecting Concepts**

87. $A = \frac{1}{2}r^2\theta$

$= \frac{1}{2}(5^2)\left(\frac{\pi}{3}\right)$

$\approx 13 \text{ in}^2$

89. $A = \frac{1}{2}r^2\theta$

$= \frac{1}{2}(120)^2 0.65$

$= 4680 \text{ cm}^2$

91. $A = \frac{1}{2}r^2\theta - \frac{1}{2}bh$

$A = \frac{1}{2}\cdot 9^2 \cdot \frac{\pi}{2} - \frac{1}{2}(9)(9)$

$\approx 63.6 - 40.5$

$= 23.1 \text{ in}^2$

93.

$$25°47' = 25° + 47'\left(\frac{1°}{60'}\right)$$

$$= 25\frac{47}{60}°$$

Convert to radians.

$$25\frac{47}{60}° \cdot \frac{\pi}{180°} = \frac{1547\pi}{10,800} \text{ radians}$$

$$s = r\theta$$

$$s = 3960\left(\frac{1547\pi}{10,800}\right)$$

$$\approx 1780$$

To the nearest 10 miles, Miami is 1780 miles north of the equator.

●   **Prepare for Section 2.2**

95. $\frac{1}{\sqrt{3}} \cdot \frac{\sqrt{3}}{\sqrt{3}} = \frac{\sqrt{3}}{3}$

96. $\frac{2}{\sqrt{2}} \cdot \frac{\sqrt{2}}{\sqrt{2}} = \frac{2\sqrt{2}}{2} = \sqrt{2}$

97. $a \div \left(\frac{a}{2}\right) = a \cdot \left(\frac{2}{a}\right) = 2$

98. $\left(\frac{a}{2}\right) \div \left(\frac{\sqrt{3}}{2}a\right) = \left(\frac{a}{2}\right) \cdot \left(\frac{2}{a\sqrt{3}}\right)$

$= \frac{1}{\sqrt{3}} \cdot \frac{\sqrt{3}}{\sqrt{3}}$

$= \frac{\sqrt{3}}{3}$

99. $\frac{\sqrt{2}}{2} = \frac{x}{5}$

$\frac{5\sqrt{2}}{2} = x$

$x \approx 3.54$

100. $\frac{\sqrt{3}}{3} = \frac{x}{18}$

$\frac{18\sqrt{3}}{3} = x$

$x = 6\sqrt{3}$

$x \approx 10.39$

Copyright © Houghton Mifflin Company. All rights reserved.

Section 2.2

1.
$$r = \sqrt{5^2 + 12^2}$$
$$r = \sqrt{25 + 144} + \sqrt{169}$$
$$r = 13$$

$$\sin\theta = \frac{y}{r} = \frac{12}{13}$$
$$\cos\theta = \frac{x}{r} = \frac{5}{13}$$
$$\tan\theta = \frac{y}{x} = \frac{12}{5}$$

$$\csc\theta = \frac{r}{y} = \frac{13}{12}$$
$$\sec\theta = \frac{r}{x} = \frac{13}{5}$$
$$\cot\theta = \frac{x}{y} = \frac{5}{12}$$

3.
$$x = \sqrt{7^2 - 4^2}$$
$$x = \sqrt{49 - 16} = \sqrt{33}$$

$$\sin\theta = \frac{y}{r} = \frac{4}{7}$$
$$\cos\theta = \frac{x}{r} = \frac{\sqrt{33}}{7}$$
$$\tan\theta = \frac{y}{x} = \frac{4}{\sqrt{33}} = \frac{4\sqrt{33}}{33}$$

$$\csc\theta = \frac{r}{y} = \frac{7}{4}$$
$$\sec\theta = \frac{r}{x} = \frac{7}{\sqrt{33}} = \frac{7\sqrt{33}}{33}$$
$$\cot\theta = \frac{x}{y} = \frac{\sqrt{33}}{4}$$

5.
$$r = \sqrt{2^2 + 5^2}$$
$$r = \sqrt{4 + 25} = \sqrt{29}$$

$$\sin\theta = \frac{y}{r} = \frac{5}{\sqrt{29}} = \frac{5\sqrt{29}}{29}$$
$$\cos\theta = \frac{x}{r} = \frac{2}{\sqrt{29}} = \frac{2\sqrt{29}}{29}$$
$$\tan\theta = \frac{y}{x} = \frac{5}{2}$$

$$\csc\theta = \frac{r}{y} = \frac{\sqrt{29}}{5}$$
$$\sec\theta = \frac{r}{x} = \frac{\sqrt{29}}{2}$$
$$\cot\theta = \frac{x}{y} = \frac{2}{5}$$

7.
$$x = \sqrt{2 + \left(\sqrt{3}\right)^2}$$
$$x = \sqrt{4 + 3} = \sqrt{7}$$

$$\sin\theta = \frac{y}{r} = \frac{\sqrt{3}}{\sqrt{7}} = \frac{\sqrt{21}}{7}$$
$$\cos\theta = \frac{x}{r} = \frac{2}{\sqrt{7}} = \frac{2\sqrt{7}}{7}$$
$$\tan\theta = \frac{y}{x} = \frac{\sqrt{3}}{2}$$

$$\csc\theta = \frac{r}{y} = \frac{\sqrt{7}}{\sqrt{3}} = \frac{\sqrt{21}}{3}$$
$$\sec\theta = \frac{r}{x} = \frac{\sqrt{7}}{2}$$
$$\cot\theta = \frac{x}{y} = \frac{2}{\sqrt{3}} = \frac{2\sqrt{3}}{3}$$

9.
$$y = \sqrt{\left(\sqrt{15}\right)^2 - \left(\sqrt{7}\right)^2}$$
$$y = \sqrt{15 - 7} = \sqrt{8}$$
$$y = 2\sqrt{2}$$

$$\sin\theta = \frac{y}{r} = \frac{2\sqrt{2}}{\sqrt{15}} = \frac{2\sqrt{30}}{15}$$
$$\cos\theta = \frac{x}{r} = \frac{\sqrt{7}}{\sqrt{15}} = \frac{\sqrt{105}}{15}$$
$$\tan\theta = \frac{y}{x} = \frac{2\sqrt{2}}{\sqrt{7}} = \frac{2\sqrt{14}}{7}$$

$$\csc\theta = \frac{r}{y} = \frac{\sqrt{15}}{2\sqrt{2}} = \frac{\sqrt{30}}{4}$$
$$\sec\theta = \frac{r}{x} = \frac{\sqrt{15}}{\sqrt{7}} = \frac{\sqrt{105}}{7}$$
$$\cot\theta = \frac{x}{y} = \frac{\sqrt{7}}{2\sqrt{2}} = \frac{\sqrt{14}}{4}$$

11.
$$\text{opposite side} = \sqrt{6^2 - 3^2}$$
$$= \sqrt{36 - 9}$$
$$= \sqrt{27} = 3\sqrt{3}$$

$$\sin\theta = \frac{\text{opp}}{\text{hyp}} = \frac{3\sqrt{3}}{6} = \frac{\sqrt{3}}{2}$$
$$\cos\theta = \frac{\text{adj}}{\text{hyp}} = \frac{3}{6} = \frac{1}{2}$$
$$\tan\theta = \frac{\text{opp}}{\text{adj}} = \frac{3\sqrt{3}}{3} = \sqrt{3}$$

$$\csc\theta = \frac{\text{hyp}}{\text{opp}} = \frac{6}{3\sqrt{3}} = \frac{2}{\sqrt{3}} = \frac{2\sqrt{3}}{3}$$
$$\sec\theta = \frac{\text{hyp}}{\text{adj}} = \frac{6}{3} = 2$$
$$\cot\theta = \frac{\text{adj}}{\text{opp}} = \frac{3}{3\sqrt{3}} = \frac{1}{\sqrt{3}} = \frac{\sqrt{3}}{3}$$

Copyright © Houghton Mifflin Company. All rights reserved.

13.

$$\text{hypotenuse} = \sqrt{5^2 + 6^2}$$
$$= \sqrt{25 + 36}$$
$$= \sqrt{61}$$

$$\sin\theta = \frac{\text{opp}}{\text{hyp}} = \frac{6}{\sqrt{61}} = \frac{6\sqrt{61}}{61} \qquad \csc\theta = \frac{\text{hyp}}{\text{opp}} = \frac{\sqrt{61}}{6}$$

$$\cos\theta = \frac{\text{adj}}{\text{hyp}} = \frac{5}{\sqrt{61}} = \frac{5\sqrt{61}}{61} \qquad \sec\theta = \frac{\text{hyp}}{\text{adj}} = \frac{\sqrt{61}}{5}$$

$$\tan\theta = \frac{\text{opp}}{\text{adj}} = \frac{6}{5} \qquad\qquad\qquad \cot\theta = \frac{\text{adj}}{\text{opp}} = \frac{5}{6}$$

For exercises 15 to 17, since $\sin\theta = \frac{y}{r} = \frac{3}{5}$, $y = 3$, $r = 5$, and $x = \sqrt{5^2 - 3^2} = 4$.

15. $\tan\theta = \frac{y}{x} = \frac{3}{4}$

17. $\cos\theta = \frac{x}{r} = \frac{4}{5}$

For exercises 18 to 20, since $\tan\theta = \frac{y}{x} = \frac{4}{3}$, $y = 4$, $x = 3$, and $r = \sqrt{3^2 + 4^2} = 5$.

19. $\cot\theta = \frac{x}{y} = \frac{3}{4}$

For exercises 21 to 23, since $\sec\beta = \frac{r}{x} = \frac{13}{12}$, $r = 13$, $x = 12$, and $y = \sqrt{13^2 - 12^2} = \sqrt{25} = 5$.

21. $\cos\beta = \frac{x}{r} = \frac{12}{13}$

23. $\csc\beta = \frac{r}{y} = \frac{13}{5}$

For exercises 24 to 26, since $\cos\theta\frac{x}{r} = \frac{2}{3}$, $x = 2$, $r = 3$, and $y = \sqrt{3^2 - 2^2} = \sqrt{9 - 4} = \sqrt{5}$.

25. $\sec\theta = \frac{r}{x} = \frac{3}{2}$

27. $\sin 45° + \cos 45° = \frac{\sqrt{2}}{2} + \frac{\sqrt{2}}{2} = \sqrt{2}$

29.
$$\sin 30° \cos 60° - \tan 45° = \frac{1}{2} \cdot \frac{1}{2} - 1$$
$$= \frac{1}{4} - 1$$
$$= -\frac{3}{4}$$

31. $\sin 30° \cos 60° + \tan 45° = \frac{1}{2} \cdot \frac{1}{2} + 1 = \frac{1}{4} + 1 = \frac{5}{4}$

33. $2 \sin 60° - \sec 45° \tan 60° = 2\left(\frac{\sqrt{3}}{2}\right) - \sqrt{2} \cdot \sqrt{3} = \sqrt{3} - \sqrt{6}$

35. $\sin\frac{\pi}{3} + \cos\frac{\pi}{6} = \frac{\sqrt{3}}{2} + \frac{\sqrt{3}}{2} = 2 \cdot \frac{\sqrt{3}}{2} = \sqrt{3}$

37. $\sin\frac{\pi}{4} + \tan\frac{\pi}{6} = \frac{\sqrt{2}}{2} + \frac{\sqrt{3}}{3} = \frac{3\sqrt{2} + 2\sqrt{3}}{6}$

39. $\sec\frac{\pi}{3}\cos\frac{\pi}{3} - \tan\frac{\pi}{6} = 2 \cdot \frac{1}{2} - \frac{\sqrt{3}}{3} = 1 - \frac{\sqrt{3}}{3} = \frac{3 - \sqrt{3}}{3}$

41. $2\csc\frac{\pi}{4} - \sec\frac{\pi}{3}\cos\frac{\pi}{6} = 2 \cdot \sqrt{2} - 2 \cdot \frac{\sqrt{3}}{2} = 2\sqrt{2} - \sqrt{3}$

43. $\tan 32° \approx 0.6249$

45. $\cos 63°20' \approx 0.4488$

47. $\cos 34.7° \approx 0.8221$

49. $\sec 5.9° \approx 1.0053$

51. $\tan\frac{\pi}{7} \approx 0.4816$

53. $\csc 1.2 \approx 1.0729$

55. $\cos 1.25 \approx 0.3153$

Copyright © Houghton Mifflin Company. All rights reserved.

57.

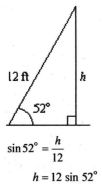

$$\sin 52° = \frac{h}{12}$$

$$h = 12 \sin 52°$$

$$h \approx 9.5 \text{ ft}$$

59.

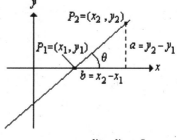

$$m = \frac{y_2 - y_1}{x_2 - x_1} = \frac{a}{b}$$

$$\tan \theta = \frac{a}{b}$$

Therefore, $\tan \theta = m$.

61.

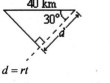

$$d = rt$$

$$t = \frac{d}{r} \qquad \cos 30° = \frac{d}{40}, \qquad d = 40 \cos 30°$$

$$t = \frac{40 \cos 30°}{10}$$

$$t \approx 3.46 \text{ h}$$

$$t \approx 3 \text{ hr } 28 \text{ min}$$

Time of closest approach: 1:00 P.M.+3 hr 28 min is
4:28 P.M.

63.

$$\cos 38° = \frac{d + 0.33}{6}$$

$$d = 6 \cos 38° + 0.33$$

$$d \approx 5.1 \text{ ft}$$

65.

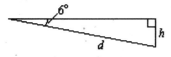

$$d = \frac{240 \text{ mi}}{\text{hr}} \left(\frac{1 \text{ hr}}{60 \text{ min}} \right) 4 \text{ min}$$

$$d = 16 \text{ mi}$$

$$\sin 6° = \frac{h}{d}$$

$$h = 16 \sin 6°$$

$$h \approx 1.7 \text{ mi}$$

67.

$$\tan 78° = \frac{h}{300}$$

$$h = 300 \tan 78°$$

$$h \approx 1411.4 \text{ ft}$$

$$h \approx 1400 \text{ ft} \quad \text{(to two significant digits)}$$

69.

$$\sin 0.056° = \frac{670,900}{d}$$

$$d = \frac{670,900}{\sin 0.056°} \text{ km}$$

$$\approx 686,000,000 \text{ km}$$

71.

$$A = \frac{1}{2} bh$$

$$= \frac{1}{2} (2a \sin \theta)(a \cos \theta)$$

$$= a^2 \sin \theta \cos \theta$$

Copyright © Houghton Mifflin Company. All rights reserved.

73.

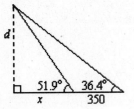

$$\tan 36.4 = \frac{d}{350 + x} \qquad \tan 51.9° = \frac{d}{x}$$

$$x = \frac{d}{\tan 51.9°}$$

$$\tan 36.4° = \frac{d}{350 + \dfrac{d}{\tan 51.9°}}$$

$$d = \frac{350 \tan 36.4°}{1 - \dfrac{\tan 36.4°}{\tan 51.9°}}$$

$$d \approx 612 \text{ ft}$$

75.

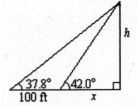

$$\tan 42° = \frac{h}{x} \qquad \tan 37.8° = \frac{h}{100 + x}$$

$$x = \frac{h}{\tan 42°} \qquad \tan 37.8° = \frac{h}{100 + \dfrac{h}{\tan 42°}}$$

$$h = \frac{100 \tan 37.8°}{1 - \dfrac{\tan 37.8°}{\tan 42.0°}}$$

$$h \approx 5.60 \times 10^2 \text{ ft}$$

$$h \approx 560 \text{ ft}$$

77. a.

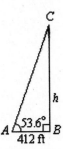

$$\tan 53.6° = \frac{h}{412}$$

$$h = 412 \tan 53.6°$$

$$h \approx 559 \text{ feet}$$

b.
$$(AC)^2 = 412^2 + 559^2$$

$$AC = \sqrt{412^2 + 559^2}$$

$$AC = \sqrt{482,225}$$

$$AC \approx 694.4 \text{ feet}$$

$$\tan 15.5° = \frac{x}{694.4}$$

$$x = 694.4 \tan 15.5°$$

$$x \approx 193 \text{ feet}$$

Copyright © Houghton Mifflin Company. All rights reserved.

79.

Consider the right triangle formed by A, B and the midpoint of AC.

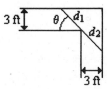

$r = \sqrt{6^2 - 3^2}$

$r = \sqrt{27}$

$r = 3\sqrt{3}$

$r \approx 5.2 \text{ m}$

81.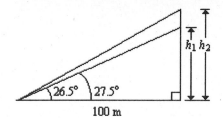

$\tan 26.5° = \dfrac{h_1}{100}$ $\tan 27.5° = \dfrac{h_2}{100}$

$h_1 = 100 \tan 26.5°$ $h_2 = 100 \tan 27.5°$

$h_1 = 49.9 \text{ m}$ $h_2 = 52.1 \text{ m}$

$49.9 \text{ m} \le h < 52.1 \text{ m}$

83.

3 ft θ d_1 d_2 3 ft

if $\theta = 45°$ $d_1 = d_2$

$\sin \theta = \dfrac{3}{d_1}$

$d_1 = \dfrac{3}{\sin 45°}$

$d = 2d_1 = \dfrac{6}{\sin 45°}$

$d \approx 8.5 \text{ ft}$

85. $-\dfrac{4}{3}$

86. $\dfrac{5}{2\sqrt{5}} \cdot \dfrac{\sqrt{5}}{\sqrt{5}} = \dfrac{5\sqrt{5}}{10} = \dfrac{\sqrt{5}}{2}$

87. $|120 - 180| = |-60| = 60$

88. $2\pi - \dfrac{9\pi}{5} = \dfrac{10\pi}{5} - \dfrac{9\pi}{5} = \dfrac{\pi}{5}$

89. $\dfrac{3}{2}\pi - \dfrac{1}{2}\pi = \dfrac{2}{2}\pi = \pi$

90. $\sqrt{(-3)^2 + (-5)^2} = \sqrt{9 + 25} = \sqrt{34}$

Copyright © Houghton Mifflin Company. All rights reserved.

Section 2.3

1.
$$x = 2, y = 3, r = \sqrt{2^2 + 3^2} = \sqrt{13}$$

$$\sin\theta = \frac{y}{r} = \frac{3}{\sqrt{13}} = \frac{3\sqrt{13}}{13} \qquad \csc\theta = \frac{\sqrt{13}}{3}$$

$$\cos\theta = \frac{x}{r} = \frac{2}{\sqrt{13}} = \frac{2\sqrt{13}}{13} \qquad \sec\theta = \frac{\sqrt{13}}{2}$$

$$\tan\theta = \frac{y}{x} = \frac{3}{2} \qquad\qquad \cot\theta = \frac{2}{3}$$

3.
$$x = -2, y = 3, r = \sqrt{(-2)^2 + (3)^2} = \sqrt{13}$$

$$\sin\theta = \frac{y}{r} = \frac{3}{\sqrt{13}} = \frac{3\sqrt{13}}{13} \qquad \csc\theta = \frac{\sqrt{13}}{3}$$

$$\cos\theta = \frac{x}{r} = \frac{-2}{\sqrt{13}} = -\frac{2\sqrt{13}}{13} \qquad \sec\theta = -\frac{\sqrt{13}}{2}$$

$$\tan\theta = \frac{y}{x} = \frac{3}{-2} = -\frac{3}{2} \qquad \cot\theta = -\frac{2}{3}$$

5.
$$x = -8, y = -5, c = \sqrt{(-8)^2 + (-5)^2} = \sqrt{89}$$

$$\sin\theta = \frac{y}{r} = \frac{-5}{\sqrt{89}} = -\frac{5\sqrt{89}}{89} \qquad\qquad \csc\theta = -\frac{\sqrt{89}}{5}$$

$$\cos\theta = \frac{x}{r} = \frac{-8}{\sqrt{89}} = -\frac{8\sqrt{89}}{89} \qquad\qquad \sec\theta = -\frac{\sqrt{89}}{8}$$

$$\tan\theta = \frac{y}{x} = \frac{-5}{-8} = \frac{5}{8} \qquad\qquad \cot\theta = \frac{8}{5}$$

7.
$$x = -5, y = 0, r = \sqrt{(-5)^2 + (0)^2} = 5$$

$$\sin\theta = \frac{y}{r} = \frac{0}{5} = 0 \qquad\qquad \csc\theta \text{ is undefined}$$

$$\cos\theta = \frac{x}{r} = \frac{-5}{5} = -1 \qquad\qquad \sec\theta = -1$$

$$\tan\theta = \frac{y}{x} = \frac{0}{-5} = 0 \qquad\qquad \cot\theta \text{ is undefined}$$

9. $\sin\theta > 0$ in quadrants I and II.
$\cos\theta > 0$ in quadrants I and IV.
quadrant I

11. $\cos\theta > 0$ in quadrants I and IV.
$\tan\theta < 0$ in quadrants II and IV.
quadrant IV

13. $\sin\theta < 0$ in quadrants III and IV.
$\cos\theta < 0$ in quadrants II and III.
quadrant III

15.
$$\sin\theta = -\frac{1}{2} = \frac{y}{r}, y = -1, r = 2, x = \pm\sqrt{2^2 - (-1)^2} = \pm\sqrt{3}, x = -\sqrt{3} \text{ in quadrant III}, \tan\theta = \frac{y}{x} = \frac{-y}{-\sqrt{3}} = \frac{\sqrt{3}}{3}$$

17.
$$\csc\theta = \sqrt{2} = \frac{r}{y}, r = \sqrt{2}, y = 1, x = \pm\sqrt{(\sqrt{2})^2 - 1^2} = \pm 1, x = -1 \text{ in quadrant II}, \cot\theta = \frac{-1}{1} = -1$$

19.
$$\theta \text{ is in quadrant IV}, \sin\theta = -\frac{1}{2} = \frac{y}{r}, y = -1, r = 2, x = \sqrt{2^2 - 1^2} = \sqrt{3}, \tan\theta = \frac{-1}{\sqrt{3}} = -\frac{\sqrt{3}}{3}$$

21.
$$\cos\theta = \frac{1}{2}, \theta \text{ is in quadrant I or IV.}$$

$$\tan\theta = \sqrt{3}, \theta \text{ is in quadrant I or III.}$$

$$\theta \text{ is in quadrant I}, x = 1, y = \sqrt{3}, r = 2$$

$$\csc\theta = \frac{r}{y} = \frac{2}{\sqrt{3}} = \frac{2\sqrt{3}}{3}$$

23.
$$\cos\theta = -\frac{1}{2}, \theta \text{ is in quadrant II or III.}$$

$$\sin\theta = \frac{\sqrt{3}}{2}, \theta \text{ is in quadrant I or II.}$$

$$\theta \text{ is in quadrant II}, x = -1, y = \sqrt{3}, r = 2$$

$$\cot\theta = \frac{x}{y} = \frac{-1}{\sqrt{3}} = -\frac{\sqrt{3}}{3}$$

Copyright © Houghton Mifflin Company. All rights reserved.

25.

$\theta = 160°$

Since $90° < \theta < 180°$,

$\theta + \theta' = 180°$

$\theta' = 20°$

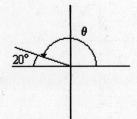

27.

$\theta = 351°$

Since $270° < \theta < 360°$,

$\theta = \theta' = 360°$

$\theta' = 9°$

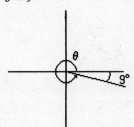

29.

$\theta = \dfrac{11\pi}{5}$

$\theta > 2\pi = \dfrac{10\pi}{5}$,

θ is coterminal with $\alpha = \dfrac{11\pi}{5} - \dfrac{10\pi}{5} = \dfrac{\pi}{5}$.

Since $0 < \alpha < \dfrac{\pi}{2}$,

$\alpha' = \alpha = \theta'$

$\theta' = \dfrac{\pi}{5}$

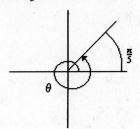

31.

$\theta = \dfrac{8}{3}$

Since $\dfrac{\pi}{2} < \theta < \pi$,

$\theta + \theta' = \pi$

$\theta' = \pi - \dfrac{8}{3}$

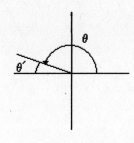

Copyright © Houghton Mifflin Company. All rights reserved.

33. $\theta = 1406° = 326° + 3 \cdot 360°$

θ is coterminal with $\alpha = 326°$.

Since $270° < \alpha < 360°$,

$\alpha + \alpha' = 360°$

$\alpha' = 34°$

$\theta' = 34°$

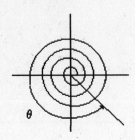

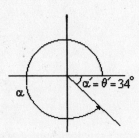

35. $\theta = -475° = 245° - 2 \cdot 360°$

θ is coterminal with $\alpha = 245°$

Since $180° < \alpha < 270°$,

$\alpha' + 180° = \alpha$

$\alpha' = 245° - 180°$

$\alpha' = 65°$

$\theta' = 65°$

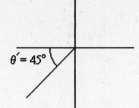

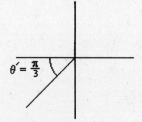

37. $\theta = 225°$ is in quadrant III.

$225° - 180° = 45°$ so $\theta' = 45°$.

Thus, $\sin 225° = -\sin 45° = -\dfrac{\sqrt{2}}{2}$.

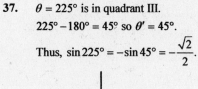

39. $\theta = 405°$ is in quadrant I.

$405° - 360° = 45°$ so $\theta' = 45°$.

Thus, $\tan 405° = \tan 45° = 1$.

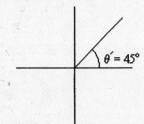

41. $\theta = \dfrac{4}{3}\pi$ is in quadrant III.

$\dfrac{4}{3}\pi - \pi = \dfrac{\pi}{3}$ so $\theta' = \dfrac{\pi}{3}$.

Thus, $\csc \dfrac{4\pi}{3} = \dfrac{1}{\sin 4\pi/3} = \dfrac{1}{-\sin \pi/3}$

$= \dfrac{1}{-\sqrt{3}/2} = -\dfrac{2}{\sqrt{3}} = -\dfrac{2\sqrt{3}}{3}$.

Copyright © Houghton Mifflin Company. All rights reserved.

43. $\theta = \dfrac{17\pi}{4} = \dfrac{16\pi}{4} + \dfrac{\pi}{4}$ is coterminal

with $\dfrac{\pi}{4}$ in quadrant I and $\theta' = \dfrac{\pi}{4}$,

so $\cos\dfrac{17\pi}{4} = \cos\dfrac{\pi}{4} = \dfrac{\sqrt{2}}{2}$.

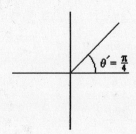

45. $\theta = 765° = 720° + 45°$ is coterminal

with $45°$ in quadrant I and $\theta' = 45°$,

so $\sec 765° = \sec 45° = \dfrac{1}{\cos 45°} = \dfrac{1}{\sqrt{2}/2} = \sqrt{2}.$

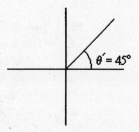

47. $\theta = 540° = 360° + 180°$ is coterminal

with $180°$, so $\cot 540° = \cot 180° = \dfrac{\cos 180°}{\sin 180°} = \dfrac{-1}{0},$

which is undefined.

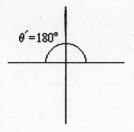

49. $\sin 127° \approx 0.798636$

51. $\cos(-116°) \approx -0.438371$

53. $\sec 578° \approx -1.26902$

55. $\sin\left(-\dfrac{\pi}{5}\right) \approx -0.587785$

57. $\csc\dfrac{9\pi}{5} \approx -1.70130$

59. $\sec(-4.45) \approx -3.85522$

61. $\sin 210° - \cos 330° \tan 330° = -\dfrac{1}{2} - \dfrac{\sqrt{3}}{2}\left(-\dfrac{\sqrt{3}}{3}\right) = -\dfrac{1}{2} + \dfrac{1}{2} = 0$

63. $\sin^2 30° + \cos^2 30° = \left(\dfrac{1}{2}\right)^\circ + \left(\dfrac{\sqrt{3}}{2}\right)^2 = \dfrac{1}{4} + \dfrac{3}{4} = 1$

65. $\sin\dfrac{3\pi}{2}\tan\dfrac{\pi}{4} - \cos\dfrac{\pi}{3} = (-1)(1) - \dfrac{1}{2} = -1 - \dfrac{1}{2} = -\dfrac{3}{2}$

67. $\sin^2\dfrac{5\pi}{4} + \cos^2\dfrac{5\pi}{4} = \left(-\dfrac{\sqrt{2}}{2}\right)^2 + \left(-\dfrac{\sqrt{2}}{2}\right)^2 = \dfrac{1}{2} + \dfrac{1}{2} = 1$

Copyright © Houghton Mifflin Company. All rights reserved.

69. $\sin\theta = \dfrac{1}{2}$, θ is in quadrant I or quadrant II

$\theta = 30°, 150°$

71. $\cos\theta = \dfrac{-\sqrt{3}}{2}$, θ is in quadrant II or quadrant III

$\theta = 150°, 120°$

73. $\csc\theta = -\sqrt{2}$
θ is in quadrant III or IV

$\theta = 225°, 315°$

75. $\tan\theta = -1$
θ is in quadrant II or IV

$\theta = \dfrac{3\pi}{4}, \dfrac{7\pi}{4}$

77. $\tan\theta = \dfrac{-\sqrt{3}}{3}$
θ is in quadrant II or IV

$\theta = \dfrac{5\pi}{6}, \dfrac{11\pi}{6}$

79. $\sin\theta = \dfrac{\sqrt{3}}{2}$
θ is in quadrant I or II

$\theta = \dfrac{\pi}{3}, \dfrac{2\pi}{3}$

81. $1 + \tan^2\theta = \sec^2\theta$

$1 + \dfrac{y^2}{x^2} = \dfrac{x^2 + y^2}{x^2}$

$= \dfrac{r^2}{x^2}$

$= \sec^2\theta$

83. $\tan\theta = \dfrac{\sin\theta}{\cos\theta}$

$= \dfrac{\dfrac{y}{r}}{\dfrac{x}{r}}$

$= \dfrac{y}{r} \cdot \dfrac{r}{x}$

$\tan\theta = \dfrac{y}{x}$

85.

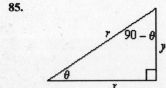

$\cos(90° - \theta) = \sin\theta$

$\dfrac{y}{r} = \dfrac{y}{r}$

87.

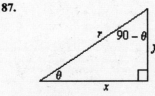

$\tan(90° - \theta) = \cot\theta$

$\dfrac{x}{y} = \dfrac{x}{y}$

89.

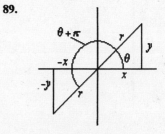

$\sin(\theta + \pi) = -\dfrac{y}{r}$

$= -\sin\theta$

91. $x^2 + y^2 = 1$
$(0)^2 + (1)^2 = 1$
Yes

92. $x^2 + y^2 = 1$

$\left(\dfrac{1}{2}\right)^2 + \left(\dfrac{\sqrt{3}}{2}\right)^2 = 1$

$\dfrac{1}{4} + \dfrac{3}{4} = 1$

Yes

93. $x^2 + y^2 = 1$

$\left(\dfrac{\sqrt{2}}{2}\right)^2 + \left(\dfrac{\sqrt{3}}{2}\right)^2 = 1$

$\dfrac{2}{4} + \dfrac{3}{4} \neq 1$

94. $C = 2\pi r$
$= 2\pi(1)$
$= 2\pi$

95. even

96. neither

Copyright © Houghton Mifflin Company. All rights reserved.

Section 2.4

1.

$t = \dfrac{\pi}{6}$

$y = \sin t \qquad\qquad x = \cos t$

$\quad = \sin\dfrac{\pi}{6} \qquad\qquad = \cos\dfrac{\pi}{6}$

$\quad = \dfrac{1}{2} \qquad\qquad\quad = \dfrac{\sqrt{3}}{2}$

The point on the unit circle corresponding to $t = \frac{\pi}{6}$ is

$\left(\dfrac{\sqrt{3}}{2}, \dfrac{1}{2}\right)$.

3.

$t = \dfrac{7\pi}{6}$

$y = \sin t \qquad\qquad x = \cos t$

$\quad = \sin\dfrac{7\pi}{6} \qquad\qquad = \cos\dfrac{7\pi}{6}$

$\quad = -\dfrac{1}{2} \qquad\qquad = -\dfrac{\sqrt{3}}{2}$

The point on the unit circle corresponding to $t = \frac{7\pi}{6}$ is

$\left(-\dfrac{\sqrt{3}}{2}, -\dfrac{1}{2}\right)$.

5.

$t = \dfrac{5\pi}{3}$

$y = \sin t \qquad\qquad x = \cos t$

$\quad = \sin\dfrac{5\pi}{3} \qquad\qquad = \cos\dfrac{5\pi}{3}$

$\quad = -\dfrac{\sqrt{3}}{2} \qquad\qquad = \dfrac{1}{2}$

The point on the unit circle corresponding to $t = \frac{5\pi}{3}$ is

$\left(\dfrac{1}{2}, -\dfrac{\sqrt{3}}{2}\right)$.

7.

$t = \dfrac{11\pi}{6}$

$y = \sin t \qquad\qquad x = \cos t$

$\quad = \sin\dfrac{11\pi}{6} \qquad\qquad = \cos\dfrac{11\pi}{6}$

$\quad = -\dfrac{1}{2} \qquad\qquad = \dfrac{\sqrt{3}}{2}$

The point on the unit circle corresponding to $t = \frac{11\pi}{6}$ is

$\left(\dfrac{\sqrt{3}}{2}, -\dfrac{1}{2}\right)$.

9.

$t = \pi$

$y = \sin t \qquad\qquad x = \cos t$

$\quad = \sin\pi \qquad\qquad = \cos\pi$

$\quad = 0 \qquad\qquad\quad = -1$

The point on the unit circle corresponding to $t = \pi$ is $(-1, 0)$.

11.

$t = -\dfrac{2\pi}{3}$

$y = \sin t \qquad\qquad x = \cos t$

$\quad = \sin\left(-\dfrac{2\pi}{3}\right) \qquad = \cos\left(-\dfrac{2\pi}{3}\right)$

$\quad = -\sin\dfrac{2\pi}{3} \qquad\quad = \cos\dfrac{2\pi}{3}$

$\quad = -\dfrac{\sqrt{3}}{2} \qquad\qquad = -\dfrac{1}{2}$

The point on the unit circle corresponding to $t = -\frac{2\pi}{3}$ is

$\left(-\dfrac{1}{2}, -\dfrac{\sqrt{3}}{2}\right)$.

13. $\tan\dfrac{11\pi}{6} = -\tan\dfrac{\pi}{6} = -\dfrac{\sqrt{3}}{3}$

15. $\cos\left(-\dfrac{2\pi}{3}\right) = -\cos\dfrac{\pi}{3} = -\dfrac{1}{2}$

17. $\csc\left(-\dfrac{\pi}{3}\right) = -\csc\dfrac{\pi}{3} = -\dfrac{2\sqrt{3}}{3}$

19. $\sin\dfrac{3\pi}{2} = -\sin\dfrac{\pi}{2} = -1$

21. $\sec\left(-\dfrac{7\pi}{6}\right) = -\sec\dfrac{\pi}{6} = -\dfrac{2\sqrt{3}}{3}$

23. $\sin 1.22 \approx 0.9391$

25. $\csc(-1.05) \approx -1.1528$

27. $\tan\dfrac{11\pi}{12} \approx -0.2679$

29. $\cos\left(-\dfrac{\pi}{5}\right) \approx 0.8090$

Copyright © Houghton Mifflin Company. All rights reserved.

31. $\sec 1.55 \approx 48.0889$

33. **a.** $\sin 2 \approx 0.9$

   b. $\cos 2 \approx -0.4$

35. **a.** $\sin 5.4 \approx -0.8$

   b. $\cos 5.4 \approx 0.6$

37. $\sin t = 0.4$ when $t = 0.4$ or 2.7

39. $\sin t = -0.3$ when $t = 3.4$ or 6.0

41. $f(-x) = -4\sin(-x)$
$$= 4\sin x$$
$$= -f(x)$$

The function defined by
$f(x) = -4\sin x$ is an odd function.

43. $G(-x) = \sin(-x) + \cos(-x)$
$$= -\sin x + \cos x$$

The function defined by
$G(x) = \sin x + \cos x$ is neither an even
nor an odd function.

45. $S(-x) = \dfrac{\sin(-x)}{-x}$
$$= -\frac{\sin x}{-x} = \frac{\sin x}{x}$$
$$= S(x)$$

The function defined by
$S(x) = \dfrac{\sin(x)}{x}$ is an even function.

47. $V(-x) = 2\sin(-x)\cos(-x)$
$$= -2\sin x \cos x$$
$$= -V(x)$$

The function defined by
$V(x) = 2\sin x \cos x$ is an odd function.

49.

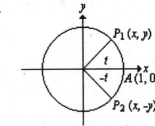

$\cos t = x$
$\cos(-t) = x$
$\cos(-t) = \cos t$

51.

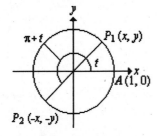

$\cos t = x$
$\cos(\pi + t) = -x$
$\cos t = -\cos(\pi + t)$

53.

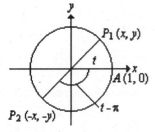

$\sin(t - \pi) = -y$
$\sin t = y$
$\sin(t - \pi) = -\sin t$

55.

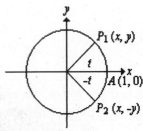

$\csc t = \dfrac{1}{y}$

$\csc(-t) = -\dfrac{1}{y}$

$\csc(-t) = -\csc t$

57. $\tan t \cos t = \dfrac{\sin t}{\cos t} \cdot \cos t$
$$= \sin t$$

59. $\dfrac{\csc t}{\cot t} = \dfrac{\dfrac{1}{\sin t}}{\dfrac{\cos t}{\sin t}}$

$$= \frac{1}{\sin t} \cdot \frac{\sin t}{\cos t}$$

$$= \frac{1}{\cos t} = \sec t$$

Copyright © Houghton Mifflin Company. All rights reserved.

61.
$$1 - \sec^2 t = 1 - \frac{1}{\cos^2 t}$$
$$= \frac{\cos^2 t - 1}{\cos^2 t}$$
$$= \frac{-\sin^2 t}{\cos^2 t} = -\tan^2 t$$

63.
$$\tan t - \frac{\sec^2 t}{\tan t} = \tan t - \frac{1 + \tan^2 t}{\tan t}$$
$$= \tan t - \frac{1}{\tan t} - \frac{\tan^2 t}{\tan t}$$
$$= \tan t - \cot t - \tan t$$
$$= -\cot t$$

65.
$$\frac{1 - \cos^2 t}{\tan^2 t} = \frac{\sin^2 t}{\frac{\sin^2 t}{\cos^2 t}}$$
$$= \sin^2 t \cdot \frac{\cos^2 t}{\sin^2 t}$$
$$= \cos^2 t$$

67.
$$\frac{1}{1 - \cos t} + \frac{1}{1 + \cos t} = \frac{1 + \cos t + 1 - \cos t}{(1 - \cos t)(1 + \cos t)}$$
$$= \frac{2}{1 - \cos^2 t}$$
$$= \frac{2}{\sin^2 t} = 2 \csc^2 t$$

69.
$$\frac{\tan t + \cot t}{\tan t} = \frac{\frac{\sin t}{\cos t} + \frac{\cos t}{\sin t}}{\frac{\sin t}{\cos t}}$$
$$= \left(\frac{\sin^2 t + \cos^2 t}{\sin t \cdot \cos t} \right) \frac{\cos t}{\sin t}$$
$$= \frac{\sin^2 t + \cos^2 t}{\sin^2 t}$$
$$= \frac{1}{\sin^2 t}$$
$$= \csc^2 t$$

71.
$$\sin^2 t \left(1 + \cot^2 t \right) = \sin^2 t \left(\csc^2 t \right)$$
$$= \sin^2 t \cdot \frac{1}{\sin^2 t}$$
$$= 1$$

73.
$$\sin^2 t + \cos^2 t = 1$$
$$\sin^2 t = 1 - \cos^2 t$$
$$\sin t = \pm\sqrt{1 - \cos^2 t}$$

Because $0 < t < \frac{\pi}{2}$, $\sin t$ is positive.

Thus, $\sin t = \sqrt{1 - \cos^2 t}$.

75.
$$\csc^2 t = 1 + \cot^2 t$$
$$\csc t = \pm\sqrt{1 + \cot^2 t}$$

Because $\frac{\pi}{2} < t < \pi$, $\csc t$ is positive.

Thus, $\csc t = \sqrt{1 + \cot^2 t}$.

77.
$$d(t) = 1970 \cos\left(\frac{\pi}{64} t \right)$$
$$d(24) = 1970 \cos\left(\frac{\pi}{64} \cdot 24 \right)$$
$$= 1970 \cos\left(\frac{3\pi}{8} \right)$$
$$\approx 750 \text{ miles}$$

79.
$$\cos t - \frac{1}{\cos t} = \frac{\cos^2 t - 1}{\cos t} = -\frac{\sin^2 t}{\cos t}$$

81.
$$\cot t + \frac{1}{\cot t} = \frac{\cot^2 t + 1}{\cot t} = \frac{\csc^2 t}{\cot t} = \frac{\frac{1}{\sin^2 t}}{\frac{\cos t}{\sin t}} = \frac{1}{\sin^2 t} \cdot \frac{\sin t}{\cos t} = \frac{1}{\sin t \cos t} = \frac{1}{\sin t} \cdot \frac{1}{\cos t} = \csc t \sec t$$

83.
$$(1 - \sin t)^2 = 1 - 2\sin t + \sin^2 t$$

85.
$$(\sin t - \cos t)^2 = \sin^2 t - 2\sin t \cos t + \cos^2 t$$
$$= 1 - 2\sin t \cos t$$

Copyright © Houghton Mifflin Company. All rights reserved.

87. $(1-\sin t)(1+\sin t) = 1 - \sin^2 t$
$$= \cos^2 t$$

89. $\dfrac{\sin t}{1+\cos t} + \dfrac{1+\cos t}{\sin t} = \dfrac{(\sin t)(\sin t) + (1+\cos t)(1+\cos t)}{\sin t(1+\cos t)}$
$$= \dfrac{\sin^2 t + 1 + 2\cos t + \cos^2 t}{\sin t(1+\cos t)}$$
$$= \dfrac{2+2\cos t}{\sin t(1+\cos t)} = \dfrac{2(1+\cos t)}{\sin t(1+\cos t)}$$
$$= \dfrac{2}{\sin t} = 2\csc t$$

91. $\cos^2 t - \sin^2 t = (\cos t - \sin t)(\cos t + \sin t)$

93. $\tan^2 t - \tan t - 6 = (\tan t + 2)(\tan t - 3)$

95. $2\sin^2 t - \sin t - 1 = (2\sin t + 1)(\sin t - 1)$

97. $\cos^4 t - \sin^4 t = \left(\cos^2 t - \sin^2 t\right)\left(\cos^2 t + \sin^2 t\right)$
$$= (\cos t - \sin t)(\cos t + \sin t)\left(\cos^2 t + \sin^2 t\right)$$
$$= (\cos t - \sin t)(\cos t + \sin t)$$

Connecting Concepts

99. $\csc t = \sqrt{2}, \quad 0 < t < \dfrac{\pi}{2}$

$\sin t = \dfrac{1}{\csc t} = \dfrac{\sqrt{2}}{2}$

$\cos^2 t + \sin^2 t = 1$

$\cos t = \pm\sqrt{1 - \sin^2 t}$

$\cos t$ is positive in quadrant I.

$\cos t = \sqrt{1 - \left(\dfrac{\sqrt{2}}{2}\right)^2} = \sqrt{\dfrac{1}{2}}$

$\cos t = \dfrac{\sqrt{2}}{2}$

101. $\sin t = \dfrac{1}{2}, \dfrac{\pi}{2} < t < \pi$

$\tan t = \dfrac{\sin t}{\cos t}$

$\tan t = \dfrac{\sin t}{\pm\sqrt{1 - \sin^2 t}}$

Because $\dfrac{\pi}{2} < t < \pi, \tan t$ is negative.

$\tan t = -\dfrac{\dfrac{1}{2}}{\sqrt{1 - \left(\dfrac{1}{2}\right)^2}}$

$\tan t = -\dfrac{\dfrac{1}{2}}{\dfrac{\sqrt{3}}{2}}$

$\tan t = -\dfrac{\sqrt{3}}{3}$

103. $\dfrac{\sin^2 t + \cos^2 t}{\sin^2 t} = \dfrac{1}{\sin^2 t}$
$$= \csc^2 t$$

105. $(\cos t - 1)(\cos t + 1) = \cos^2 t - 1$
$$= -\left(1 - \cos^2 t\right)$$
$$= -\sin^2 t$$

Prepare for Section 2.5

107. $\sin \dfrac{3\pi}{4} \approx 0.7$

108. $\cos \dfrac{5\pi}{4} \approx -0.7$

109. Reflect the graph of $y = f(x)$ across the x-axis to produce $y = -f(x)$.

110. Contract each point on the graph of $y = f(x)$ toward the y-axis by a factor of $\dfrac{1}{2}$.

111. $\dfrac{2\pi}{\frac{1}{3}} = \dfrac{2\pi}{1} \cdot \dfrac{3}{1} = 6\pi$

112. $\dfrac{2\pi}{\frac{2}{5}} = \dfrac{2\pi}{1} \cdot \dfrac{5}{2} = 5\pi$

Copyright © Houghton Mifflin Company. All rights reserved.

Section 2.5

1. $y = 2\sin x$
$a = 2, p = 2\pi$

3. $y = \sin 2x$
$a = 1, p = \dfrac{2\pi}{2} = \pi$

5. $y = \dfrac{1}{2}\sin 2\pi x$
$a = \dfrac{1}{2}, p = \dfrac{2\pi}{2\pi} = 1$

7. $y = -2\sin\dfrac{x}{2}$
$a = |-2| = 2, p = \dfrac{2\pi}{1/2} = 4\pi$

9. $y = \dfrac{1}{2}\cos x$
$a = \dfrac{1}{2}, p = 2\pi$

11. $y = \cos\dfrac{x}{4}$
$a = 1, p = \dfrac{2\pi}{1/4} = 8\pi$

13. $y = 2\cos\dfrac{\pi x}{3}$
$a = 2, p = \dfrac{2\pi}{\pi/3} = 6$

15. $y = -3\cos\dfrac{2x}{3}$
$a = |-3| = 3, p = \dfrac{2\pi}{2/3} = 3\pi$

17. $y = \dfrac{1}{2}\sin x, a = \dfrac{1}{2}, p = 2\pi$

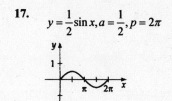

19. $y = 3\cos x, a = 3, p = 2\pi$

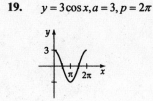

21. $y = -\dfrac{7}{2}\cos x, \quad a = \left|-\dfrac{7}{2}\right| = \dfrac{7}{2}, \quad p = 2\pi$

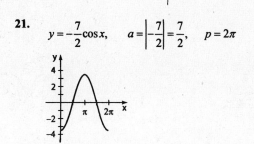

23. $y = -4\sin x, \quad a = |-4| = 4, \quad p = 2\pi$

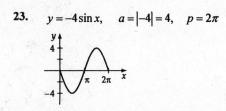

25. $y = \cos 3x, \quad a = 1, \quad p = \dfrac{2\pi}{3}$

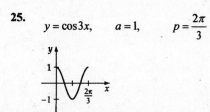

27. $y = \sin\dfrac{3x}{2}, \quad a = 1, \quad p = \dfrac{4\pi}{3}$

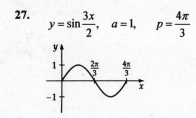

29. $y = \cos\dfrac{\pi}{2}x, \quad a = 1, \quad p = 4$

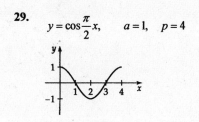

31. $y = \sin 2\pi x, \quad a = 1, \quad p = 1$

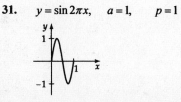

33. $y = 4\cos\dfrac{x}{2}, \quad a = 4, \quad p = \dfrac{2\pi}{1/2} = 4\pi$

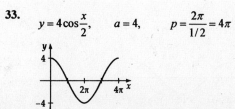

Copyright © Houghton Mifflin Company. All rights reserved.

35. $y = -2\cos\dfrac{x}{3}, \qquad a = |-2| = 2, \qquad p = \dfrac{2\pi}{1/3} = 6\pi$

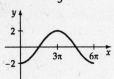

37. $y = 2\sin\pi x, \qquad a = 2, \qquad p = \dfrac{2\pi}{\pi} = 2$

39. $y = \dfrac{3}{2}\cos\dfrac{\pi x}{2}, \qquad a = \dfrac{3}{2}, \qquad p = \dfrac{2\pi}{\pi/2} = 4$

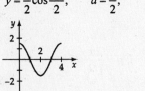

41. $y = 4\sin\dfrac{2\pi}{3}x, a = |4| = 4, p = \dfrac{2\pi}{2\pi/3} = 3$

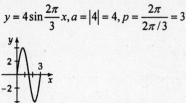

43. $y = 2\cos 2x, a = 2, p = \dfrac{2\pi}{2} = \pi$

45. $y = -2\sin 1.5x, a = |-2| = 2, p = \dfrac{2\pi}{1.5} = \dfrac{4\pi}{3}$

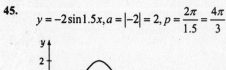

47. $y = \left|2\sin\dfrac{x}{2}\right|$

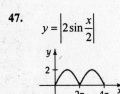

49. $y = \left|-2\cos 3x\right|$

51. $y = -\left|2\sin\dfrac{x}{3}\right|$

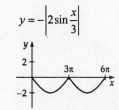

53. $y = -\left|3\cos\pi x\right|$

55. $\dfrac{2\pi}{b} = \pi, b = 2, a = 1$

$y = \cos 2x$

57. $\dfrac{2x}{b} = 3\pi, b = \dfrac{2}{3}, a = 2$

$y = 2\sin\dfrac{2}{3}x$

59. $\dfrac{2\pi}{b} = 2, b = \pi, a = 2$

$y = -2\cos\pi x$

61. $f(x) = 2\sin\dfrac{2x}{3}, a = 2, p = \dfrac{2\pi}{2/3} = 3\pi$

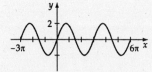

63. $f(x) = 2\cos\dfrac{x}{2}, a = 2, p = \dfrac{2\pi}{1/2} = 4\pi, g(x) = 2\cos x, a = 2, p = 2\pi$

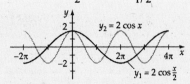

Copyright © Houghton Mifflin Company. All rights reserved.

65. $y = \cos^2 x$

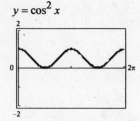

67. $y = \cos|x|$

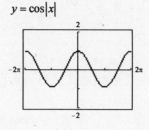

69. $y = \dfrac{1}{2} x \sin x$

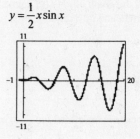

71. $y = -x \cos x$

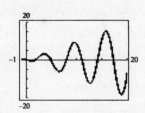

73. $y = e^{\sin x}$

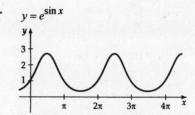

The maximum value of $e^{\sin x}$ is e.

The minimum value of $e^{\sin x}$ is $\dfrac{1}{e} \approx 0.3679$.

The function defined by $y = e^{\sin x}$ is periodic with a period of 2π.

75.

$a = 60, p = \dfrac{2\pi}{B} = 20, B = \dfrac{\pi}{10}$

$f(t) = A \cos Bt$

$f(t) = 60 \cos \dfrac{\pi}{10} t$

77. $a = 2$

$p = \dfrac{2\pi}{b} = 3\pi$

$b = \dfrac{2}{3}$

$y = 2 \sin \dfrac{2}{3} x$

79. $a = 4$

$p = \dfrac{2\pi}{b} = 2$

$b = \pi$

$y = 4 \sin \pi x$

81. $a = 3$

$p = \dfrac{2\pi}{b} = \dfrac{\pi}{2}$

$b = 4$

$y = 3 \cos 4x$

Connecting Concepts

83. $a = 3$

$p = \dfrac{2\pi}{b} = 2.5$

$b = \dfrac{4\pi}{5}$

$y = 3 \cos \dfrac{4\pi}{5} x$

Prepare for Section 2.6

85. $\tan \dfrac{\pi}{3} \approx 1.7$

86. $\cot \dfrac{\pi}{3} \approx 0.6$

87. Stretch each point on the graph of $y = f(x)$ away from the x-axis by a factor of 2 to produce $y = 2 f(x)$.

Copyright © Houghton Mifflin Company. All rights reserved.

88. Shift the graph of $y = f(x)$ 2 units to the right and up 3 units.

89. $\dfrac{\pi}{\frac{1}{2}} = \dfrac{\pi}{1} \cdot \dfrac{2}{1} = 2\pi$

90. $\dfrac{\pi}{|-3/4|} = \dfrac{\pi}{1} \cdot \left|-\dfrac{4}{3}\right| = \dfrac{\pi}{1} \cdot \dfrac{4}{3} = \dfrac{4}{3}\pi$

Section 2.6

1. $y = \tan x$ is undefined for $\dfrac{\pi}{2} + k\pi$, k an integer.

3. $y = \sec x$ is undefined for $\dfrac{\pi}{2} + k\pi$, k an integer.

5. $p = 2\pi$

7. $p = \pi$

9. $p = \dfrac{\pi}{1/2} = 2\pi$

11. $p = \dfrac{2\pi}{3}$

13. $p = \dfrac{\pi}{3}$

15. $p = \dfrac{2\pi}{1/4} = 8\pi$

17. $p = \dfrac{\pi}{\pi} = 1$

19. $p = \dfrac{2\pi}{\pi/2} = 4$

21. $y = 3\tan x$, $p = \pi$

23. $y = \dfrac{3}{2}\cot x$, $p = \pi$

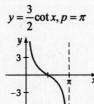

25. $y(x) = 2\sec x$, $p = 2\pi$

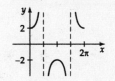

27. $y = \dfrac{1}{2}\csc x$, $p = 2\pi$

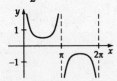

29. $y = 2\tan\dfrac{x}{2}$, $p = \dfrac{\pi}{1/2} = 2\pi$

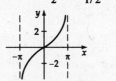

31. $y = -3\cot\dfrac{x}{2}$, $p = \dfrac{\pi}{1/2} = 2\pi$

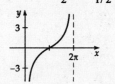

33. $y = -2\csc\dfrac{x}{3}$, $p = \dfrac{2\pi}{1/3} = 6\pi$

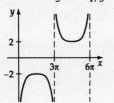

35. $y = \dfrac{1}{2}\sec 2x$, $p = \dfrac{2\pi}{2} = \pi$

37. $y = -2\sec \pi x$, $p = \dfrac{2\pi}{\pi} = 2$

39. $y = 3\tan 2\pi x$, $p = \dfrac{\pi}{2\pi} = \dfrac{1}{2}$

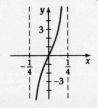

41. $y = 2\csc 3x$, $p = \dfrac{2\pi}{3}$

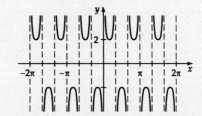

Copyright © Houghton Mifflin Company. All rights reserved.

43. $y = 3 \sec \pi x, p = \dfrac{2\pi}{\pi} = 2$

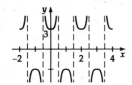

45. $y = 2 \cot 2x, p = \dfrac{\pi}{2}$

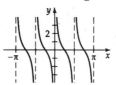

47. $y = 3 \tan \pi x, p = \dfrac{\pi}{\pi} = 1$

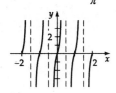

49. $\dfrac{\pi}{b} = \dfrac{2\pi}{3}, b = \dfrac{3}{2}$

$y = \cot \dfrac{3}{2}x$

51. $\dfrac{2\pi}{b} = 3\pi, b = \dfrac{2}{3}$

$y = \csc \dfrac{2}{3}x$

53. $\dfrac{2\pi}{b} = \dfrac{8\pi}{3}, b = \dfrac{3}{4}$

$y = \sec \dfrac{3}{4}x$

55. $y = \tan |x|$

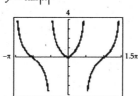

57. $y = |\csc x|$

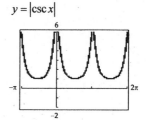

59.

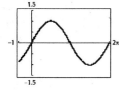

61. $y = \tan x, x = \tan y$

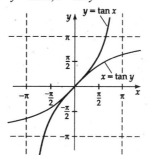

• •

Connecting Concepts

63. $\dfrac{\pi}{b} = \dfrac{\pi}{3}, b = 3$

$y = \tan 3x$

65. $\dfrac{2\pi}{b} = \dfrac{3\pi}{4}, b = \dfrac{8}{3}$

$y = \sec \dfrac{8}{3}x$

67. $\dfrac{\pi}{b} = 2, b = \dfrac{\pi}{2}$

$y = \cot \dfrac{\pi}{2}x$

69. $\dfrac{2\pi}{b} = 1.5, b = \dfrac{4}{3}\pi$

$y = \csc \dfrac{4\pi}{3}x$

Copyright © Houghton Mifflin Company. All rights reserved.

71. $y = 2\sin 2x$
amplitude $= 2$, period $= \pi$

72. $y = \dfrac{2}{3}\cos\dfrac{x}{3}$

amplitude $= \dfrac{2}{3}$, period $= 6\pi$

73. $y = -4\sin 2\pi x$
amplitude $= 4$, period $= 1$

74. maximum at 2

75. minimum at -3

76. $f(x) = \cos x$ is symmetric to y-axis.

Section 2.7

1. $a = 2, p = 2\pi$, phase shift $= \dfrac{\pi}{2}$

3. $a = 1, p = \dfrac{2\pi}{2} = \pi$, phase shift $= \dfrac{\pi/4}{2} = \dfrac{\pi}{8}$

5. $a = |-4| = 4, p = \dfrac{2\pi}{2/3} = 3\pi$, phase shift $= \dfrac{-\pi/6}{2/3} = -\dfrac{\pi}{4}$

7. $a = \dfrac{5}{4}, p = \dfrac{2\pi}{3}$, phase shift $= \dfrac{2\pi}{3}$

9. $p = \dfrac{\pi}{2}$, phase shift $= \dfrac{\pi/4}{2} = \dfrac{\pi}{8}$

11. $p = \dfrac{2\pi}{1/3} = 6\pi$, phase shift $= \dfrac{-\pi}{1/3} = -3\pi$

13. $p = \dfrac{2\pi}{2} = \pi$, phase shift $= \dfrac{\pi/8}{2} = \dfrac{\pi}{16}$

15. $p = \dfrac{\pi}{1/4} = 4\pi$, phase shift $= \dfrac{-3\pi}{1/4} = -12\pi$

17. $y = \sin\left(x - \dfrac{\pi}{2}\right)$

$0 \le x - \dfrac{\pi}{2} \le 2\pi$

$\dfrac{\pi}{2} \le x \le \dfrac{5\pi}{2}$

period $= 2\pi$, phase shift $\dfrac{\pi}{2}$

19. $y = \cos\left(\dfrac{x}{2} + \dfrac{\pi}{3}\right)$

$0 \le \dfrac{x}{2} + \dfrac{\pi}{3} \le 2\pi$

$-\dfrac{\pi}{3} \le \dfrac{x}{2} \le \dfrac{5\pi}{3}$

$-\dfrac{2\pi}{3} \le x \le \dfrac{10\pi}{3}$

period $= 4\pi$, phase shift $= -\dfrac{2\pi}{3}$

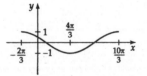

21. $y = \tan\left(x + \dfrac{\pi}{4}\right)$

$-\dfrac{\pi}{2} < x + \dfrac{\pi}{4} < \dfrac{\pi}{2}$

$-\dfrac{3\pi}{4} < x < \dfrac{\pi}{4}$

period $= \pi$, phase shift $= -\dfrac{\pi}{4}$

23. $y = 2\cot\left(\dfrac{x}{2} - \dfrac{\pi}{8}\right)$

$0 < \dfrac{x}{2} - \dfrac{\pi}{8} < \pi$

$\dfrac{\pi}{8} < \dfrac{x}{2} < \dfrac{9\pi}{8}$

$\dfrac{\pi}{4} < x < \dfrac{9\pi}{4}$

period $= 2\pi$, phase shift $= \dfrac{\pi}{4}$

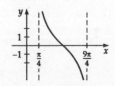

25. $y = \sec\left(x + \dfrac{\pi}{4}\right)$

$0 \le x + \dfrac{\pi}{4} \le 2\pi$

$-\dfrac{\pi}{4} \le x \le \dfrac{7\pi}{4}$

period $= 2\pi$, phase shift $= -\dfrac{\pi}{4}$

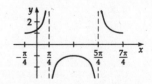

Copyright © Houghton Mifflin Company. All rights reserved.

27.
$$y = \csc\left(\frac{x}{3} - \frac{\pi}{2}\right)$$

$$0 \le \frac{x}{3} - \frac{\pi}{2} \le 2\pi$$

$$\frac{\pi}{2} \le \frac{x}{3} \le \frac{5\pi}{2}$$

$$\frac{3\pi}{2} \le x \le \frac{15\pi}{2}$$

period $= 6\pi$, phase shift $= \dfrac{3\pi}{2}$

29.
$$y = -2\sin\left(\frac{x}{3} - \frac{2\pi}{3}\right)$$

$$0 \le \frac{x}{3} - \frac{2\pi}{3} \le 2\pi$$

$$\frac{2\pi}{3} \le \frac{x}{3} \le \frac{8\pi}{3}$$

$$2\pi \le x \le 8\pi$$

period $= 6\pi$, phase shift $= 2\pi$

31.
$$y = -3\cos\left(3x + \frac{\pi}{4}\right)$$

$$0 \le 3x + \frac{\pi}{4} \le 2\pi$$

$$-\frac{\pi}{4} \le 3x \le \frac{7\pi}{4}$$

$$-\frac{\pi}{12} \le x \le \frac{7\pi}{12}$$

period $= \dfrac{2\pi}{3}$, phase shift $= -\dfrac{\pi}{12}$

33. $y = \sin x + 1, p = 2\pi$

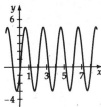

35. $y = -\cos x - 2, p = 2\pi$

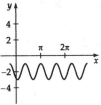

37. $y = \sin 2x - 2, p = \pi$

39. $y = 4\cos(\pi x - 2) + 1, p = 2$

phase shift $= \dfrac{2}{\pi}$

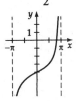

41. $y = -\sin(\pi x + 1) - 2, p = 2$

phase shift $= -\dfrac{1}{\pi}$

43.
$$y = \sin\left(x - \frac{\pi}{2}\right) - \frac{1}{2}, p = 2\pi,$$

phase shift $= \dfrac{\pi}{2}$

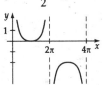

45. $y = \tan\dfrac{x}{2} - 4, p = 2\pi$

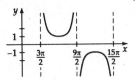

47. $y = \sec 2x - 2, p = \pi$

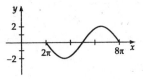

49. $y = \csc\dfrac{x}{2} - 1, p = 4\pi$

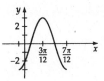

Copyright © Houghton Mifflin Company. All rights reserved.

51. **a.**   phase shift: $-\dfrac{c}{b} = -\dfrac{-1.25\pi}{\pi/6} = 1.25(6) = 7.5$ months

period: $\dfrac{2\pi}{b} = \dfrac{2\pi}{\pi/6} = 2(6) = 12$ months

b.   First graph $y_1 = 4.1\cos\left(\dfrac{\pi}{6}t\right)$. Because the phase shift

is 7.5 months, shift the graph of y_1 7.5 units to the
right to produce the graph of y_2. Now shift the graph
of y_2 upward 7 units to product the graph of S.

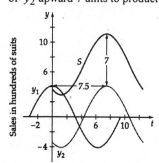

c.   7.5 months after January 1 is the middle of August.

53. $y = x - \sin x$

55. $y = x + \sin 2x$

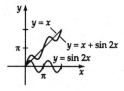

57. $y = \sin x + \cos x$

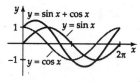

59.

sine curve, $a = 1, \dfrac{2\pi}{b} = \pi, b = 2$

phase shift $= -\dfrac{c}{b} = \dfrac{\pi}{6}, c = -\dfrac{\pi}{3}$

$y = \sin\left(2x - \dfrac{\pi}{3}\right)$

61.

cosecant curve, $\dfrac{2\pi}{b} = 4\pi, b = \dfrac{1}{2}$

phase shift $= -\dfrac{c}{b} = 2\pi, c = -\pi$

$y = \csc\left(\dfrac{x}{2} - \pi\right)$

63.

secant curve, $\dfrac{2\pi}{b} = 2\pi, b = 1$

phase shift $= -\dfrac{c}{b} = \dfrac{\pi}{2}, c = -\dfrac{\pi}{2}$

$y = \sec\left(x - \dfrac{\pi}{2}\right)$

65. $y = 2.3\sin 2\pi t + 1.25t + 315$

$t = 20$ between 1972 and 1992

$y = 2.3\sin 2\pi(20) + 1.25(20) + 315$

$y \approx 340$ ppm

difference $\approx 340 - 315$

≈ 25 ppm

Copyright © Houghton Mifflin Company. All rights reserved.

67. 5 rpm $a = 7$

$p = 0.2$ min $p = \dfrac{2\pi}{b}$

$0.2 = \dfrac{2\pi}{b}$

$b = 10\pi$

$s = 7\cos 10\pi t + 5$

69. Change 6 rpm to radians/second.

$6 \text{ rpm} = \dfrac{6 \cdot 2\pi \text{ radians}}{1 \text{ minute}} \cdot \dfrac{1 \text{ minute}}{60 \text{ sec}}$

$= \dfrac{\pi}{5}$ radians/sec

in t seconds, θ increased by $\dfrac{\pi}{5}t$ radians.

$\tan\dfrac{\pi}{5}t = \dfrac{s}{400}$

$400\tan\dfrac{\pi}{5}t = s$ for $0 \le t < 2.5$ or $7.5 < t \le 10$

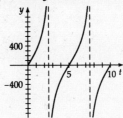

71. amplitude $A = 3$

$k = 9$

$24 = 2$ cycles

$12 = 1$ cycle

$\dfrac{2\pi}{B} = 12 \Rightarrow B = \dfrac{\pi}{6}$

$y = 3\cos\dfrac{\pi}{6}t + 9$

At 6:00 P.M., $t = 12$.

$y = 3\cos\dfrac{\pi}{6}\cdot 12 + 9$

$y = 3\cos 2\pi + 9$

$y = 3 + 9$

$y = 12$ ft

73. $y = \sin x - \cos\dfrac{x}{2}$

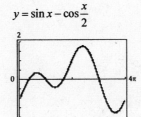

75. $y = 2\cos x + \sin\dfrac{x}{2}$

77. $y = \dfrac{x}{2}\sin x$

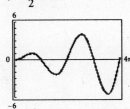

79. $y = x\sin\dfrac{x}{2}$

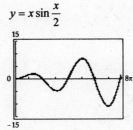

81. $y = x\sin\left(x + \dfrac{\pi}{2}\right)$

83.

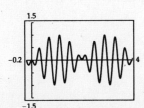

85.

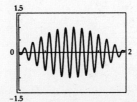

Copyright © Houghton Mifflin Company. All rights reserved.

87. sine function, $a = 2, p = \pi$

$\dfrac{2\pi}{b} = \pi, b = 2$

phase shift $= -\dfrac{c}{b} = \dfrac{\pi}{3}, c = -\dfrac{2\pi}{3}$

$y = 2\sin\left(2x - \dfrac{2\pi}{3}\right)$

89. tangent function, $p = 2\pi$

$\dfrac{\pi}{b} = 2\pi \Rightarrow b = \dfrac{1}{2}$

phase shift $= -\dfrac{c}{b} = \dfrac{\pi}{2} \Rightarrow c = -\dfrac{\pi}{4}$

$y = \tan\left(\dfrac{x}{2} - \dfrac{\pi}{4}\right)$

91. secant function, $p = 4\pi$

$\dfrac{2\pi}{b} = 4\pi, b = \dfrac{1}{2}$

phase shift $= -\dfrac{c}{b} = \dfrac{3\pi}{4}, c = -\dfrac{3\pi}{8}$

$y = \sec\left(\dfrac{x}{2} - \dfrac{3\pi}{8}\right)$

93. $g(x) + h(x) = \sin^2 x + \cos^2 x = 1$

95. $g[h(x)] = (\cos x)^2 + 2$

$\qquad = \cos^2 x + 2$

97. $\dfrac{\sin x}{x} \to 1$ as $x \to 0$

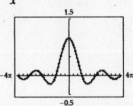

99. $y = |x|\sin x$

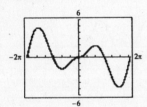

101. $\dfrac{3}{2\pi}$

102. $\dfrac{5}{2}$

103. $4\cos 2\pi(0) = 4\cos 0 = 4$

104. $\sqrt{\dfrac{18}{2}} = \sqrt{9} = 3$

105. $4\cos\left(\sqrt{\dfrac{16}{4}} \cdot 2\pi\right) = 4\cos(4 \cdot 2\pi)$

$\qquad\qquad = 4\cos 8\pi = 4$

106. $y = 4\cos \pi x$

Section 2.8

1. $y = 2\sin 2t$

amplitude $= 2$

$p = \dfrac{2\pi}{2} = \pi$

frequency $= \dfrac{1}{p} = \dfrac{1}{\pi}$

3. $y = 3\cos\dfrac{2t}{3}$

amplitude $= 3$

$p = \dfrac{2\pi}{2/3} = 3\pi$

frequency $= \dfrac{1}{p} = \dfrac{1}{3\pi}$

5. $y = 4\cos \pi t$

amplitude $= 4$

$p = \dfrac{2\pi}{\pi} = 2$

frequency $= \dfrac{1}{p} = \dfrac{1}{2}$

Copyright © Houghton Mifflin Company. All rights reserved.

7. $y = \dfrac{3}{4}\sin\dfrac{\pi t}{2}$

amplitude $= \dfrac{3}{4}$

$p = \dfrac{2\pi}{\pi/2} = 4$

frequency $= \dfrac{1}{p} = \dfrac{1}{4}$

9. $a = 4$

$p = \dfrac{1}{\text{frequency}} = \dfrac{1}{1.5} = \dfrac{2}{3}$

$\dfrac{2\pi}{b} = \dfrac{2}{3} \Rightarrow b = 3\pi$

$y = 4\cos 3\pi t$

11. $p = 1.5$

amplitude $= \dfrac{3}{2}$

$\dfrac{2\pi}{b} = 1.5 \Rightarrow b = \dfrac{4\pi}{3}$

$y = \dfrac{3}{2}\cos\dfrac{4\pi}{3}t$

13. amplitude $= 2$

$p = \pi$

$\dfrac{2\pi}{b} = \pi \Rightarrow b = 2$

$y = 2\sin 2t$

15. amplitude $= 1$

$p = 2$

$\dfrac{2\pi}{b} = 2 \Rightarrow b = \pi$

$y = \sin \pi t$

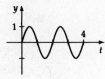

17. amplitude $= 2$

frequency $= 1$

$p = \dfrac{1}{\text{frequency}} = \dfrac{1}{1} = 1$

$\dfrac{2\pi}{b} = 1 \Rightarrow b = 2\pi$

$y = 2\sin 2\pi t$

19. amplitude $= \dfrac{1}{2}$

frequency $= \dfrac{2}{\pi} \Rightarrow p = \dfrac{\pi}{2}$

$\dfrac{2\pi}{b} = \dfrac{\pi}{2} \Rightarrow b = 4$

$y = \dfrac{1}{2}\cos 4t$

21. amplitude $= 2.5$

frequency $= 0.5 \Rightarrow p = 2$

$\dfrac{2\pi}{b} = 2 \Rightarrow b = \pi$

$y = 2.5\cos \pi t$

23. amplitude $= \dfrac{1}{2}$

$p = 3$

$\dfrac{2\pi}{b} = 3 \Rightarrow b = \dfrac{2}{3}\pi$

$y = \dfrac{1}{2}\cos\dfrac{2\pi}{3}t$

25. amplitude $= 4$

$p = \dfrac{\pi}{2}$

$\dfrac{2\pi}{b} = \dfrac{\pi}{2} \Rightarrow b = 4$

$y = 4\cos 4t$

27. amplitude $= 2$ feet, $a = -2$

frequency $= f = \dfrac{1}{2\pi}\sqrt{\dfrac{k}{m}} = \dfrac{1}{2\pi}\sqrt{\dfrac{8}{32}} = \dfrac{1}{2\pi}\cdot\dfrac{1}{2} = \dfrac{1}{4\pi}$

period $= p = \dfrac{1}{f} = 4\pi$

$y = a\cos 2\pi ft = -2\cos 2\pi\left(\dfrac{1}{4\pi}\right)t$

$\quad = -2\cos\dfrac{1}{2}t$

Copyright © Houghton Mifflin Company. All rights reserved.

29. a. The pseudoperiod is $\frac{2\pi}{2} = \pi$. There are $\frac{10}{\pi} \approx 3$ complete oscillations of length π in $0 \le t \le 10$.

   b. $|f(t)| < 0.01$ for all $t > 59.8$(nearest tenth).

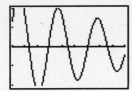

Xmin = 56, Xmax = 65, Xscl = 1,
Ymin = −0.01, Ymax = 0.01, Yscl = 0.005

31. a. The pseudoperiod is $\frac{2\pi}{2\pi} = 1$. There are 10 complete oscillations of length 1 in $0 \le t \le 10$.

   b. $|f(t)| < 0.01$ for all $t > 71.0$ (nearest tenth).

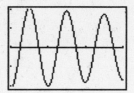

Xmin = 70, Xmax = 73, Xscl = 1
Ymin = −0.01, Ymax = 0.01, Yscl = 0.005

33. a. The pseudoperiod is $\frac{2\pi}{2\pi} = 1$. There are 10 complete oscillations of length 1 in $0 \le t \le 10$.

   b. $|f(t)| < 0.01$ for all $t > 9.1$ (nearest tenth).

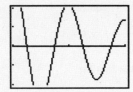

Xmin = 8, Xmax = 10, Xscl = 1,
Ymin = −0.01, Ymax = 0.01, Yscl = 0.005

35. a. The pseudoperiod is $\frac{2\pi}{2\pi} = 1$. There are 10 complete oscillations of length 1 in $0 \le t \le 10$.

   b. $|f(t)| < 0.01$ for all $t > 6.1$(nearest tenth).

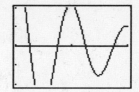

Xmin = 5, Xmax = 7, Xscl = 1,
Ymin = −0.01, Ymax = .01, Yscl = 0.005

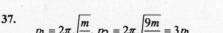

Connecting Concepts

37.

$$p_1 = 2\pi\sqrt{\frac{m}{k}}, \quad p_2 = 2\pi\sqrt{\frac{9m}{k}} = 3p_1$$

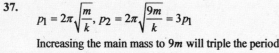

Increasing the main mass to $9m$ will triple the period.

39. yes

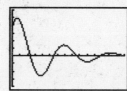

Xmin = 0, Xmax = 15, Xscl = 1,
Ymin = −1, Ymax = 1.5, Yscl = 0.25

41. yes

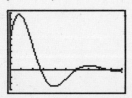

Xmin = 0, Xmax = 10, Xscl = 1,
Ymin = −3, Ymax = 9, Yscl = 1

Copyright © Houghton Mifflin Company. All rights reserved.

1. False; the initial side must be along the positive x-axis.

2. True

3. True

4. False; in the third quadrant, $\cos\theta < 0$ and $\tan\theta > 0$.

5. False; $\sec^2\theta - \tan^2\theta = 1$ is an identity.

6. False; the tangent function has no amplitude.

7. False; the period is 2π.

8. True

9. False; $\sin 45° + \cos(90° - 45°) = \sin 45° + \cos 45°$
$$= \frac{\sqrt{2}}{2} + \frac{\sqrt{2}}{2} = \sqrt{2}.$$

10. False; $\sin\left(\dfrac{\pi}{2} + \dfrac{\pi}{2}\right) = \sin\pi$ $\qquad \sin\dfrac{\pi}{2} + \sin\dfrac{\pi}{2} = 1 + 1$
$$= 0 \qquad\qquad\qquad = 2$$

11. False; $\sin^2\dfrac{\pi}{6} = \left(\sin\dfrac{\pi}{6}\right)^2$ $\qquad \sin\left(\dfrac{\pi}{6}\right)^2 = \sin\dfrac{\pi^2}{36}$
$$= \left(\dfrac{1}{2}\right)^2 = \dfrac{1}{4} \qquad\qquad \approx 0.2707.$$

12. False; the phase shift is $\dfrac{\pi/3}{2} = \dfrac{\pi}{6}$.

13. True

14. False; 1 rad $\approx 57.3°$.

15. False, the graph lies on or between the graphs of $y = 2^{-x}$ and $y = -2^{-x}$.

16. False; $|f(t)| \to 0$ as $t \to \infty$.

1. complement: $90° - 65° = 25°$ [2.1]
   supplement: $180° - 65° = 115°$

2. $\theta = 980° = 260° + 2\cdot 360°$ [2.3]
   θ is coterminal with $\alpha = 260°$ and
   $\theta' = \alpha'$.
   Since $180° < \alpha < 270°$,
   $180° + \alpha' = \alpha$
   $180° + a' = 260°$
   $\alpha' = 80°$
   $\theta = 80°$

3. $2 = 2\left(\dfrac{180°}{\pi}\right)$ [2.1]
   $= 114.59°$

4. $315° = 315°\left(\dfrac{\pi}{180°}\right)$ [2.1]
   $= \dfrac{7\pi}{4}$

5. $s = r\theta = 3(75°)\left(\dfrac{\pi}{180°}\right)$ [2.1]
   $= 3.93\text{ m}$

6. $\theta = \dfrac{s}{r} = \dfrac{12}{40}$ [2.1]
   $= 0.3$

7. $w = \dfrac{V}{r} = \dfrac{50}{16}\cdot\dfrac{63360}{3600}$ [2.1]
   $\approx 55\text{ rad/sec}$

For exercises 8 to 11, $\csc\theta = \dfrac{3}{2} = \dfrac{r}{y}$, $r = 3$, $y = 2$, and $x = \sqrt{3^2 - 2^2} = \sqrt{5}$.

8. $\cos\theta = \dfrac{x}{r} = \dfrac{\sqrt{5}}{3}$ [2.2]

9. $\cot\theta = \dfrac{x}{y} = \dfrac{\sqrt{5}}{2}$ [2.2]

10. $\sin\theta = \dfrac{y}{r} = \dfrac{2}{3}$ [2.2]

11. $\sec\theta = \dfrac{r}{x} = \dfrac{3}{\sqrt{5}} = \dfrac{3\sqrt{5}}{5}$ [2.2]

Copyright © Houghton Mifflin Company. All rights reserved.

12.

(1, -3)

$x = 1, y = -3, r = \sqrt{1^2 + (-3)^2} = \sqrt{10}$

$\sin\theta = -\dfrac{3}{\sqrt{10}} = -\dfrac{3\sqrt{10}}{10}$ $\cos\theta = \dfrac{1}{\sqrt{10}} = \dfrac{\sqrt{10}}{10}$ $\tan\theta = \dfrac{-3}{1} = -3$

$\csc\theta = -\dfrac{\sqrt{10}}{3}$ $\sec\theta = \sqrt{10}$ $\cot\theta = -\dfrac{1}{3}$

13.

a. $\sec 150° = \dfrac{2}{-\sqrt{3}} = -\dfrac{2\sqrt{3}}{3}$

b. $\tan\left(-\dfrac{3\pi}{4}\right) = 1$

c. $\cot(-225°) = -1$

d. $\cos\dfrac{2\pi}{3} = -\dfrac{1}{2}$ [2.3]

14.

a. $\cos 123° \approx -0.5446$

b. $\cot 4.22 \approx 0.5365$

c. $\sec 612° \approx -3.2361$

d. $\tan\dfrac{2\pi}{5} \approx 3.0777$ [2.3]

15.

$\cos\phi = -\dfrac{\sqrt{3}}{2} = \dfrac{x}{r}, x = -\sqrt{3}, r = 2, y = -\sqrt{2^2 - \left(-\sqrt{3}\right)^2} = -1$ [2.3]

a. $\sin\phi = \dfrac{y}{r} = -\dfrac{1}{2}$

b. $\tan\phi = \dfrac{y}{x} = \dfrac{-1}{-\sqrt{3}} = \dfrac{\sqrt{3}}{3}$

16.

$\tan\phi = -\dfrac{\sqrt{3}}{3} = \dfrac{y}{x}, y = \sqrt{3}, x = -3, r = \sqrt{(-3)^2 + \left(\sqrt{3}\right)^2} = 2\sqrt{3}$ [2.3]

a. $\sec\phi = \dfrac{r}{x} = \dfrac{2\sqrt{3}}{-3} = -\dfrac{2\sqrt{3}}{3}$

b. $\csc\phi = \dfrac{r}{y} = \dfrac{2\sqrt{3}}{\sqrt{3}} = 2$

17.

$\sin\phi = -\dfrac{\sqrt{2}}{2}, y = -\sqrt{2}, r = 2, x = -\sqrt{2^2 - \left(-\sqrt{2}\right)^2} = \sqrt{2}$ [2.3]

a. $\cos\phi = \dfrac{x}{r} = \dfrac{\sqrt{2}}{2}$

b. $\cot\phi = \dfrac{x}{y} = \dfrac{\sqrt{2}}{-\sqrt{2}} = -1$

Copyright © Houghton Mifflin Company. All rights reserved.

18. **a.** $w(\pi) = (-1, 0)$ [2.4]

b. $w\left(-\dfrac{\pi}{3}\right) = \left(\dfrac{1}{2}, -\dfrac{\sqrt{3}}{2}\right)$

c. $w\left(\dfrac{5\pi}{4}\right) = \left(-\dfrac{\sqrt{2}}{2}, -\dfrac{\sqrt{2}}{2}\right)$

19. $f(x) = \sin(x)\tan(x)$ [2.4]

$f(-x) = \sin(-x)\tan(-x) = (-\sin x)(-\tan x)$

$\qquad = \sin x \tan x$

$\qquad = f(x)$

The function defined by $f(x) = \sin(x)\tan(x)$ is an even function.

20.

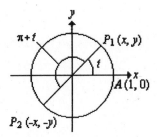

$\cos(\pi + t) = -x$

$\qquad \cos t = x$

$\cos(\pi + t) = -\cos t$

21.

$\tan(-t) = -\dfrac{y}{x}$

$\tan t = \dfrac{y}{x}$

$\tan(-t) = -\tan t$

22. $1 + \dfrac{\sin^2\phi}{\cos^2\phi} = 1 + \tan^2\phi$ [2.4]

$\qquad\qquad = \sec^2\phi$

23. $\dfrac{\tan\phi + 1}{\cot\phi + 1} = \dfrac{\dfrac{\sin\phi}{\cos\phi} + 1}{\dfrac{\cos\phi}{\sin\phi} + 1}$ [2.4]

$\qquad = \dfrac{\dfrac{\sin\phi + \cos\phi}{\cos\phi}}{\dfrac{\cos\phi + \sin\phi}{\sin\phi}}$

$\qquad = \dfrac{\sin\phi(\sin\phi + \cos\phi)}{\cos\phi(\cos\phi + \sin\phi)}$

$\qquad = \tan\phi$

24. $\dfrac{\cos^2\phi + \sin^2\phi}{\csc\phi} = \dfrac{1}{\csc\phi}$ [2.4]

$\qquad\qquad = \sin\phi$

25. $\sin^2\phi(\tan^2\phi + 1) = \sin^2\phi\sec^2\phi$ [2.4]

$\qquad\qquad = \dfrac{\sin^2\phi}{\cos^2\phi}$

$\qquad\qquad = \tan^2\phi$

26. $1 + \dfrac{1}{\tan^2\phi} = \dfrac{\tan^2\phi + 1}{\tan^2\phi}$ [2.4]

$\qquad = \dfrac{\sec^2\phi}{\tan^2\phi}$

$\qquad = \dfrac{\dfrac{1}{\cos^2\phi}}{\dfrac{\sin^2\phi}{\cos^2\phi}}$

$\qquad = \dfrac{1}{\sin^2\phi}$

$\qquad = \csc^2\phi$

27. $\dfrac{\cos^2\phi}{1-\sin^2\phi} - 1 = \dfrac{1-\sin^2\phi}{1-\sin^2\phi} - 1$ [2.4]

$\qquad = 1 - 1$

$\qquad = 0$

Copyright © Houghton Mifflin Company. All rights reserved.

28. $y = 3\cos(2x - \pi)$ [2.5]

$a = |3| = 3$; period $= \dfrac{2\pi}{b} = \dfrac{2\pi}{2} = \pi$

phase shift $= -\dfrac{c}{b} = -\dfrac{-\pi}{2} = \dfrac{\pi}{2}$

29. $y = 2\tan 3x$ [2.6]

no amplitude; period $= \dfrac{\pi}{b} = \dfrac{\pi}{3}$

phase shift $= 0$

30. $y = -2\sin\left(3x + \dfrac{\pi}{3}\right)$ [2.5]

$a = |-2| = 2$; period $= \dfrac{2\pi}{b} = \dfrac{2\pi}{3}$

phase shift $= -\dfrac{c}{b} = -\dfrac{\pi/3}{3} = -\dfrac{\pi}{9}$

31. $y = \cos\left(2x - \dfrac{2\pi}{3}\right) + 2$ [2.5]

$a = |1| = 1$; period $= \dfrac{2\pi}{b} = \dfrac{2\pi}{2} = \pi$

phase shift $= -\dfrac{c}{b} = -\dfrac{-2\pi/3}{2} = \dfrac{\pi}{3}$

32. $y = -4\sec\left(4x - \dfrac{3\pi}{2}\right)$ [2.6]

no amplitude; period $= \dfrac{2\pi}{b} = \dfrac{2\pi}{4} = \dfrac{\pi}{2}$

phase shift $= -\dfrac{c}{b} = -\dfrac{-3\pi/2}{4} = \dfrac{3\pi}{8}$

33. $y = 2\csc\left(x - \dfrac{\pi}{4}\right) - 3$ [2.6]

no amplitude; period $= \dfrac{2\pi}{b} = \dfrac{2\pi}{1} = 2\pi$

phase shift $= -\dfrac{c}{b} = -\dfrac{-\pi/4}{1} = \dfrac{\pi}{4}$

34. $y = 2\cos \pi x, \; p = \dfrac{2\pi}{\pi} = 2$

35. $y = -\sin \dfrac{2x}{3}, \; p = \dfrac{2\pi}{2/3} = 3\pi$

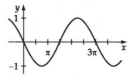

36. $y = 2\sin \dfrac{3x}{2}, \; p = \dfrac{2\pi}{3/2} = \dfrac{4\pi}{3}$

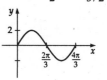

37. $y = \cos\left(x - \dfrac{\pi}{2}\right), \; p = 2\pi,$

phase shift $= \dfrac{\pi}{2}$

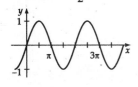

38. $y = \dfrac{1}{2}\sin\left(2x + \dfrac{\pi}{4}\right), \; p = \dfrac{2\pi}{2} = \pi,$

phase shift $-\dfrac{\pi}{8}$

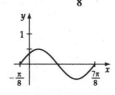

39. $y = 3\cos 3(x - \pi), \; p = \dfrac{2\pi}{3},$

phase shift $= \pi$

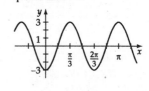

40. $y = -\tan \dfrac{x}{2}, \; p = \dfrac{\pi}{1/2} = 2\pi$

41. $y = 2\cot 2x, \; p = \dfrac{\pi}{2}$

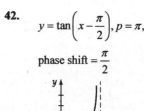

42. $y = \tan\left(x - \dfrac{\pi}{2}\right), \; p = \pi,$

phase shift $= \dfrac{\pi}{2}$

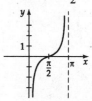

Copyright © Houghton Mifflin Company. All rights reserved.

43. $y = -\cot\left(2x + \dfrac{\pi}{4}\right),\ p = \dfrac{\pi}{2},$

phase shift $= -\dfrac{\pi}{8}$

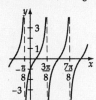

44. $y = -2\csc\left(2x - \dfrac{\pi}{3}\right),\ p = \dfrac{2\pi}{2} = \pi,$

phase shift $= \dfrac{\pi}{6}$

45. $y = 3\sec\left(x + \dfrac{\pi}{4}\right),\ p = 2\pi,$

phase shift $= -\dfrac{\pi}{4}$

46. $y = 3\sin 2x - 3$

47. $y = 2\cos 3x + 3$

48. $y = -\cos\left(3x + \dfrac{\pi}{2}\right) + 2$

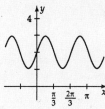

49. $y = 3\sin\left(4x - \dfrac{2\pi}{3}\right) - 3$

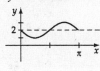

50. $y = 2 - \sin 2x$

51. $y = \sin x - \sqrt{3}\cos x$

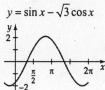

52.

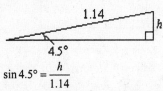

$\sin 4.5° = \dfrac{h}{1.14}$

$h = 1.14\sin 4.5° \approx 0.089$ mi

[2.2]

53.

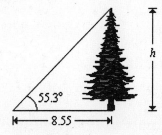

$\tan 55.3° = \dfrac{h}{8.55}$

$h = 8.55\tan 55.3° \approx 12.3$ feet

[2.2]

Copyright © Houghton Mifflin Company. All rights reserved.

54.

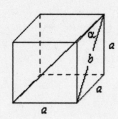

$$b = \sqrt{a^2 + a^2}$$
$$= a\sqrt{2}$$

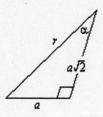

$$r^2 = a^2 + (a\sqrt{2})^2$$
$$= a^2 + 2a^2$$
$$= 3a^2$$

$$\sin\alpha = \frac{a}{r} = \frac{a}{a\sqrt{3}} = \frac{1}{\sqrt{3}} = \frac{\sqrt{3}}{3}$$

[2.2]

55.

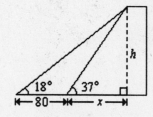

(1) $\cot 18° = \frac{80 + x}{h} = \frac{80}{h} + \frac{x}{h}$

(2) $\cot 37° = \frac{x}{h}$

Substitute for $\frac{x}{h}$ in equation (1).

$$\cot 18° = \frac{80}{h} + \cot 37°$$

Solve for h. $\frac{80}{h} = \cot 18° - \cot 37°$

$$\frac{h}{80} = \frac{1}{\cot 18° - \cot 37°}$$

$$h = \frac{80}{\cot 18° - \cot 37°} \approx 46 \text{ ft}$$

[2.2]

56. $y = 2.5\sin 50t$ [2.8]

amplitude $= 2.5$

$$p = \frac{2\pi}{b} = \frac{2\pi}{50} = \frac{\pi}{25}$$

frequency $= \frac{1}{p} = \frac{25}{\pi}$

57.   amplitude $= 0.5$ [2.8]

$$f = \frac{1}{2\pi}\sqrt{\frac{k}{m}} = \frac{1}{2\pi}\sqrt{\frac{20}{5}} = \frac{1}{\pi}$$

$p = \pi$

$$y = -0.5\cos 2\pi ft = -0.5\cos 2\pi\left(\frac{1}{\pi}\right)t$$

$$y = -0.5\cos 2t$$

58. $\left|f(t)\right| < 0.01$ for all $t > 7.2$ [2.8]

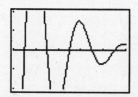

Xmin $= 5$, Xmax $= 10$, Xcsl $= 1$
Ymin $= -.01$, Ymax $= .01$, Yscl $= .005$

··

Chapter Test

1. $150° = 150°\left(\frac{\pi}{180°}\right)$ [2.1]

$$= \frac{5\pi}{6}$$

2. $\pi - \frac{11}{12}\pi = \frac{\pi}{12}$ [2.1]

3. $s = r\theta$ [2.1]

$$s = 10(75°)\left(\frac{\pi}{180°}\right)$$

$$s \approx 13.1 \text{ cm}$$

4. $w = 6\frac{rev}{sec}$ [2.1]

$$w = 6\frac{rev}{sec}\left(\frac{2\pi \ rad}{rev}\right)$$

$$w = 12\pi \ rad/sec$$

5. $v = rw$ [2.1]
$$= 8 \cdot 10$$
$$= 80 \text{ cm/sec}$$

6.

$$r = \sqrt{7^2 + 3^2}$$
$$r = \sqrt{58}$$

$$\sec\theta = \frac{\sqrt{58}}{7}$$ [2.2]

Copyright © Houghton Mifflin Company. All rights reserved.

7. $\csc 67° \approx 1.0864$ [2.2]

8.
$$\tan\frac{\pi}{6}\cos\frac{\pi}{3}-\sin\frac{\pi}{2}=\frac{1}{\sqrt{3}}\cdot\frac{1}{2}-1 \quad [2.3]$$
$$=\frac{1}{2\sqrt{3}}-1$$
$$=\frac{\sqrt{3}}{6}-1$$
$$=\frac{\sqrt{3}-6}{6}$$

9.
$$t=\frac{11\pi}{6} \quad [2.4]$$
$$x=\cos t \qquad y=\sin t$$
$$=\frac{\sqrt{3}}{2} \qquad =-\frac{1}{2}$$
$$P(x,y)=P\left(\frac{\sqrt{3}}{2},-\frac{1}{2}\right)$$

10.
$$\frac{\sec^2 t-1}{\sec^2 t}=\frac{\dfrac{1}{\cos^2 t}-1}{\dfrac{1}{\cos^2 t}} \quad [2.4]$$
$$=\frac{\dfrac{1-\cos^2 t}{\cos^2 t}}{\dfrac{1}{\cos^2 t}}$$
$$=1-\cos^2 t$$
$$=\sin^2 t$$

11.
$$\text{period}=\frac{\pi}{b}=\frac{\pi}{3} \quad [2.6]$$

12.
$$a=\left|-3\right|=3; \quad \text{period}=\frac{2\pi}{b}=\frac{2\pi}{2}=\pi \quad [2.7]$$
$$\text{phase shift}=-\frac{\pi}{4}$$

13.
$$\text{period}=\frac{\pi}{\pi/3}=3 \quad [2.7]$$
$$\text{phase shift}=-\frac{c}{b}=-\frac{\pi/6}{\pi/3}=-\frac{1}{2}$$

14.
$$y=3\cos\frac{1}{2}x, \, p=4\pi$$

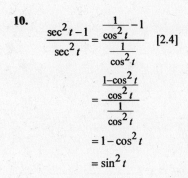

15.
$$y=-2\sec\frac{1}{2}x, \qquad p=4\pi$$

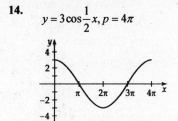

16. Shift the graph [of $y=2\sin(2x)$] [2.7]
$\frac{\pi}{4}$ units to the right and down 1 unit.

17.
$$y=2-\sin\frac{x}{2}$$

18.
$$y=\sin x-\cos 2x$$

Copyright © Houghton Mifflin Company. All rights reserved.

19.

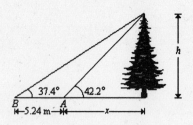

$$\tan 42.2° = \frac{h}{x}$$

$$x = \frac{h}{\tan 42.2°}$$

$$= h \cot 422.2°$$

$$\tan 37.4° = \frac{h}{5.24x}$$

$$= \frac{h}{5.24 + h \cot 42.2°}$$

Solve for h.

$$\tan 37.4° = \frac{h}{5.24 + h \cot 42.2°}$$

$$\tan 37.4°(5.24 + h \cot 42.2°) = h$$

$$5.24 \tan 37.4° + h \tan 37.4° \cot 42.2° = h$$

$$h - h \tan 37.4° \cot 42.2° = 5.24 \tan 37.4°$$

$$h(1 - \tan 37.4° \cot 42.2°) = 5.24 \tan 37.4°$$

$$h = \frac{5.24 \tan 37.4°}{1 - \tan 37.4° \cot 42.2°}$$

$$h \approx 25.5 \text{ meters}$$

The height of the tree is approximately 25.5 meters. [2.2]

20.

$$p = 5 \qquad 5 = \frac{2\pi}{b}, b = \frac{2\pi}{5}$$
$$a = 13$$

$$y = 13 \cos \frac{2\pi}{5} t \text{ or } y = 13 \sin \frac{2\pi}{5} t \quad [2.8]$$

● ●

1.

$$d = \sqrt{(4 - (-3))^2 + (1 - 2)^2} = \sqrt{49 + 1} = \sqrt{50} \quad [1.1]$$

2.

$$c^2 = a^2 + b^2 \qquad [1.1]$$

$$1^2 = \left(\frac{1}{2}\right)^2 + b^2$$

$$\frac{3}{4} = b^2$$

$$\frac{\sqrt{3}}{2} = b$$

3. Intercepts: $(-3, 0), (3, 0), (-9, 0)$ [1.2]

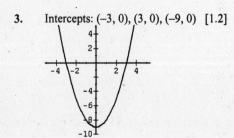

4.

$$f(-x) = \frac{-x}{(-x)^2 + 1} = \frac{-x}{x^2 + 1} = -f(x) \quad [1.4]$$

Odd function

Copyright © Houghton Mifflin Company. All rights reserved.

5.
$$f(x) = \frac{x}{2x-3} \qquad [1.5]$$
$$x = \frac{y}{2y-3}$$
$$x(2y-3) = 2xy - 3x = y$$
$$2xy - y = y(2x-1) = 3x$$
$$y = \frac{3x}{2x-1}$$
$$f^{-1}(x) = \frac{3x}{2x-1}$$

6. Domain: $(-\infty, 4) \cup (4, \infty)$ [1.3]

7.
$$x^2 + x - 6 = 0$$
$$(x+3)(x-2) = 0$$
$$x+3 = 0 \qquad x-2 = 0$$
$$x = -3 \qquad x = 2$$
The solutions are -3 and 2.   [1.1]

8. Shift the graph of $y = f(x)$ horizontally 3 units to the right. [1.4]

9. Reflect the graph of $y = f(x)$ across the y-axis. [1.4[

10.
$$300° = 300°\left(\frac{\pi}{180°}\right) = \frac{5\pi}{3} \quad [2.1]$$

11.
$$\frac{5\pi}{4} = \frac{5\pi}{4}\left(\frac{180°}{\pi}\right) = 225° \quad [2.1]$$

12.
$$f\left(\frac{\pi}{3}\right) = \sin\left(\frac{\pi}{3} + \frac{\pi}{6}\right) = \sin\left(\frac{\pi}{2}\right) = 1 \quad [2.3]$$

13.
$$f\left(\frac{\pi}{3}\right) = \sin\left(\frac{\pi}{3}\right) + \sin\left(\frac{\pi}{6}\right) = \frac{\sqrt{3}}{2} + \frac{1}{2} = \frac{\sqrt{3}+1}{2} \quad [2.2]$$

14.
$$\cos^2 45° + \sin^2 60° + = \left(\frac{\sqrt{2}}{2}\right)^2 + \left(\frac{\sqrt{3}}{2}\right)^2 = \frac{2}{4} + \frac{3}{4} = \frac{5}{4} \quad [2.2]$$

15. negative [2.3]

16.
$$\theta = 210° \quad [2.3]$$
Since $180° < \theta < 270°$,
$$\theta' + 180° = \theta$$
$$\theta' = 30°$$

17.
$$\theta = \frac{2\pi}{3} \quad [2.3]$$
Since $\frac{\pi}{2} < \theta < \pi$,
$$\theta + \theta' = \pi$$
$$\theta' = \frac{\pi}{3}$$

18. Domain: $(-\infty, \infty)$ [2.4]

19. Range: $[-1, 1]$ [2.4]

20.
$$\tan\theta = \frac{\text{opp}}{\text{adj}} = \frac{3}{4}$$
$$\text{hypotenuse} = \sqrt{3^2 + 4^2}$$
$$= \sqrt{9+16}$$
$$= \sqrt{25}$$
$$= 5$$
$$\sin\theta = \frac{\text{opp}}{\text{hyp}} = \frac{3}{5} \quad [2.2]$$

Copyright © Houghton Mifflin Company. All rights reserved.

Chapter 3
Trigonometric Identities and Equations

Section 3.1

1. Not an identity. If $x = \pi/4$, the left side is 2 and the right side is 1.

3. Not an identity. If $x = 0°$, the left side is $\sqrt{3}/2$ and the right side is $(2+\sqrt{3})/2$.

5. Not an identity. If $x = 0$, the left side is -1 and the right side is 1.

7. Not an identity. If $x = 0$, the left side is -1 and the right side is 1.

9. Not an identity. If $x = \pi/4$, the left side is 2 and the right side is 1.

11. $\tan x \csc x \cos x = \dfrac{\sin x}{\cos x} \cdot \dfrac{1}{\sin x} \cdot \cos x = 1$

13. $\dfrac{4\sin^2 x - 1}{2\sin x + 1} = \dfrac{(2\sin x - 1)(2\sin x + 1)}{2\sin x + 1} = 2\sin x - 1$

15.
$$(\sin x - \cos x)(\sin x + \cos x) = \sin^2 x - \cos^2 x$$
$$= 1 - \cos^2 x - \cos^2 x$$
$$= 1 - 2\cos^2 x$$

17.
$$\dfrac{1}{\sin x} - \dfrac{1}{\cos x} = \dfrac{\cos x}{\sin x \cos x} - \dfrac{\sin x}{\sin x \cos x}$$
$$= \dfrac{\cos x - \sin x}{\sin x \cos x}$$

19.
$$\dfrac{\cos x}{1 - \sin x} = \dfrac{\cos x(1 + \sin x)}{(1 - \sin x)(1 + \sin x)}$$
$$= \dfrac{\cos x(1 + \sin x)}{1 - \sin^2 x}$$
$$= \dfrac{\cos x(1 + \sin x)}{\cos^2 x}$$
$$= \dfrac{(1 + \sin x)}{\cos x} = \dfrac{1}{\cos x} + \dfrac{\sin x}{\cos x}$$
$$= \sec x + \tan x$$

21.
$$\dfrac{1 - \tan^4 x}{\sec^2 x} = \dfrac{(1 + \tan^2 x)(1 - \tan^2 x)}{\sec^2 x}$$
$$= \dfrac{\sec^2 x(1 - \tan^2 x)}{\sec^2 x}$$
$$= 1 - \tan^2 x$$

23.
$$\dfrac{1 + \tan^3 x}{1 + \tan x} = \dfrac{(1 + \tan x)(1 - \tan x + \tan^2 x)}{1 + \tan x}$$
$$= 1 - \tan x + \tan^2 x$$

25.
$$\dfrac{\sin x - 2 + \frac{1}{\sin x}}{\sin x - \frac{1}{\sin x}} = \dfrac{\sin x - 2 + \frac{1}{\sin x}}{\sin x - \frac{1}{\sin x}} \cdot \dfrac{\sin x}{\sin x}$$
$$= \dfrac{\sin^2 x - 2\sin x + 1}{\sin^2 x - 1}$$
$$= \dfrac{(\sin x - 1)(\sin x - 1)}{(\sin x - 1)(\sin x + 1)}$$
$$= \dfrac{\sin x - 1}{\sin x + 1}$$

27.
$$(\sin x + \cos x)^2 = \sin^2 x + 2\sin x \cos x + \cos^2 x$$
$$= \sin^2 x + \cos^2 x + 2\sin x \cos x$$
$$= 1 + 2\sin x \cos x$$

29.
$$\dfrac{\cos x}{1 + \sin x} = \dfrac{\cos x(1 - \sin x)}{(1 + \sin x)(1 - \sin x)}$$
$$= \dfrac{\cos x(1 - \sin x)}{1 - \sin^2 x}$$
$$= \dfrac{\cos x(1 - \sin x)}{\cos^2 x}$$
$$= \dfrac{1 - \sin x}{\cos x}$$
$$= \dfrac{1}{\cos x} - \dfrac{\sin x}{\cos x}$$
$$= \sec x - \tan x$$

Copyright © Houghton Mifflin Company. All rights reserved.

31.
$$\frac{\cot x + \tan x}{\sec x} = \frac{\dfrac{\cos x}{\sin x} + \dfrac{\sin x}{\cos x}}{\dfrac{1}{\cos x}}$$

$$= \frac{\dfrac{\cos x}{\sin x} + \dfrac{\sin x}{\cos x}}{\dfrac{1}{\cos x}} \cdot \frac{\sin x \cos x}{\sin x \cos x}$$

$$= \frac{\cos^2 x + \sin^2 x}{\sin x}$$

$$= \frac{1}{\sin x}$$

$$= \csc x$$

33.
$$\frac{\cos x \tan x + 2\cos x - \tan x - 2}{\tan x + 2} = \frac{\cos x(\tan x + 2) - (\tan x + 2)}{\tan x + 2}$$

$$= \frac{(\tan x + 2)(\cos x - 1)}{\tan x + 2}$$

$$= \cos x - 1$$

35.
$$\frac{1 - \sin x}{\cos x} = \frac{1}{\cos x} - \frac{\sin x}{\cos x} = \sec x - \tan x$$

37.
$$\sin^2 x - \cos^2 x = \sin^2 x - (1 - \sin^2 x)$$

$$= \sin^2 x - 1 + \sin^2 x$$

$$= 2\sin^2 x - 1$$

39.
$$\frac{1}{\sin^2 x} + \frac{1}{\cos^2 x} = \frac{\cos^2 x + \sin^2 x}{\sin^2 x \cos^2 x}$$

$$= \frac{1}{\sin^2 x \cos^2 x}$$

$$= \csc^2 x \sec^2 x$$

41.
$$\sec x - \cos x = \frac{1}{\cos x} - \cos x$$

$$= \frac{1 - \cos^2 x}{\cos x}$$

$$= \frac{\sin^2 x}{\cos x}$$

$$= \sin x \tan x$$

43.
$$\frac{\dfrac{1}{\sin x} + 1}{\dfrac{1}{\sin x} - 1} = \frac{\dfrac{1}{\sin x} + 1}{\dfrac{1}{\sin x} - 1} \cdot \frac{\sin x}{\sin x}$$

$$= \frac{1 + \sin x}{1 - \sin x}$$

$$= \frac{(1 + \sin x)}{1 - \sin x} \cdot \frac{(1 + \sin x)}{1 + \sin x}$$

$$= \frac{1 + 2\sin x + \sin^2 x}{1 - \sin^2 x}$$

$$= \frac{1 + 2\sin x + \sin^2 x}{\cos^2 x}$$

$$= \frac{1}{\cos^2 x} + \frac{2\sin x}{\cos^2 x} + \frac{\sin^2 x}{\cos^2 x}$$

$$= \sec^2 x + 2\tan x \sec x + \tan^2 x$$

45.
$$\sin^4 x - \cos^4 x = (\sin^2 x + \cos^2 x)(\sin^2 x - \cos^2 x)$$

$$= 1(\sin^2 x - \cos^2 x)$$

$$= \sin^2 x - (1 - \sin^2 x)$$

$$= \sin^2 x - 1 + \sin^2 x$$

$$= 2\sin^2 x - 1$$

47.
$$\frac{1}{1 - \cos x} = \frac{1}{1 - \cos x} \cdot \frac{1 + \cos x}{1 + \cos x}$$

$$= \frac{1 + \cos x}{1 - \cos^2 x}$$

$$= \frac{1 + \cos x}{\sin^2 x}$$

49.
$$\frac{\sin x}{1 - \sin x} - \frac{\cos x}{1 - \sin x} = \frac{\sin x - \cos x}{1 - \sin x}$$

$$= \frac{\dfrac{\sin x}{\sin x} - \dfrac{\cos x}{\sin x}}{\dfrac{1}{\sin x} - \dfrac{\sin x}{\sin x}}$$

$$= \frac{1 - \cot x}{\csc x - 1}$$

Copyright © Houghton Mifflin Company. All rights reserved.

51.
$$\frac{1}{1+\cos x} - \frac{1}{1-\cos x} = \frac{(1-\cos x)-(1+\cos x)}{(1+\cos x)(1-\cos x)}$$
$$= \frac{1-\cos x-1-\cos x}{1-\cos^2 x}$$
$$= \frac{-2\cos x}{\sin^2 x}$$
$$= -2\cot x \csc x$$

53.
$$\frac{\frac{1}{\sin x}+\csc x}{\frac{1}{\sin x}-\sin x} = \frac{\frac{1}{\sin x}+\csc x}{\frac{1}{\sin x}-\sin x} \cdot \frac{\sin x}{\sin x}$$
$$= \frac{1+1}{1-\sin^2 x}$$
$$= \frac{2}{\cos^2 x}$$

55.
$$\frac{\cot x}{1+\csc x} + \frac{1+\csc x}{\cot x} = \frac{\frac{\cos x}{\sin x}}{1+\frac{1}{\sin x}} + \frac{1+\frac{1}{\sin x}}{\frac{\cos x}{\sin x}}$$
$$= \frac{\cos x}{\sin x+1} + \frac{\sin x+1}{\cos x}$$
$$= \frac{\cos^2 x + (\sin x+1)^2}{\cos x(\sin x+1)}$$
$$= \frac{\cos^2 x + \sin^2 x + 2\sin x + 1}{\cos x(\sin x+1)}$$
$$= \frac{1+2\sin x+1}{\cos x(\sin x+1)}$$
$$= \frac{2(1+\sin x)}{\cos x(\sin x+1)}$$
$$= 2\sec x$$

57.
$$\sqrt{\frac{1+\sin x}{1-\sin x}} = \sqrt{\frac{1+\sin x}{1-\sin x} \cdot \frac{1+\sin x}{1+\sin x}}$$
$$= \sqrt{\frac{(1+\sin x)^2}{1-\sin^2 x}}$$
$$= \sqrt{\frac{(1+\sin x)^2}{\cos^2 x}}$$
$$= \frac{1+\sin x}{\cos x}, \quad \cos x > 0$$

59.
$$\frac{\sin^3 x + \cos^3 x}{\sin x + \cos x} = \frac{(\sin x+\cos x)(\sin^2 x-\sin x\cos x+\cos^2 x)}{\sin x+\cos x}$$
$$= \sin^2 x - \sin x\cos x + \cos^2 x$$
$$= 1 - \sin x\cos x$$

61.
$$\frac{\sec x-1}{\sec x+1} - \frac{\sec x+1}{\sec x-1} = \frac{(\sec x-1)^2-(\sec x+1)^2}{(\sec x-1)(\sec x+1)}$$
$$= \frac{\sec^2 x-2\sec x+1-\sec^2 x-2\sec x-1}{\sec^2 x-1}$$
$$= \frac{-4\sec x}{\tan^2 x}$$
$$= \frac{-4}{\cos x} \cdot \frac{\cos^2 x}{\sin^2 x}$$
$$= -4\csc x\cot x$$

63.
$$\frac{1+\sin x}{\cos x} - \frac{\cos x}{1-\sin x} = \frac{(1+\sin x)(1-\sin x)-\cos x(\cos x)}{\cos x(1-\sin x)}$$
$$= \frac{1-\sin^2 x-\cos^2 x}{\cos x(1-\sin x)}$$
$$= \frac{\cos^2 x-\cos^2 x}{\cos x(1-\sin x)}$$
$$= 0$$

65.
$$\frac{\sec x+\tan x}{\sec x-\tan x} = \frac{\frac{1}{\cos x}+\frac{\sin x}{\cos x}}{\frac{1}{\cos x}-\frac{\sin x}{\cos x}} \cdot \frac{\cos x}{\cos x}$$
$$= \frac{1+\sin x}{1-\sin x}$$
$$= \frac{1+\sin x}{1-\sin x} \cdot \frac{1+\sin x}{1+\sin x}$$
$$= \frac{(1+\sin x)^2}{1-\sin^2 x}$$
$$= \frac{(1+\sin x)^2}{\cos^2 x}$$

67.
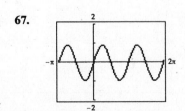

69.
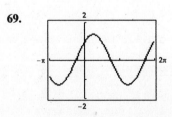

71.
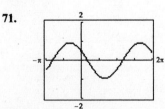

Copyright © Houghton Mifflin Company. All rights reserved.

73.

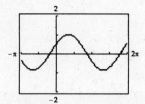

• •

75.
$$\frac{1-\sin x+\cos x}{1+\sin x+\cos x}=\frac{1-\sin x+\cos x}{1+\sin x+\cos x}\cdot\frac{1+\sin x+\cos x}{1+\sin x-\cos x}$$

$$=\frac{1-\sin^2 x+2\sin x\cos x-\cos^2 x}{1+2\sin x+\sin^2 x-\cos^2 x}$$

$$=\frac{1-(\sin^2 x+\cos^2 x)+2\sin x\cos x}{1+2\sin x+\sin^2 x-(1-\sin^2 x)}$$

$$=\frac{1-1+2\sin x\cos x}{1+2\sin x+\sin^2 x-1+\sin^2 x}$$

$$=\frac{2\sin x\cos x}{2\sin x+2\sin^2 x}=\frac{2\sin x\cos x}{2\sin x(1+\sin x)}$$

$$=\frac{\cos x}{1+\sin x}$$

77.
$$\frac{2\sin^4 x+2\sin^2 x\cos^2 x-3\sin^2 x-3\cos^2 x}{2\sin^2 x}=\frac{2\sin^2 x(\sin^2 x+\cos^2 x)-3(\sin^2 x+\cos^2 x)}{2\sin^2 x}$$

$$=\frac{(2\sin^2 x-3)(\sin^2 x+\cos^2 x)}{2\sin^2 x}$$

$$=\frac{2\sin^2 x-3}{2\sin^2 x}$$

$$=\frac{2\sin^2 x}{2\sin^2 x}-\frac{3}{2\sin^2 x}$$

$$=1-\frac{3}{2}\csc^2 x$$

79.
$$\frac{\sin x(\tan x+1)-2\tan x\cos x}{\sin x-\cos x}=\frac{\sin x\tan x+\sin x-2\dfrac{\sin x}{\cos x}\cos x}{\sin x-\cos x}$$

$$=\frac{\sin x\tan x+\sin x-2\sin x}{\sin x-\cos x}$$

$$=\frac{\sin x(\tan x-1)}{\sin x-\cos x}$$

$$=\frac{\dfrac{\sin x(\tan x-1)}{\cos x}}{\dfrac{\sin x}{\cos x}-\dfrac{\cos x}{\cos x}}$$

$$=\frac{\tan x(\tan x-1)}{\tan x-1}$$

$$=\tan x$$

Copyright © Houghton Mifflin Company. All rights reserved.

81. $\sin^4 x + \cos^4 x = \sin^4 x + 2\sin^2 x \cos^2 x + \cos^4 x - 2\sin^2 x \cos^2 x$

$\qquad\qquad\qquad = (\sin^2 x + \cos^2 x)^2 - 2\sin^2 x \cos^2 x$

$\qquad\qquad\qquad = 1 - 2\sin^2 x \cos^2 x$

83. $\cos\left(\dfrac{\pi}{2} - \dfrac{\pi}{6}\right) = \cos\left(\dfrac{\pi}{3}\right) = \dfrac{1}{2}$

$\cos\left(\dfrac{\pi}{2}\right)\cos\left(\dfrac{\pi}{6}\right) + \sin\left(\dfrac{\pi}{2}\right)\sin\left(\dfrac{\pi}{6}\right) = 0 \cdot \dfrac{\sqrt{3}}{2} + 1 \cdot \dfrac{1}{2} = \dfrac{1}{2}$

Both functional values equal $\dfrac{1}{2}$.

84. $\sin\left(\dfrac{\pi}{2} + \dfrac{\pi}{3}\right) = \sin\left(\dfrac{5\pi}{6}\right) = \dfrac{1}{2}$

$\sin\left(\dfrac{\pi}{2}\right)\cos\left(\dfrac{\pi}{3}\right) + \cos\left(\dfrac{\pi}{2}\right)\sin\left(\dfrac{\pi}{3}\right) = 1 \cdot \dfrac{1}{2} + 0 \cdot \dfrac{\sqrt{3}}{2} = \dfrac{1}{2}$

Both functional values equal to $\dfrac{1}{2}$.

85. $\sin(90° - 30°) = \sin(60°) = \dfrac{\sqrt{3}}{2} = \cos(30°)$

$\sin(90° - 45°) = \sin(45°) = \dfrac{\sqrt{2}}{2} = \cos(45°)$

$\sin(90° - 120°) = \sin(-30°) = -\dfrac{1}{2} = \cos(120°)$

For each of the given values of θ, the functional values are equal.

86. $\tan\left(\dfrac{\pi}{2} - \dfrac{\pi}{6}\right) = \tan\left(\dfrac{\pi}{3}\right) = \cot\left(\dfrac{\pi}{6}\right)$

$\tan\left(\dfrac{\pi}{2} - \dfrac{\pi}{4}\right) = \tan\left(\dfrac{\pi}{4}\right) = \cot\left(\dfrac{\pi}{4}\right)$

$\tan\left(\dfrac{\pi}{2} - \dfrac{4\pi}{3}\right) = \tan\left(-\dfrac{5\pi}{6}\right) = \cot\left(\dfrac{4\pi}{3}\right)$

For each of the given values of θ, the functional values are equal.

87. $\tan\left(\dfrac{\pi}{3} - \dfrac{\pi}{6}\right) = \tan\left(\dfrac{\pi}{6}\right) = \dfrac{\sqrt{3}}{3}$

$\dfrac{\tan\left(\dfrac{\pi}{3}\right) - \tan\left(\dfrac{\pi}{6}\right)}{1 + \tan\left(\dfrac{\pi}{3}\right)\tan\left(\dfrac{\pi}{6}\right)} = \dfrac{\sqrt{3} - \dfrac{\sqrt{3}}{3}}{1 + \sqrt{3} \cdot \dfrac{\sqrt{3}}{3}} = \dfrac{\dfrac{3\sqrt{3}}{3} - \dfrac{\sqrt{3}}{3}}{1 + 1} = \dfrac{\dfrac{2\sqrt{3}}{3}}{2} = \dfrac{\sqrt{3}}{3}$

Both functional values equal $\dfrac{\sqrt{3}}{3}$.

88. For k is any integer, the value of $(2k + 1)$ will result in odd integers.

Thus $\sin[(2k+1)\pi]$ will be 0.

Section 3.2

1. $\sin(45° + 30°) = \sin 45° \cos 30° + \cos 45° \sin 30°$

$\qquad\qquad\qquad = \dfrac{\sqrt{2}}{2} \cdot \dfrac{\sqrt{3}}{2} + \dfrac{\sqrt{2}}{2} \cdot \dfrac{1}{2}$

$\qquad\qquad\qquad = \dfrac{\sqrt{6}}{4} + \dfrac{\sqrt{2}}{4} = \dfrac{\sqrt{6} + \sqrt{2}}{4}$

3. $\cos(45° - 30°) = \cos 45° \cos 30° + \sin 45° \sin 30°$

$\qquad\qquad\qquad = \dfrac{\sqrt{2}}{2} \cdot \dfrac{\sqrt{3}}{2} + \dfrac{\sqrt{2}}{2} \cdot \dfrac{1}{2}$

$\qquad\qquad\qquad = \dfrac{\sqrt{6}}{4} + \dfrac{\sqrt{2}}{4} = \dfrac{\sqrt{6} + \sqrt{2}}{4}$

5. $\tan(45° - 30°) = \dfrac{\tan 45° - \tan 30°}{1 + \tan 45° \tan 30°}$

$\qquad\qquad = \dfrac{1 - \dfrac{\sqrt{3}}{3}}{1 + 1\left(\dfrac{\sqrt{3}}{3}\right)} = \dfrac{\dfrac{3 - \sqrt{3}}{3}}{\dfrac{3 + \sqrt{3}}{3}} = \dfrac{3 - \sqrt{3}}{3 + \sqrt{3}} = 2 - \sqrt{3}$

7. $\sin\left(\dfrac{5\pi}{4} - \dfrac{\pi}{6}\right) = \sin\dfrac{5\pi}{4}\cos\dfrac{\pi}{6} - \cos\dfrac{5\pi}{4}\sin\dfrac{\pi}{6}$

$\qquad\qquad = -\dfrac{\sqrt{2}}{2} \cdot \dfrac{\sqrt{3}}{2} - \left(-\dfrac{\sqrt{2}}{2}\right) \cdot \dfrac{1}{2}$

$\qquad\qquad = -\dfrac{\sqrt{6}}{4} + \dfrac{\sqrt{2}}{4} = \dfrac{-\sqrt{6} + \sqrt{2}}{4}$

Copyright © Houghton Mifflin Company. All rights reserved.

9.
$$\cos\left(\frac{3\pi}{4}+\frac{\pi}{6}\right)=\cos\frac{3\pi}{4}\cos\frac{\pi}{6}-\sin\frac{3\pi}{4}\sin\frac{\pi}{6}$$
$$=-\frac{\sqrt{2}}{2}\cdot\frac{\sqrt{3}}{2}-\frac{\sqrt{2}}{2}\cdot\frac{1}{2}$$
$$=-\frac{\sqrt{6}}{4}-\frac{\sqrt{2}}{4}=-\frac{\sqrt{6}+\sqrt{2}}{4}$$

11.
$$\tan\left(\frac{\pi}{6}+\frac{\pi}{4}\right)=\frac{\tan\frac{\pi}{6}+\tan\frac{\pi}{4}}{1-\tan\frac{\pi}{6}\tan\frac{\pi}{4}}$$
$$=\frac{\frac{\sqrt{3}}{3}+1}{1-\frac{\sqrt{3}}{3}\cdot1}=\frac{\frac{\sqrt{3}+3}{3}}{\frac{3-\sqrt{3}}{3}}$$
$$=\frac{\sqrt{3}+3}{3-\sqrt{3}}\cdot\frac{3+\sqrt{3}}{3+\sqrt{3}}=\frac{9+6\sqrt{3}+3}{9-3}$$
$$=\frac{12+6\sqrt{3}}{6}=2+\sqrt{3}$$

13. $\cos 212°\cos 122° + \sin 212°\sin 122° = \cos(212° - 122°) = \cos 90° = 0$

15. $\sin\frac{5\pi}{12}\cos\frac{\pi}{4}-\cos\frac{5\pi}{12}\sin\frac{\pi}{4}=\sin\left(\frac{5\pi}{12}-\frac{\pi}{4}\right)=\sin\frac{\pi}{6}=\frac{1}{2}$

17. $\dfrac{\tan\frac{7\pi}{12}-\tan\frac{\pi}{4}}{1+\tan\frac{7\pi}{12}\tan\frac{\pi}{4}}=\tan\left(\frac{7\pi}{12}-\frac{\pi}{4}\right)=\tan\frac{\pi}{3}=\sqrt{3}$

19. $\sin 42° = \cos(90° - 42°)$
     $= \cos 48°$

21. $\tan 15° = \cot(90° - 15°)$
     $= \cot 75°$

23. $\sec 25° = \csc(90° - 25°)$
     $= \csc 65°$

25. $\sin 7x\cos 2x - \cos 7x\sin 2x = \sin(7x - 2x) = \sin 5x$

27. $\cos x\cos 2x + \sin x\sin 2x = \cos(x - 2x) = \cos(-x) = \cos x$

29. $\sin 7x\cos 3x - \cos 7x\sin 3x = \sin(7x - 3x) = \sin 4x$

31. $\cos 4x\cos(-2x) - \sin 4x\sin(-2x) = \cos 4x\cos 2x + \sin 4x\sin 2x$
$$= \cos(4x - 2x)$$
$$= \cos 2x$$

33. $\sin\frac{x}{3}\cos\frac{2x}{3}+\cos\frac{x}{3}\sin\frac{2x}{3}=\sin\left(\frac{x}{3}+\frac{2x}{3}\right)=\sin x$

35. $\dfrac{\tan 3x + \tan 4x}{1-\tan 3x\tan 4x}=\tan(3x+4x)=\tan 7x$

Copyright © Houghton Mifflin Company. All rights reserved.

37.

$\tan\alpha = -\dfrac{4}{3}, \ \sin\alpha = \dfrac{4}{5}, \ \cos\alpha = -\dfrac{3}{5},$

$\tan\beta = \dfrac{15}{8}, \ \sin\beta = -\dfrac{15}{17}, \ \cos\beta = -\dfrac{8}{17}$

a. $\sin(\alpha - \beta) = \sin\alpha\cos\beta - \cos\alpha\sin\beta$

$\qquad = \left(\dfrac{4}{5}\right)\left(-\dfrac{8}{17}\right) - \left(-\dfrac{3}{5}\right)\left(-\dfrac{15}{17}\right)$

$\qquad = -\dfrac{32}{85} - \dfrac{45}{85} = -\dfrac{77}{85}$

b. $\cos(\alpha + \beta) = \cos\alpha\cos\beta - \sin\alpha\sin\beta$

$\qquad = \left(-\dfrac{3}{5}\right)\left(-\dfrac{8}{17}\right) - \left(\dfrac{4}{5}\right)\left(-\dfrac{15}{17}\right)$

$\qquad = \dfrac{24}{85} + \dfrac{60}{85} = \dfrac{84}{85}$

c. $\tan(\alpha - \beta) = \dfrac{\tan\alpha - \tan\beta}{1 + \tan\alpha\tan\beta}$

$\qquad = \dfrac{-\dfrac{4}{3} - \dfrac{15}{8}}{1 + \left(-\dfrac{4}{3}\right)\left(\dfrac{15}{8}\right)} = \dfrac{-\dfrac{4}{3} - \dfrac{15}{8}}{1 - \dfrac{60}{24}} \cdot \dfrac{24}{24}$

$\qquad = \dfrac{-32 - 45}{24 - 60} = \dfrac{77}{36}$

39.

$\sin\alpha = \dfrac{3}{5}, \ \cos\alpha = \dfrac{4}{5}, \ \tan\alpha = \dfrac{3}{4},$

$\cos\beta = -\dfrac{5}{13}, \ \sin\beta = \dfrac{12}{13}, \ \tan\beta = -\dfrac{12}{5}$

a. $\sin(\alpha - \beta) = \sin\alpha\cos\beta - \cos\alpha\sin\beta$

$\qquad = \dfrac{3}{5}\left(-\dfrac{5}{13}\right) - \dfrac{4}{5}\left(\dfrac{12}{13}\right)$

$\qquad = -\dfrac{15}{65} - \dfrac{48}{65}$

$\qquad = -\dfrac{63}{65}$

b. $\cos(\alpha + \beta) = \cos\alpha\cos\beta - \sin\alpha\sin\beta$

$\qquad = \dfrac{4}{5}\left(-\dfrac{5}{13}\right) - \dfrac{3}{5}\left(\dfrac{12}{13}\right)$

$\qquad = -\dfrac{20}{65} - \dfrac{36}{65}$

$\qquad = -\dfrac{56}{65}$

c. $\tan(\alpha - \beta) = \dfrac{\tan\alpha - \tan\beta}{1 + \tan\alpha\tan\beta}$

$\qquad = \dfrac{\dfrac{3}{4} - \left(-\dfrac{12}{5}\right)}{1 + \left(\dfrac{3}{4}\right)\left(-\dfrac{12}{5}\right)}$

$\qquad = \dfrac{\dfrac{3}{4} + \dfrac{12}{5}}{1 - \dfrac{36}{20}} \cdot \dfrac{20}{20}$

$\qquad = \dfrac{15 + 48}{20 - 36} = -\dfrac{63}{16}$

41.

$\sin\alpha = -\dfrac{4}{5}, \ \cos\alpha = -\dfrac{3}{5}, \ \tan\alpha = \dfrac{4}{3},$

$\cos\beta = -\dfrac{12}{13}, \ \sin\beta = \dfrac{5}{13}, \ \tan\beta = -\dfrac{5}{12}$

a. $\sin(\alpha - \beta) = \sin\alpha\cos\beta - \cos\alpha\sin\beta$

$\qquad = \left(-\dfrac{4}{5}\right)\left(-\dfrac{12}{13}\right) - \left(-\dfrac{3}{5}\right)\dfrac{5}{13}$

$\qquad = \dfrac{48}{65} + \dfrac{15}{65} = \dfrac{63}{65}$

b. $\cos(\alpha + \beta) = \cos\alpha\cos\beta - \sin\alpha\sin\beta$

$\qquad = \left(-\dfrac{3}{5}\right)\left(-\dfrac{12}{13}\right) - \left(-\dfrac{4}{5}\right)\dfrac{5}{13}$

$\qquad = \dfrac{36}{65} + \dfrac{20}{65} = \dfrac{56}{65}$

c. $\tan(\alpha + \beta) = \dfrac{\tan\alpha + \tan\beta}{1 - \tan\alpha\tan\beta}$

$\qquad = \dfrac{\dfrac{4}{3} + \left(-\dfrac{5}{12}\right)}{1 - \left(\dfrac{4}{3}\right)\left(-\dfrac{5}{12}\right)}$

$\qquad = \dfrac{\dfrac{4}{3} - \dfrac{5}{12}}{1 + \dfrac{29}{36}} = \dfrac{\dfrac{4}{3} - \dfrac{5}{12}}{1 + \dfrac{20}{36}} \cdot \dfrac{36}{36}$

$\qquad = \dfrac{48 - 15}{36 + 20} = \dfrac{33}{56}$

43.

$\cos\alpha = \dfrac{15}{17}, \ \sin\alpha = \dfrac{8}{17}, \ \tan\alpha = \dfrac{8}{15},$

$\sin\beta = -\dfrac{3}{5}, \ \cos\beta = -\dfrac{4}{5}, \ \tan\beta = \dfrac{3}{4}$

a. $\sin(\alpha + \beta) = \sin\alpha\cos\beta + \cos\alpha\sin\beta$

$\qquad = \dfrac{8}{17}\left(-\dfrac{4}{5}\right) + \dfrac{15}{17}\left(-\dfrac{3}{5}\right)$

$\qquad = -\dfrac{32}{85} - \dfrac{45}{85} = -\dfrac{77}{85}$

b. $\cos(\alpha - \beta) = \cos\alpha\cos\beta + \sin\alpha\sin\beta$

$\qquad = \dfrac{15}{17}\left(-\dfrac{4}{5}\right) + \dfrac{8}{17}\left(-\dfrac{3}{5}\right)$

$\qquad = -\dfrac{60}{85} - \dfrac{24}{85} = -\dfrac{84}{85}$

c. $\tan(\alpha - \beta) = \dfrac{\tan\alpha - \tan\beta}{1 + \tan\alpha\tan\beta}$

$\qquad = \dfrac{\dfrac{8}{15} - \dfrac{3}{4}}{1 + \dfrac{8}{15}\left(\dfrac{3}{4}\right)} = \dfrac{\dfrac{8}{15} - \dfrac{3}{4}}{1 + \dfrac{24}{60}} = \dfrac{\dfrac{8}{15} - \dfrac{3}{4}}{1 + \dfrac{24}{60}} \cdot \dfrac{60}{60}$

$\qquad = \dfrac{32 - 45}{60 + 24} = -\dfrac{13}{84}$

Copyright © Houghton Mifflin Company. All rights reserved.

45.

$$\cos\alpha = -\frac{3}{5}, \quad \sin\alpha = -\frac{4}{5}, \quad \tan\alpha = \frac{4}{3},$$

$$\sin\beta = \frac{5}{13}, \quad \cos\beta = \frac{12}{13}, \quad \tan\beta = \frac{5}{12}$$

a. $\sin(\alpha - \beta) = \sin\alpha\cos\beta - \cos\alpha\sin\beta$

$$= \left(-\frac{4}{5}\right)\frac{12}{13} - \left(-\frac{3}{5}\right)\left(\frac{5}{13}\right)$$

$$= -\frac{48}{65} + \frac{15}{65} = -\frac{33}{65}$$

b. $\cos(\alpha + \beta) = \cos\alpha\cos\beta - \sin\alpha\sin\beta$

$$= \left(-\frac{3}{5}\right)\frac{12}{13} - \left(-\frac{4}{5}\right)\frac{5}{13}$$

$$= -\frac{36}{65} + \frac{20}{65} = -\frac{16}{65}$$

c. $\tan(\alpha + \beta) = \dfrac{\tan\alpha + \tan\beta}{1 - \tan\alpha\tan\beta}$

$$= \frac{\frac{4}{3} + \frac{5}{12}}{1 - \frac{4}{3}\left(\frac{5}{12}\right)} = \frac{\frac{4}{3} + \frac{5}{12}}{1 - \frac{20}{36}}$$

$$= \frac{\frac{4}{3} + \frac{5}{12}}{1 - \frac{20}{36}} \cdot \frac{36}{36} = \frac{48 + 15}{36 - 20} = \frac{63}{16}$$

47.

$$\sin\alpha = \frac{3}{5}, \quad \cos\alpha = \frac{4}{5}, \quad \tan\alpha = \frac{3}{4},$$

$$\tan\beta = \frac{5}{12}, \quad \sin\beta = -\frac{5}{13}, \quad \cos\beta = -\frac{12}{13}$$

a. $\sin(\alpha + \beta) = \sin\alpha\cos\beta + \cos\alpha\sin\beta$

$$= \left(\frac{3}{5}\right)\left(-\frac{12}{13}\right) + \left(\frac{4}{5}\right)\left(-\frac{5}{13}\right)$$

$$= -\frac{36}{65} - \frac{20}{65} = -\frac{56}{65}$$

b. $\cos(\alpha - \beta) = \cos\alpha\cos\beta + \sin\alpha\sin\beta$

$$= \left(\frac{4}{5}\right)\left(-\frac{12}{13}\right) + \left(\frac{3}{5}\right)\left(-\frac{5}{13}\right)$$

$$= -\frac{48}{65} - \frac{15}{65} = -\frac{63}{65}$$

c. $\tan(\alpha - \beta) = \dfrac{\tan\alpha - \tan\beta}{1 + \tan\alpha\tan\beta}$

$$= \frac{\frac{3}{4} - \frac{5}{12}}{1 + \left(\frac{3}{4}\right)\left(\frac{5}{12}\right)} = \frac{\frac{3}{4} - \frac{5}{12}}{1 + \frac{15}{48}} \cdot \frac{48}{48}$$

$$= \frac{36 - 20}{48 + 15} = \frac{16}{63}$$

49.

$$\cos\left(\frac{\pi}{2} - \theta\right) = \cos\frac{\pi}{2}\cos\theta + \sin\frac{\pi}{2}\sin\theta$$

$$= 0 \cdot \cos\theta + 1 \cdot \sin\theta$$

$$= \sin\theta$$

51.

$$\sin\left(\theta + \frac{\pi}{2}\right) = \sin\theta\cos\frac{\pi}{2} + \cos\theta\sin\frac{\pi}{2}$$

$$= \sin\theta(0) + \cos\theta(1)$$

$$= \cos\theta$$

53.

$$\tan\left(\theta + \frac{\pi}{4}\right) = \frac{\tan\theta + \tan\frac{\pi}{4}}{1 - \tan\theta\tan\frac{\pi}{4}}$$

$$= \frac{\tan\theta + 1}{1 - \tan\theta}$$

55.

$$\cos\left(\frac{3\pi}{2} - \theta\right) = \cos\frac{3\pi}{2}\cos\theta + \sin\frac{3\pi}{2}\sin\theta$$

$$= 0(\cos\theta) + (-1)\sin\theta$$

$$= -\sin\theta$$

57.

$$\cot\left(\frac{\pi}{2} - \theta\right) = \frac{\cos(\pi/2 - \theta)}{\sin(\pi/2 - \theta)}$$

$$= \frac{\left(\cos\frac{\pi}{2}\right)\cos\theta + \left(\sin\frac{\pi}{2}\right)\sin\theta}{\left(\sin\frac{\pi}{2}\right)\cos\theta - \left(\cos\frac{\pi}{2}\right)\sin\theta}$$

$$= \frac{(0)\cos\theta + (1)\sin\theta}{(1)\cos\theta - (0)\sin\theta}$$

$$= \frac{\sin\theta}{\cos\theta}$$

$$= \tan\theta$$

59.

$$\csc(\pi - \theta) = \frac{1}{\sin(\pi - \theta)}$$

$$= \frac{1}{\sin\pi\cos\theta - \cos\pi\sin\theta}$$

$$= \frac{1}{(0)\cos\theta - (-1)\sin\theta}$$

$$= \frac{1}{\sin\theta}$$

$$= \csc\theta$$

61.

$$\sin 6x\cos 2x - \cos 6x\sin 2x = \sin(6x - 2x)$$

$$= \sin 4x$$

$$= \sin(2x + 2x)$$

$$= \sin 2x\cos 2x + \cos 2x\sin 2x$$

$$= 2\sin 2x\cos 2x$$

63.

$$\cos(\alpha + \beta) + \cos(\alpha - \beta) = \cos\alpha\cos\beta - \sin\alpha\sin\beta + \cos\alpha\cos\beta + \sin\alpha\sin\beta$$

$$= 2\cos\alpha\cos\beta$$

Copyright © Houghton Mifflin Company. All rights reserved.

65. $\sin(\alpha + \beta) + \sin(\alpha - \beta) = \sin\alpha\cos\beta + \cos\alpha\sin\beta + \sin\alpha\cos\beta - \cos\alpha\sin\beta$
$$= 2\sin\alpha\cos\beta$$

67. $\dfrac{\cos(\alpha - \beta)}{\sin(\alpha + \beta)} = \dfrac{\cos\alpha\cos\beta + \sin\alpha\sin\beta}{\sin\alpha\cos\beta + \cos\alpha\sin\beta}$

$$= \dfrac{\dfrac{\cos\alpha\cos\beta}{\sin\alpha\cos\beta} + \dfrac{\sin\alpha\sin\beta}{\sin\alpha\cos\beta}}{\dfrac{\sin\alpha\cos\beta}{\sin\alpha\cos\beta} + \dfrac{\cos\alpha\sin\beta}{\sin\alpha\cos\beta}}$$

$$= \dfrac{\cot\alpha + \tan\beta}{1 + \cot\alpha\tan\beta}$$

69. $\sin\left(\dfrac{\pi}{2} + \alpha - \beta\right) = \sin\left[\dfrac{\pi}{2} + (\alpha - \beta)\right]$

$$= \sin\dfrac{\pi}{2}\cos(\alpha - \beta) + \cos\dfrac{\pi}{2}\sin(\alpha - \beta)$$
$$= (1)\cos(\alpha - \beta) + (0)\sin(\alpha - \beta)$$
$$= \cos(\alpha - \beta)$$
$$= \cos\alpha\cos\beta + \sin\alpha\sin\beta$$

71. $\sin 3x = \sin(2x + x)$
$$= \sin 2x\cos x + \cos 2x\sin x$$
$$= \sin(x + x)\cos x + \cos(x + x)\sin x$$
$$= (\sin x\cos x + \cos x\sin x)\cos x + (\cos x\cos x - \sin x\sin x)\sin x$$
$$= 2\sin x\cos^2 x + \sin x\cos^2 x - \sin^3 x$$
$$= 3\sin x\cos^2 x - \sin^3 x$$
$$= 3\sin x(1 - \sin^2 x) - \sin^3 x$$
$$= 3\sin x - 3\sin^3 x - \sin^3 x$$
$$= 3\sin x - 4\sin^3 x$$

73. $\cos(\theta + 3\pi) = \cos\theta\cos 3\pi - \sin\theta\sin 3\pi$
$$= (\cos\theta)(-1) - (\sin\theta)(0)$$
$$= -\cos\theta$$

75. $\tan(\theta + \pi) = \dfrac{\tan\theta + \tan\pi}{1 - \tan\theta\tan\pi}$
$$= \dfrac{\tan\theta + 0}{1 - (\tan\theta)(0)}$$
$$= \tan\theta$$

77. $\sin(\theta + 2k\pi) = \sin\theta\cos(2k\pi) + \cos\theta\sin 2k\pi$
$$= (\sin\theta)(1) + (\cos\theta)(0)$$
$$= \sin\theta$$

79. $y = \sin\left(\dfrac{\pi}{2} - x\right)$ and $y = \cos x$ both have the following graph.

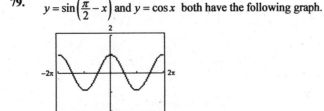

81. $y = \sin 7x\,\cos 2x - \cos 7x\,\sin 2x$ and $y = \sin 5x$ both have the following graph.

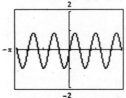

Connecting Concepts

83. $\sin(x - y)\cdot\sin(x + y) = (\sin x\cos y - \cos x\sin y)(\sin x\cos y + \cos x\sin y)$
$$= \sin^2 x\cos^2 y - \cos^2 x\sin^2 y$$

85. $\cos(x + y + z) = \cos[x + (y + z)]$
$$= \cos x\cos(y + z) - \sin x\sin(y + z)$$
$$= \cos x[\cos y\cos z - \sin y\sin z] - \sin x[\sin y\cos z + \cos y\sin z]$$
$$= \cos x\cos y\cos z - \cos x\sin y\sin z - \sin x\sin y\cos z - \sin x\cos y\sin z$$

Copyright © Houghton Mifflin Company. All rights reserved.

87.
$$\frac{\cos(x-y)}{\cos x \sin y} = \frac{\cos x \cos y + \sin x \sin y}{\cos x \sin y}$$
$$= \frac{\cos x \cos y}{\cos x \sin y} + \frac{\sin x \sin y}{\cos x \sin y}$$
$$= \cot y + \tan x$$

89.
$$\frac{\cos(x+h)-\cos x}{h} = \frac{\cos x \cos h - \sin x \sin h - \cos x}{h}$$
$$= \frac{\cos x(\cos h - 1)}{h} - \frac{\sin x \sin h}{h}$$
$$= \cos x \frac{\cos h - 1}{h} - \sin x \frac{\sin h}{h}$$

●●●

91.
$$\sin 2\alpha = \sin(\alpha + \alpha)$$
$$= \sin \alpha \cos \alpha + \cos \alpha \sin \alpha$$
$$= 2 \sin \alpha \cos \alpha$$

92.
$$\cos 2\alpha = \cos(\alpha + \alpha)$$
$$= \cos \alpha \cos \alpha - \sin \alpha \sin \alpha$$
$$= \cos^2 \alpha - \sin^2 \alpha$$

93.
$$\tan 2\alpha = \tan(\alpha + \alpha)$$
$$= \frac{\tan \alpha + \tan \alpha}{1 - \tan \alpha \tan \alpha}$$
$$= \frac{2 \tan \alpha}{1 - \tan^2 \alpha}$$

94.
$$\tan\left(\frac{60°}{2}\right) = \tan(30°) = \frac{\sqrt{3}}{3} \qquad \frac{\sin(60°)}{1+\cos(60°)} = \frac{\frac{\sqrt{3}}{2}}{1+\frac{1}{2}} = \frac{\frac{\sqrt{3}}{2}}{\frac{3}{2}} = \frac{\sqrt{3}}{3}$$

$$\tan\left(\frac{90°}{2}\right) = \tan(45°) = 1 \qquad \frac{\sin(90°)}{1+\cos(90°)} = \frac{1}{1+0} = 1$$

$$\tan\left(\frac{120°}{2}\right) = \tan(60°) = \sqrt{3} \qquad \frac{\sin(120°)}{1+\cos(120°)} = \frac{\frac{\sqrt{3}}{2}}{1-\frac{1}{2}} = \frac{\frac{\sqrt{3}}{2}}{\frac{1}{2}} = \sqrt{3}$$

For each of the given values of α, the functional values are equal.

95. Let $\alpha = 45°$; then the left side of the equation is 1, and the right side of the equation is $\sqrt{2}$.

96. Let $\alpha = 60°$; then the left side of the equation is $\frac{\sqrt{3}}{2}$, and the right side of the equation is $\frac{1}{4}$.

Section 3.3

1.
$$2\sin 2\alpha \cos 2\alpha = \sin 2(2\alpha)$$
$$= \sin 4\alpha$$

3.
$$1 - 2\sin^2 5\beta = \cos 2(5\beta)$$
$$= \cos 10\beta$$

5.
$$\cos^2 3\alpha - \sin^2 3\alpha = \cos 2(3\alpha)$$
$$= \cos 6\alpha$$

7.
$$\frac{2\tan 3\alpha}{1 - \tan^2 3\alpha} = \tan 2(3\alpha)$$
$$= \tan 6\alpha$$

9.
$$\sin 75° = \sin \frac{150°}{2}$$
$$= +\sqrt{\frac{1-\cos 150°}{2}}$$
$$= \sqrt{\frac{1-(-\sqrt{3}/2)}{2}}$$
$$= \sqrt{\frac{2+\sqrt{3}}{4}}$$
$$= \frac{\sqrt{2+\sqrt{3}}}{2}$$

11.
$$\tan 67.5° = \tan \frac{135°}{2}$$
$$= \frac{1-\cos 135°}{\sin 135°}$$
$$= \frac{1-(-\sqrt{2}/2)}{\sqrt{2}/2}$$
$$= \frac{2+\sqrt{2}}{\sqrt{2}} \cdot \frac{\sqrt{2}}{\sqrt{2}}$$
$$= \frac{2\sqrt{2}+2}{2}$$
$$= \sqrt{2}+1$$

Copyright © Houghton Mifflin Company. All rights reserved.

13.
$$\cos 157.5° = \cos\frac{315°}{2}$$
$$= -\sqrt{\frac{1+\cos 315°}{2}}$$
$$= -\sqrt{\frac{1+(\sqrt{2}/2)}{2}}$$
$$= -\sqrt{\frac{2+\sqrt{2}}{4}}$$
$$= -\frac{\sqrt{2+\sqrt{2}}}{2}$$

15.
$$\sin 22.5° = \sin\frac{45°}{2}$$
$$= +\sqrt{\frac{1-\cos 45°}{2}}$$
$$= \sqrt{\frac{1-\sqrt{2}/2}{2}}$$
$$= \sqrt{\frac{2-\sqrt{2}}{4}}$$
$$= \frac{\sqrt{2-\sqrt{2}}}{2}$$

17.
$$\sin\frac{7\pi}{8} = \sin\frac{7\pi/4}{2}$$
$$= +\sqrt{\frac{1-\cos 7\pi/4}{2}}$$
$$= \sqrt{\frac{1-\sqrt{2}/2}{2}}$$
$$= \sqrt{\frac{2-\sqrt{2}}{4}}$$
$$= \frac{\sqrt{2-\sqrt{2}}}{2}$$

19.
$$\cos\frac{5\pi}{12} = \cos\frac{5\pi/6}{2}$$
$$= +\sqrt{\frac{1+\cos 5\pi/6}{2}}$$
$$= \sqrt{\frac{1-\sqrt{3}/2}{2}}$$
$$= \sqrt{\frac{2-\sqrt{3}}{4}}$$
$$= \frac{\sqrt{2-\sqrt{3}}}{2}$$

21.
$$\tan\frac{7\pi}{12} = \tan\frac{7\pi/6}{2}$$
$$= \frac{1-\cos 7\pi/6}{\sin 7\pi/6}$$
$$= \frac{1-(-\sqrt{3}/2)}{-1/2}$$
$$= \frac{2+\sqrt{3}}{-1}$$
$$= -2-\sqrt{3}$$

23.
$$\cos\frac{\pi}{12} = \cos\frac{\pi/6}{2}$$
$$= +\sqrt{\frac{1+\cos\pi/6}{2}}$$
$$= \sqrt{\frac{1+\sqrt{3}/2}{2}}$$
$$= \sqrt{\frac{2+\sqrt{3}}{4}}$$
$$= \frac{\sqrt{2+\sqrt{3}}}{2}$$

25.
$$\cos\theta = -\frac{4}{5}, \quad \sin\theta = \sqrt{1-\left(\frac{4}{5}\right)^2} = \frac{3}{5}, \quad \tan\theta = \frac{\sin\theta}{\cos\theta} = \frac{3/5}{-4/5} = -\frac{3}{4}$$

$$\sin 2\theta = 2\sin\theta\cos\theta$$
$$= 2\left(\frac{3}{5}\right)\left(-\frac{4}{5}\right)$$
$$= -\frac{24}{25}$$

$$\cos 2\theta = \cos^2\theta - \sin^2\theta$$
$$= \left(-\frac{4}{5}\right)^2 - \left(\frac{3}{5}\right)^2$$
$$= \frac{16}{25} - \frac{9}{25}$$
$$= \frac{7}{25}$$

$$\tan 2\theta = \frac{2\tan\theta}{1-\tan^2\theta}$$
$$= \frac{2\left(-\frac{3}{4}\right)}{1-\left(-\frac{3}{4}\right)^2} = \frac{-\frac{6}{4}}{1-\frac{9}{16}}$$
$$= \frac{-\frac{6}{4}}{1-\frac{9}{16}}\cdot\frac{16}{16} = \frac{-24}{16-9}$$
$$= -\frac{24}{7}$$

27.
$$\sin\theta = \frac{8}{17}, \quad \cos\theta = -\sqrt{1-\left(\frac{8}{17}\right)^2} = -\frac{15}{17}, \quad \tan\theta = \frac{\sin\theta}{\cos\theta} = \frac{8/17}{-15/17} = -\frac{8}{15}$$

$$\sin 2\theta = 2\sin\theta\cos\theta$$
$$= 2\left(\frac{8}{17}\right)\left(-\frac{15}{17}\right)$$
$$= -\frac{240}{289}$$

$$\cos 2\theta = \cos^2\theta - \sin^2\theta$$
$$= \left(-\frac{15}{17}\right)^2 - \left(\frac{8}{17}\right)^2$$
$$= \frac{225}{289} - \frac{64}{289}$$
$$= \frac{161}{289}$$

$$\tan 2\theta = \frac{2\tan\theta}{1-\tan^2\theta}$$
$$= \frac{2\left(-\frac{8}{15}\right)}{1-\left(-\frac{8}{15}\right)^2} = \frac{-\frac{16}{15}}{1-\frac{64}{225}}$$
$$= \frac{-\frac{16}{15}}{1-\frac{64}{225}}\cdot\frac{225}{225} = \frac{-240}{225-64}$$
$$= -\frac{240}{161}$$

Copyright © Houghton Mifflin Company. All rights reserved.

29. $\tan\theta = -\dfrac{24}{7}, \; r = \sqrt{24^2 + 7^2} = \sqrt{576 + 49} = \sqrt{625} = 25, \; \sin\theta = -24/25, \; \cos\theta = 7/25$

$\begin{aligned} \sin 2\theta &= 2\sin\theta\cos\theta \\ &= 2\left(-\frac{24}{25}\right)\left(\frac{7}{25}\right) \\ &= -\frac{336}{625} \end{aligned}$

$\begin{aligned} \cos 2\theta &= \cos^2\theta - \sin^2\theta \\ &= \left(\frac{7}{25}\right)^2 - \left(-\frac{24}{25}\right)^2 \\ &= \frac{49}{625} - \frac{576}{625} \\ &= -\frac{527}{625} \end{aligned}$

$\begin{aligned} \tan 2\theta &= \frac{2\tan\theta}{1-\tan^2\theta} \\ &= \frac{2\left(-\frac{24}{7}\right)}{1-\left(-\frac{24}{7}\right)^2} \\ &= \frac{-\frac{48}{7}}{1-\frac{576}{49}} \cdot \frac{49}{49} \\ &= \frac{-336}{49-576} = \frac{336}{527} \end{aligned}$

31. $\sin\theta = \dfrac{15}{17}, \; \cos\theta = \sqrt{1-\left(\frac{15}{17}\right)^2} = \sqrt{1-\frac{225}{289}} = \sqrt{\frac{289-225}{289}} = \sqrt{\frac{64}{289}} = \frac{8}{17}, \; \tan\theta = \dfrac{15/17}{8/17} = \dfrac{15}{8}$

$\begin{aligned} \sin 2\theta &= 2\sin\theta\cos\theta \\ &= 2\left(\frac{15}{17}\right)\left(\frac{8}{17}\right) \\ &= \frac{240}{289} \end{aligned}$

$\begin{aligned} \cos 2\theta &= \cos^2\theta - \sin^2\theta \\ &= \left(\frac{8}{17}\right)^2 - \left(\frac{15}{17}\right)^2 \\ &= \frac{64}{289} - \frac{225}{289} \\ &= -\frac{161}{289} \end{aligned}$

$\begin{aligned} \tan 2\theta &= \frac{2\tan\theta}{1-\tan^2\theta} \\ &= \frac{2\left(\frac{15}{8}\right)}{1-\left(\frac{15}{8}\right)^2} = \frac{\frac{15}{4}}{1-\frac{225}{64}} \\ &= \frac{\frac{15}{4}}{1-\frac{225}{64}} \cdot \frac{64}{64} = \frac{240}{64-225} \\ &= -\frac{240}{161} \end{aligned}$

33. $\cos\theta = \dfrac{40}{41}, \; \sin\theta = -\sqrt{1-\left(\frac{40}{41}\right)^2} = -\sqrt{1-\frac{1600}{1681}} = -\frac{9}{41}, \; \tan\theta = \dfrac{-9/41}{40/41} = -\dfrac{9}{40}$

$\begin{aligned} \sin 2\theta &= 2\sin\theta\cos\theta \\ &= 2\left(-\frac{9}{41}\right)\left(\frac{40}{41}\right) \\ &= -\frac{720}{1681} \end{aligned}$

$\begin{aligned} \cos 2\theta &= \cos^2\theta - \sin^2\theta \\ &= \left(\frac{40}{41}\right)^2 - \left(-\frac{9}{41}\right)^2 \\ &= \frac{1600}{1681} - \frac{81}{1681} \\ &= \frac{1519}{1681} \end{aligned}$

$\begin{aligned} \tan 2\theta &= \frac{2\tan\theta}{1-\tan^2\theta} \\ &= \frac{2\left(-\frac{9}{40}\right)}{1-\left(-\frac{9}{40}\right)^2} = \frac{-\frac{9}{20}}{1-\frac{81}{1600}} \\ &= \frac{-\frac{9}{20}}{1-\frac{81}{1600}} \cdot \frac{1600}{1600} \\ &= \frac{-720}{1600-81} = -\frac{720}{1519} \end{aligned}$

35. $\tan\theta = \dfrac{15}{8}, \; r = \sqrt{15^2 + 8^2} = \sqrt{225 + 64} = 17, \; \sin\theta = -\dfrac{15}{17}, \; \cos\theta = -\dfrac{8}{17}$

$\begin{aligned} \sin 2\theta &= 2\sin\theta\cos\theta \\ &= 2\left(-\frac{15}{17}\right)\left(-\frac{8}{17}\right) \\ &= \frac{240}{289} \end{aligned}$

$\begin{aligned} \cos 2\theta &= \cos^2\theta - \sin^2\theta \\ &= \left(-\frac{8}{17}\right)^2 - \left(-\frac{15}{17}\right)^2 \\ &= \frac{64}{289} - \frac{225}{289} \\ &= -\frac{161}{289} \end{aligned}$

$\begin{aligned} \tan 2\theta &= \frac{2\tan\theta}{1-\tan^2\theta} \\ &= \frac{2\left(\frac{15}{8}\right)}{1-\left(\frac{15}{8}\right)^2} = \frac{\frac{15}{4}}{1-\frac{225}{64}} \\ &= \frac{\frac{15}{4}}{1-\frac{225}{64}} \cdot \frac{64}{64} = \frac{120}{64-225} \\ &= -\frac{120}{161} \end{aligned}$

Copyright © Houghton Mifflin Company. All rights reserved.

37.

$$\sin\alpha = \frac{5}{13}, \quad \cos\alpha = -\sqrt{1-\left(\frac{5}{13}\right)^2} = -\sqrt{1-\frac{25}{169}} = -\frac{12}{13}.$$

$$\sin\frac{\alpha}{2} = \sqrt{\frac{1-\cos\alpha}{2}} \qquad\qquad \cos\frac{\alpha}{2} = \sqrt{\frac{1+\cos\alpha}{2}} \qquad\qquad \tan\frac{\alpha}{2} = \frac{1-\cos\alpha}{\sin\alpha}$$

$$= \sqrt{\frac{1-(-12/13)}{2}} \qquad\qquad = \sqrt{\frac{1-12/13}{2}} \qquad\qquad = \frac{1+\frac{12}{13}}{\frac{5}{13}}$$

$$= \sqrt{\frac{13+12}{26}} \qquad\qquad = \sqrt{\frac{13-12}{26}} \qquad\qquad = \frac{13+12}{5}$$

$$= \sqrt{\frac{25}{26}} = \frac{5}{\sqrt{26}} \qquad\qquad = \sqrt{\frac{1}{26}} = \frac{1}{\sqrt{26}} \qquad\qquad = 5$$

$$= \frac{5\sqrt{26}}{26} \qquad\qquad = \frac{\sqrt{26}}{26}$$

39.

$$\cos\alpha = -\frac{8}{17}, \quad \sin\alpha = -\sqrt{1-\left(-\frac{8}{17}\right)^2} = -\sqrt{1-\frac{64}{289}} = -\frac{15}{17}$$

$$\sin\frac{\alpha}{2} = \sqrt{\frac{1-\cos\alpha}{2}} \qquad\qquad \cos\frac{\alpha}{2} = -\sqrt{\frac{1+\cos\alpha}{2}} \qquad\qquad \tan\frac{\alpha}{2} = \frac{1-\cos\alpha}{\sin\alpha}$$

$$= \sqrt{\frac{1-(-8/17)}{2}} \qquad\qquad = -\sqrt{\frac{1-8/17}{2}} \qquad\qquad = \frac{1+\frac{18}{17}}{-\frac{15}{17}}$$

$$= \sqrt{\frac{17+8}{34}} \qquad\qquad = -\sqrt{\frac{17-8}{34}} \qquad\qquad = \frac{17+8}{-15}$$

$$= \sqrt{\frac{25}{34}} \qquad\qquad = -\sqrt{\frac{9}{34}} \qquad\qquad = -\frac{5}{3}$$

$$= \frac{5\sqrt{34}}{34} \qquad\qquad = -\frac{3\sqrt{34}}{34}$$

41.

$$\tan\alpha = \frac{4}{3}, \quad r = \sqrt{3^2+4^2} = \sqrt{25} = 5, \quad \sin\alpha = \frac{4}{5}, \quad \cos = \frac{3}{5}$$

$$\sin\frac{\alpha}{2} = \sqrt{\frac{1-\cos\alpha}{2}} \qquad\qquad \cos\frac{\alpha}{2} = \sqrt{\frac{1+\cos\alpha}{2}} \qquad\qquad \tan\frac{\alpha}{2} = \frac{1-\cos\alpha}{\sin\alpha}$$

$$= \sqrt{\frac{1-3/5}{2}} \qquad\qquad = \sqrt{\frac{1+3/5}{2}} \qquad\qquad = \frac{1-\frac{3}{5}}{\frac{4}{5}}$$

$$= \sqrt{\frac{5-3}{10}} \qquad\qquad = \sqrt{\frac{5+3}{10}} \qquad\qquad = \frac{5-3}{4}$$

$$= \sqrt{\frac{1}{5}} \qquad\qquad = \sqrt{\frac{4}{5}} \qquad\qquad = \frac{1}{2}$$

$$= \frac{\sqrt{5}}{5} \qquad\qquad = \frac{2\sqrt{5}}{5}$$

43.

$$\cos\alpha = \frac{24}{25}, \quad \sin\alpha = -\sqrt{1-\left(\frac{24}{25}\right)^2} = -\sqrt{1-\frac{576}{625}} = -\frac{7}{25}$$

$$\sin\frac{\alpha}{2} = \sqrt{\frac{1-\cos\alpha}{2}} \qquad\qquad \cos\frac{\alpha}{2} = -\sqrt{\frac{1+\cos\alpha}{2}} \qquad\qquad \tan\frac{\alpha}{2} = \frac{1-\cos\alpha}{\sin\alpha}$$

$$= \sqrt{\frac{1-24/25}{2}} \qquad\qquad = -\sqrt{\frac{1+24/25}{2}} \qquad\qquad = \frac{1-\frac{24}{25}}{-\frac{7}{25}} = \frac{25-24}{-7}$$

$$= \sqrt{\frac{25-24}{50}} = \sqrt{\frac{1}{50}} \qquad\qquad = -\sqrt{\frac{25+24}{50}} = -\sqrt{\frac{49}{50}} \qquad\qquad = -\frac{1}{7}$$

$$= \frac{\sqrt{2}}{10} \qquad\qquad = -\frac{7\sqrt{2}}{10}$$

Copyright © Houghton Mifflin Company. All rights reserved.

45.
$$\sec\alpha = \frac{17}{15}, \ \cos\alpha = \frac{15}{17}, \ \sin\alpha = \sqrt{1 - \left(\frac{15}{17}\right)^2} = \sqrt{1 - \frac{225}{289}} = \frac{8}{17}$$

$$\sin\frac{\alpha}{2} = \sqrt{\frac{1 - \cos\alpha}{2}} \qquad\qquad \cos\frac{\alpha}{2} = \sqrt{\frac{1 + \cos\alpha}{2}} \qquad\qquad \tan\frac{\alpha}{2} = \frac{1 - \cos\alpha}{\sin\alpha}$$

$$= \sqrt{\frac{1 - 15/17}{2}} \qquad\qquad\quad = \sqrt{\frac{1 + 15/17}{2}} \qquad\qquad\quad = \frac{1 - \frac{15}{17}}{\frac{8}{17}} = \frac{17 - 15}{8}$$

$$= \sqrt{\frac{17 - 15}{34}} = \sqrt{\frac{1}{17}} \qquad\qquad = \sqrt{\frac{17 + 15}{34}} = \sqrt{\frac{16}{17}} \qquad\qquad = \frac{1}{4}$$

$$= \frac{\sqrt{17}}{17} \qquad\qquad\qquad\qquad = \frac{4}{\sqrt{17}}$$

$$\qquad\qquad\qquad\qquad\qquad\qquad\qquad = \frac{4\sqrt{17}}{17}$$

47.
$$\cot\alpha = \frac{8}{15}, \ r = \sqrt{8^2 + 15^2} = \sqrt{64 + 225} = 17, \ \sin\alpha = -\frac{15}{17}, \ \cos\alpha = -\frac{8}{17}$$

$$\sin\frac{\alpha}{2} = \sqrt{\frac{1 - \cos\alpha}{2}} \qquad\qquad \cos\frac{\alpha}{2} = -\sqrt{\frac{1 + \cos\alpha}{2}} \qquad\qquad \tan\frac{\alpha}{2} = \frac{1 - \cos\alpha}{\sin\alpha}$$

$$= \sqrt{\frac{1 - (-8/17)}{2}} \qquad\qquad = -\sqrt{\frac{1 + (-8/17)}{2}} \qquad\qquad = \frac{1 - \left(-\frac{8}{17}\right)}{-\frac{15}{17}}$$

$$= \sqrt{\frac{17 + 8}{34}} = \sqrt{\frac{25}{34}} \qquad\quad = -\sqrt{\frac{17 - 8}{34}} = -\sqrt{\frac{9}{34}} \qquad\quad = \frac{17 + 8}{-15} = \frac{25}{-15}$$

$$= \frac{5\sqrt{34}}{34} \qquad\qquad\qquad\quad = -\frac{3\sqrt{34}}{34} \qquad\qquad\qquad = -\frac{5}{3}$$

49.
$$\sin 3x \cos 3x = \frac{1}{2}(2\sin 3x \cos 3x)$$
$$= \frac{1}{2}\sin 2(3x)$$
$$= \frac{1}{2}\sin 6x$$

51.
$$\sin^2 x + \cos 2x = \sin^2 x + \cos^2 x - \sin^2 x$$
$$= \cos^2 x$$

53.
$$\frac{1 + \cos 2x}{\sin 2x} = \frac{1 + 2\cos^2 x - 1}{2\sin x \cos x}$$
$$= \frac{2\cos^2 x}{2\sin x \cos x}$$
$$= \cot x$$

55.
$$\frac{\sin 2x}{1 - \sin^2 x} = \frac{2\sin x \cos x}{\cos^2 x}$$
$$= 2\tan x$$

57.
$$\frac{\cos 2x}{\cos^2 x} = \frac{\cos^2 x - \sin^2 x}{\cos^2 x}$$
$$= \frac{\cos^2 x}{\cos^2 x} - \frac{\sin^2 x}{\cos^2 x}$$
$$= 1 - \tan^2 x$$

59.
$$\sin 2x - \tan x = 2\sin x \cos x - \frac{\sin x}{\cos x}$$
$$= \frac{2\sin x \cos^2 x - \sin x}{\cos x}$$
$$= \frac{\sin x(2\cos^2 x - 1)}{\cos x}$$
$$= \tan x \cos 2x$$

61.
$$\cos^4 x - \sin^4 x = (\cos^2 x + \sin^2 x)(\cos^2 x - \sin^2 x)$$
$$= \cos^2 x - \sin^2 x$$
$$= \cos 2x$$

Copyright © Houghton Mifflin Company. All rights reserved.

63.
$$\cos^2 x - 2\sin^2 x \cos^2 x - \sin^2 x + 2\sin^4 x = \cos^2 x(1 - 2\sin^2 x) - \sin^2 x(1 - 2\sin^2 x)$$
$$= (1 - 2\sin^2 x)(\cos^2 x - \sin^2 x)$$
$$= \cos 2x \cos 2x$$
$$= \cos^2 2x$$

65.
$$\cos 4x = \cos 2(2x)$$
$$= 2\cos^2 2x - 1$$
$$= 2(2\cos^2 x - 1)^2 - 1$$
$$= 2(4\cos^4 x - 4\cos^2 x + 1) - 1$$
$$= 8\cos^4 x - 8\cos^2 x + 1$$

67.
$$\cos 3x - \cos x = \cos(2x + x) - \cos x$$
$$= \cos 2x \cos x - \sin 2x \sin x - \cos x$$
$$= (2\cos^2 x - 1)\cos x - 2\sin x \cos x \cdot \sin x - \cos x$$
$$= 2\cos^3 x - \cos x - 2\sin^2 x \cos x - \cos x$$
$$= 2\cos^3 x - 2\cos x - 2\sin^2 x \cos x$$
$$= 2\cos^3 x - 2\cos x - 2(1 - \cos^2 x)\cos x$$
$$= 2\cos^3 x - 2\cos x - 2\cos x + 2\cos^3 x$$
$$= 4\cos^3 x - 4\cos x$$

69.
$$\sin^3 x + \cos^3 x = (\sin x + \cos x)(\sin^2 x - \sin x \cos x + \cos^2 x)$$
$$= (\sin x + \cos x)\left(\sin^2 x + \cos^2 x - \frac{2\sin x \cos x}{2}\right)$$
$$= (\sin x + \cos x)\left(1 - \frac{1}{2}\sin 2x\right)$$

71.
$$\sin^2 \frac{x}{2} = \left[\pm\sqrt{\frac{1 - \cos x}{2}}\right]^2$$
$$= \frac{1 - \cos x}{2}$$
$$= \frac{1 - \cos x}{2} \cdot \frac{\sec x}{\sec x}$$
$$= \frac{\sec x - 1}{2\sec x}$$

73.
$$\tan \frac{x}{2} = \frac{1 - \cos x}{\sin x}$$
$$= \frac{1}{\sin x} - \frac{\cos x}{\sin x}$$
$$= \csc x - \cot x$$

75.
$$2\sin \frac{x}{2}\cos \frac{x}{2} = \sin 2\left(\frac{x}{2}\right)$$
$$= \sin x$$

77.
$$\left(\cos \frac{x}{2} + \sin \frac{x}{2}\right)^2 = \cos^2 \frac{x}{2} + 2\sin \frac{x}{2}\cos \frac{x}{2} + \sin^2 \frac{x}{2}$$
$$= \cos^2 \frac{x}{2} + \sin^2 \frac{x}{2} + \sin 2\left(\frac{x}{2}\right)$$
$$= 1 + \sin x$$

79.
$$\sin^2 \frac{x}{2}\sec x = \left(\pm\sqrt{\frac{1 - \cos x}{2}}\right)^2 \sec x$$
$$= \frac{1 - \cos x}{2} \cdot \sec x$$
$$= \frac{1}{2}(\sec x - 1)$$

81.
$$\cos^2 \frac{x}{2} - \cos x = \left(\pm\sqrt{\frac{1 + \cos x}{2}}\right)^2 - \cos x$$
$$= \frac{1 + \cos x}{2} - \cos x$$
$$= \frac{1 + \cos x - 2\cos x}{2}$$
$$= \frac{1 - \cos x}{2}$$
$$= \sin^2 \frac{x}{2}$$

Copyright © Houghton Mifflin Company. All rights reserved.

83.
$$\sin^2\frac{x}{2} - \cos^2\frac{x}{2} = -\left(\cos^2\frac{x}{2} - \sin^2\frac{x}{2}\right)$$
$$= -\cos 2\left(\frac{x}{2}\right)$$
$$= -\cos x$$

85.
$$\sin 2x - \cos x = 2\sin x \cos x - \cos x$$
$$= (\cos x)(2\sin x - 1)$$

87.
$$\tan 2x = \frac{2\tan x}{1 - \tan^2 x}$$
$$= \frac{\dfrac{2\tan x}{\tan x}}{\dfrac{1}{\tan x} - \dfrac{\tan^2 x}{\tan x}}$$
$$= \frac{2}{\cot x - \tan x}$$

89.
$$\frac{\sin^2 x + 1 - \cos^2 x}{\sin x(1 + \cos x)} = \frac{1 - \cos^2 x + 1 - \cos^2 x}{\sin x(1 + \cos x)}$$
$$= \frac{2(1 - \cos^2 x)}{\sin x(1 + \cos x)}$$
$$= \frac{2(1 - \cos x)(1 + \cos x)}{\sin x(1 + \cos x)}$$
$$= \frac{2(1 - \cos x)}{\sin x}$$
$$= 2\tan\frac{x}{2}$$

91.
$$\csc 2x = \frac{1}{\sin 2x}$$
$$= \frac{1}{2\sin x \cos x}$$
$$= \frac{1}{2}\csc x \sec x$$

93.
$$\cos\frac{x}{5} = \cos 2\left(\frac{x}{10}\right)$$
$$= 1 - 2\sin^2\frac{x}{10}$$

Connecting Concepts

95. $y = \sin^2 x + \cos 2x$ and $y = \cos^2 x$ both have the following graph.

97. $y = 2\sin\frac{x}{2}\cos\frac{x}{2}$ and $y = \sin x$ both have the following graph.

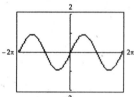

99.
$$\frac{\sin^3 x + \cos^3 x}{\sin x + \cos x} = \frac{(\sin x + \cos x)(\sin^2 x - \sin x \cos x + \cos^2 x)}{\sin x + \cos x}$$
$$= \sin^2 x + \cos^2 x - \frac{2\sin x \cos x}{2}$$
$$= 1 - \frac{1}{2}\sin 2x$$

101.
$$\sin\frac{x}{2} - \cos\frac{x}{2} = \sqrt{\left(\sin\frac{x}{2} + \cos\frac{x}{2}\right)^2}$$
$$= \sqrt{\sin^2\frac{x}{2} - 2\sin\frac{x}{2}\cos\frac{x}{2} + \cos^2\frac{x}{2}}$$
$$= \sqrt{\sin^2\frac{x}{2} + \cos^2\frac{x}{2} - 2\sin\frac{x}{2}\cos\frac{x}{2}}$$
$$= \sqrt{1 - \sin x},\ 0° \le x \le 90°$$

103. $x + y = 90°,\ y = 90° - x$
$$\sin(x - y) = \sin(x - (90° - x))$$
$$= \sin(x - 90° + x)$$
$$= \sin(2x - 90°)$$
$$= \sin 2x \cos 90° - \cos 2x \sin 90°$$
$$= \sin 2x(0) - \cos 2x(1)$$
$$= -\cos 2x$$

105. $x + y = 180°,\ y = 180° - x$
$$\sin(x - y) = \sin(x - 180° + x)$$
$$= \sin(2x - 180°)$$
$$= \sin 2x \cos 180° - \cos 2x \sin 180°$$
$$= \sin 2x(-1) - \cos 2x(0)$$
$$= -\sin 2x$$

Copyright © Houghton Mifflin Company. All rights reserved.

107.

$\dfrac{1}{2}\big[\sin(\alpha+\beta)+\sin(\alpha-\beta)\big]$

$=\dfrac{1}{2}\big[\sin\alpha\cos\beta+\cos\alpha\sin\beta+\sin\alpha\cos\beta-\cos\alpha\sin\beta\big]$

$=\sin\alpha\cos\beta$

108.

$\dfrac{1}{2}\big[\cos(\alpha+\beta)+\cos(\alpha-\beta)\big]$

$=\dfrac{1}{2}\big[\cos\alpha\cos\beta-\sin\alpha\sin\beta+\cos\alpha\cos\beta-\sin\alpha\sin\beta\big]$

$=\cos\alpha\cos\beta$

109. $\sin\pi-\sin\dfrac{\pi}{6}=0-\dfrac{1}{2}=-\dfrac{1}{2}$

$2\cos\left(\dfrac{\pi+\dfrac{\pi}{6}}{\dfrac{2}{1}}\right)\sin\left(\dfrac{\pi+\dfrac{\pi}{6}}{2}\right)=2\cos\left(\dfrac{7\pi}{6}\right)\cos\left(\dfrac{5\pi}{6}\right)$

$2\left(-\sqrt{\dfrac{1+\cos\left(\dfrac{7\pi}{6}\right)}{2}}\right)+\left(\sqrt{\dfrac{1-\cos\left(\dfrac{5\pi}{6}\right)}{2}}\right)$

$=-2\sqrt{\dfrac{1-\dfrac{\sqrt{3}}{2}}{2}}\sqrt{\dfrac{1+\dfrac{\sqrt{3}}{2}}{2}}$

$=-\sqrt{1-\dfrac{3}{4}}=-\sqrt{\dfrac{1}{4}}$

$=-\dfrac{1}{2}$

Both functional values equal $-\dfrac{1}{2}$.

110.

$\sqrt{2}\sin\left(x+\dfrac{\pi}{4}\right)=\sqrt{2}\left[\sin x\cos\left(\dfrac{\pi}{4}\right)+\cos x\sin\left(\dfrac{\pi}{4}\right)\right]$

$=\sqrt{2}\left[(\sin x)\left(\dfrac{\sqrt{2}}{2}\right)+(\cos x)\left(\dfrac{\sqrt{2}}{2}\right)\right]$

$=\sin x\cos x$

111. Answers will vary.

112. $\sqrt{(-1)^2+\left(\sqrt{3}\right)^2}=\sqrt{1+3}=\sqrt{4}=2$

Section 3.4

1.

$2\sin x\cos 2x=2\cdot\dfrac{1}{2}\big[\sin(x+2x)+\sin(x-2x)\big]$

$=\sin 3x+\sin(-x)$

$=\sin 3x-\sin x$

3.

$\cos 6x\sin 2x=\dfrac{1}{2}\big[\sin(6x+2x)-\sin(6x-2x)\big]$

$=\dfrac{1}{2}\big[\sin 8x-\sin 4x\big]$

5.

$2\sin 5x\cos 3x=\sin(5x+3x)+\sin(5x-3x)$

$=\sin 8x+\sin 2x$

7.

$\sin x\cos 5x=\dfrac{1}{2}\big[\cos(x-5x)-\cos(x+5x)\big]$

$=\dfrac{1}{2}\big[\cos(-4x)-\cos 6x\big]$

$=\dfrac{1}{2}(\cos 4x-\cos 6x)$

9.

$\cos 75°\cos 15°=\dfrac{1}{2}\big[\cos(75°+15°)+\cos(75°-15°)\big]$

$=\dfrac{1}{2}(\cos 90°+\cos 60°)$

$=\dfrac{1}{2}\left(0+\dfrac{1}{2}\right)$

$=\dfrac{1}{4}$

11.

$\cos 157.5°\sin 22.5°=\dfrac{1}{2}\big[\sin(157.5°+22.5°)-\sin(157.5°-22.5°)\big]$

$=\dfrac{1}{2}(\sin 180°-\sin 135°)$

$=\dfrac{1}{2}\left(0-\dfrac{\sqrt{2}}{2}\right)$

$=-\dfrac{\sqrt{2}}{4}$

Copyright © Houghton Mifflin Company. All rights reserved.

13.

$$\sin\frac{13\pi}{12}\cos\frac{\pi}{12} = \frac{1}{2}\left[\sin\left(\frac{13\pi}{12}+\frac{\pi}{12}\right)+\sin\left(\frac{13\pi}{12}-\frac{\pi}{12}\right)\right]$$

$$= \frac{1}{2}\left(\sin\frac{7\pi}{6}+\sin\pi\right)$$

$$= \frac{1}{2}\left(-\frac{1}{2}+0\right)$$

$$= -\frac{1}{4}$$

15.

$$\sin\frac{\pi}{12}\cos\frac{7\pi}{12} = \frac{1}{2}\left[\sin\left(\frac{\pi}{12}+\frac{7\pi}{12}\right)+\sin\left(\frac{\pi}{12}-\frac{7\pi}{12}\right)\right]$$

$$= \frac{1}{2}\left[\sin\frac{2\pi}{3}+\sin\left(-\frac{\pi}{2}\right)\right]$$

$$= \frac{1}{2}\left(\sin\frac{2\pi}{3}-\sin\frac{\pi}{2}\right)$$

$$= \frac{1}{2}\left(\frac{\sqrt{3}}{2}-1\right)$$

$$= \frac{\sqrt{3}-2}{4}$$

17.

$$\sin 4\theta + \sin 2\theta = 2\sin\frac{4\theta+2\theta}{2}\cos\frac{4\theta-2\theta}{2}$$

$$= 2\sin 3\theta\cos\theta$$

19.

$$\cos 3\theta + \cos\theta = 2\cos\frac{3\theta+\theta}{2}\cos\frac{3\theta-\theta}{2}$$

$$= 2\cos 2\theta\cos\theta$$

21.

$$\cos 6\theta - \cos 2\theta = -2\sin\frac{6\theta+2\theta}{2}\sin\frac{6\theta-2\theta}{2}$$

$$= -2\sin 4\theta\sin 2\theta$$

23.

$$\cos\theta + \cos 7\theta = 2\cos\frac{\theta+7\theta}{2}\cos\frac{\theta-7\theta}{2}$$

$$= 2\cos 4\theta\cos(-3\theta)$$

$$= 2\cos 4\theta\cos 3\theta$$

25.

$$\sin 5\theta + \sin 9\theta = 2\sin\frac{5\theta+9\theta}{2}\cos\frac{5\theta-9\theta}{2}$$

$$= 2\sin 7\theta\cos(-2\theta)$$

$$= 2\sin 7\theta\cos 2\theta$$

27.

$$\cos 2\theta - \cos\theta = -2\sin\frac{2\theta+\theta}{2}\sin\frac{2\theta-\theta}{2}$$

$$= -2\sin\frac{3}{2}\theta\sin\frac{1}{2}\theta$$

29.

$$\cos\frac{\theta}{2} - \cos\theta = -2\sin\frac{\frac{\theta}{2}+\theta}{2}\sin\frac{\frac{\theta}{2}-\theta}{2}$$

$$= -2\sin\frac{3}{4}\theta\sin\left(-\frac{1}{4}\theta\right)$$

$$= 2\sin\frac{3}{4}\theta\sin\frac{1}{4}\theta$$

31.

$$\sin\frac{\theta}{2} - \sin\frac{\theta}{3} = 2\cos\frac{\frac{\theta}{2}+\frac{\theta}{3}}{2}\sin\frac{\frac{\theta}{2}-\frac{\theta}{3}}{2}$$

$$= 2\cos\frac{5}{12}\theta\sin\frac{1}{12}\theta$$

33.

$$\cos(\alpha+\beta) + \cos(\alpha-\beta) = \cos\alpha\cos\beta - \sin\alpha\sin\beta + \cos\alpha\cos\beta + \sin\alpha\sin\beta$$

$$= 2\cos\alpha\cos\beta$$

35.

$$2\cos 3x\sin x = 2\cdot\frac{1}{2}\left[\sin(3x+x)-\sin(3x-x)\right]$$

$$= \sin 4x - \sin 2x$$

$$= 2\sin 2x\cos 2x - \sin 2x$$

$$= \sin 2x(2\cos 2x - 1)$$

$$= 2\sin x\cos x\left[2(1-2\sin^2 x)-1\right]$$

$$= 4\sin x\cos x - 8\sin^3 x\cos x - 2\sin x\cos x$$

$$= 2\sin x\cos x - 8\cos x\sin^3 x$$

37.

$$2\cos 5x\cos 7x = 2\cdot\frac{1}{2}\left[\cos(5x+7x)+\cos(5x-7x)\right]$$

$$= \cos 12x + \cos(-2x)$$

$$= \cos 12x + \cos 2x$$

$$= \cos^2 6x - \sin^2 6x + 2\cos^2 x - 1$$

39.

$$\sin 3x - \sin x = 2\cos\frac{3x+x}{2}\sin\frac{3x-x}{2}$$

$$= 2\cos 2x\sin x$$

$$= 2(1-2\sin^2 x)\sin x$$

$$= 2\sin x - 4\sin^3 x$$

41.

$$\sin 2x + \sin 4x = 2\cos\frac{2x+4}{2}\cos\frac{2x-x4}{2}$$

$$= 2\cos 3x\cos(-x)$$

$$= 2\cos 3x\cos x$$

$$= 2\cos x\sin(2x+x)$$

$$= 2\cos x(\sin 2x\cos x + \cos 2x\sin x)$$

$$= 2\cos x[(2\sin x\cos x)\cos x + (2\cos^2 x - 1)\sin x]$$

$$= 2\cos x\sin x(4\cos^2 x - 1)$$

Copyright © Houghton Mifflin Company. All rights reserved.

43.

$$\frac{\sin 3x - \sin x}{\cos 3x - \cos x} = \frac{2\cos\frac{3x+x}{2}\sin\frac{3x-x}{2}}{-2\sin\frac{3x+x}{2}\sin\frac{3x-x}{2}}$$

$$= -\frac{\cos 2x}{\sin 2x}$$

$$= -\cot 2x$$

45.

$$\frac{\sin 5x + \sin 3x}{4\sin x \cos^3 x - 4\sin^3 x \cos x} = \frac{2\sin\frac{5x+3x}{2}\cos\frac{5x-3x}{2}}{4\sin x \cos x(\cos^2 x - \sin^2 x)}$$

$$= \frac{\sin 4x \cos x}{2\sin x \cos x \cos 2x}$$

$$= \frac{2\sin 2x \cos 2x \cos x}{\sin 2x \cos 2x}$$

$$= 2\cos x$$

47.

$$\sin(x+y)\cos(x-y) = \frac{1}{2}\Big[\sin(x+y+x-y) + \sin(x+y-x+y)\Big]$$

$$= \frac{1}{2}\Big[\sin 2x + \sin 2y\Big]$$

$$= \frac{1}{2}\Big[2\sin x \cos x + 2\sin y \cos y\Big]$$

$$= \sin x \cos x + \sin y \cos y$$

49. $a = -1,\ b = -1,\ k = \sqrt{(-1)^2 + (-1)^2} = \sqrt{2},$

α is a third quadrant angle.

$$\sin\beta = \left|\frac{-1}{\sqrt{2}}\right| = \frac{1}{\sqrt{2}}$$

$$\beta = 45°$$

$$\alpha = -180° + 45° = -135°$$

$$y = \sqrt{2}\sin(x - 135°)$$

51. $a = \dfrac{1}{2},\ b = -\dfrac{\sqrt{3}}{2},\ k = \sqrt{\left(\dfrac{1}{2}\right)^2 + \left(-\dfrac{\sqrt{3}}{2}\right)^2} = 1,$

α is a fourth quadrant angle.

$$\sin\beta = \left|\frac{-\frac{\sqrt{3}}{2}}{1}\right| = \frac{\sqrt{3}}{2}$$

$$\beta = 60°$$

$$\alpha = -60°$$

$$y = \sin(x - 60°)$$

53. $a = \dfrac{1}{2},\ b = -\dfrac{1}{2},\ k = \sqrt{\left(\dfrac{1}{2}\right)^2 + \left(\dfrac{1}{2}\right)^2} = \dfrac{\sqrt{2}}{2},$

α is a fourth quadrant angle.

$$\sin\beta = \left|\frac{-1/2}{\sqrt{2}/2}\right| = \frac{\sqrt{2}}{2}$$

$$\beta = 45°$$

$$\alpha = -45°$$

$$y = \frac{\sqrt{2}}{2}\sin(x - 45°)$$

55. $a = -3,\ b = 3,\ k = \sqrt{(-3)^2 + 3^2} = 3\sqrt{2},$

α is a second quadrant angle.

$$\sin\beta = \left|\frac{3}{3\sqrt{2}}\right| = \frac{1}{\sqrt{2}} = \frac{\sqrt{2}}{2}$$

$$\beta = 45°$$

$$\alpha = 180° - 45° = 135°$$

$$y = 3\sqrt{2}\sin(x + 135°)$$

57. $a = \pi,\ b = -\pi,\ k = \sqrt{\pi^2 + (-\pi)^2} = \pi\sqrt{2},$

α is a fourth quadrant angle.

$$\sin\beta = \left|\frac{-\pi}{\pi\sqrt{2}}\right| = \frac{1}{\sqrt{2}} = \frac{\sqrt{2}}{2}$$

$$\beta = 45°$$

$$\alpha = -45°$$

$$y = \pi\sqrt{2}\sin(x - 45°)$$

59. $a = -1,\ b = 1,\ k = \sqrt{(-1)^2 + 1^2} = \sqrt{2},$

α is a second quadrant angle.

$$\sin\beta = \left|\frac{1}{\sqrt{2}}\right| = \frac{1}{\sqrt{2}} = \frac{\sqrt{2}}{2}$$

$$\beta = \frac{\pi}{4}$$

$$\alpha = \pi - \frac{\pi}{4} = \frac{3\pi}{4}$$

$$y = \sqrt{2}\sin\left(x + \frac{3\pi}{4}\right)$$

Copyright © Houghton Mifflin Company. All rights reserved.

61.
$$a = \frac{\sqrt{3}}{2},\ b = \frac{1}{2},\ k = \sqrt{\left(\frac{\sqrt{3}}{2}\right)^2 + \left(\frac{1}{2}\right)^2} = 1,$$

α is a first quadrant angle.

$$\sin\beta = \left|\frac{1/2}{1}\right| = \frac{1}{2}$$

$$\beta = \frac{\pi}{6}$$

$$\alpha = \frac{\pi}{6}$$

$$y = \sin\left(x + \frac{\pi}{6}\right)$$

63.
$$a = -10,\ b = 10\sqrt{3},\ k = \sqrt{(-10)^2 + (10\sqrt{3})^2} = 20,$$

α is a second quadrant angle.

$$\sin\beta = \frac{10\sqrt{3}}{20} = \frac{\sqrt{3}}{2}$$

$$\beta = \frac{\pi}{3}$$

$$\alpha = \pi - \frac{\pi}{3} = \frac{2\pi}{3}$$

$$y = 20\sin\left(x + \frac{2\pi}{3}\right)$$

65.
$$a = -5,\ b = 5,\ k = \sqrt{(-5)^2 + 5^2} = 5\sqrt{2},$$

α is a second quadrant angle.

$$\sin\beta = \left|\frac{5}{5\sqrt{2}}\right| = \frac{\sqrt{2}}{2}$$

$$\beta = \frac{\pi}{4}$$

$$\alpha = \pi - \frac{\pi}{4} = \frac{3\pi}{4}$$

$$y = 5\sqrt{2}\sin\left(x + \frac{3\pi}{4}\right)$$

67.
$$y = -\sin x - \sqrt{3}\cos x$$

$$y = 2\sin\left(x - \frac{2\pi}{3}\right)$$

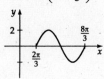

69.
$$y = 2\sin x + 2\cos x$$

$$y = 2\sqrt{2}\sin\left(x + \frac{\pi}{4}\right)$$

71.
$$y = -\sqrt{3}\sin x - \cos x$$

$$y = 2\sin\left(x - \frac{5\pi}{6}\right)$$

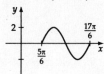

73.
$$y = -5\sin x + 5\sqrt{3}\cos x$$

$$y = 10\sin\left(x + \frac{2\pi}{3}\right)$$

75.
$$y = 6\sqrt{3}\sin x - 6\cos x$$

$$y = 12\sin\left(x - \frac{\pi}{6}\right)$$

77. **a.** $p(t) = \cos(2\pi \cdot 1336t) + \cos(2\pi \cdot 770t)$

b. $p(t) = 2\cos\left(\dfrac{2\pi \cdot 1336t + 2\pi \cdot 770t}{2}\right)\cos\left(\dfrac{2\pi \cdot 1336t - 2\pi \cdot 770t}{2}\right)$

$ = 2\cos(2106\pi t)\cos(556\pi t)$

c. $\dfrac{1336 + 770}{2} = \dfrac{2106}{2} = 1053$ cycles per second

79.

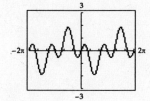

81.

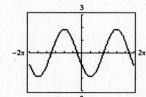

83.

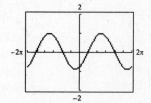

Copyright © Houghton Mifflin Company. All rights reserved.

85. Let $x = \alpha + \beta$ and $y = \alpha - \beta$.

$x + y = \alpha + \beta + \alpha - \beta$ and $x - y = \alpha + \beta - (\alpha - \beta)$

$x + y = 2\alpha$ $\qquad\qquad\qquad$ $x - y = 2\beta$

$\alpha = \dfrac{x+y}{2}$ $\qquad\qquad\qquad$ $2\beta = \dfrac{x-y}{2}$

$\cos(\alpha - \beta) + \cos(\alpha + \beta) = 2\cos\alpha\cos\beta$

$\cos\left[\dfrac{x+y}{2} - \dfrac{x-y}{2}\right] + \cos\left[\dfrac{x+y}{2} + \dfrac{x-y}{2}\right] = 2\cos\dfrac{x+y}{2}\cos\dfrac{x-y}{2}$

$\cos y + \cos x = 2\cos\dfrac{x+y}{2}\cos\dfrac{x-y}{2}$

87. $x + y = 180°$

$\qquad y = 180° - x$

$\sin x + \sin y = \sin x + \sin(180° - x)$

$\qquad\qquad\qquad = \sin x + \sin 180° \cos x - \cos 180° \sin x$

$\qquad\qquad\qquad = \sin x + 0(\cos x) - (-1)\sin x$

$\qquad\qquad\qquad = 2\sin x$

89.

$\sin 2x + \sin 4x + \sin 6x = 2\sin\dfrac{2x+4x}{2}\cos\dfrac{2x-4x}{2} + 2\sin 3x\cos 3x$

$\qquad\qquad\qquad\qquad = 2\sin 3x\cos x + 2\sin 3x\cos 3x$

$\qquad\qquad\qquad\qquad = 2\sin 3x(\cos x + \cos 3x)$

$\qquad\qquad\qquad\qquad = 2\sin 3x\left(2\cos\dfrac{x+3x}{2}\cos\dfrac{x-3x}{2}\right)$

$\qquad\qquad\qquad\qquad = 4\sin 3x\cos 2x\cos x$

91.

$\dfrac{\cos 10x + \cos 8x}{\sin 10x - \sin 8x} = \dfrac{2\cos\frac{10x+8x}{2}\cos\frac{10x-8x}{2}}{2\cos\frac{10x+8x}{2}\sin\frac{10x-8x}{2}}$

$\qquad\qquad\qquad\quad = \dfrac{2\cos 9x\cos x}{2\cos 9x\sin x}$

$\qquad\qquad\qquad\quad = \cot x$

93. $\dfrac{\sin 2x + \sin 4x + \sin 6x}{\cos 2x + \cos 4x + \cos 6x} = \dfrac{\sin 2x + \sin 6x + \sin 4x}{\cos 2x + \cos 6x + \cos 4x}$

$\qquad\qquad\qquad\qquad = \dfrac{2\sin\dfrac{2x+6x}{2}\cos\dfrac{2x-6x}{2} + \sin 4x}{2\cos\dfrac{2x+6x}{2}\cos\dfrac{2x-6x}{2} + \cos 4x}$

$\qquad\qquad\qquad\qquad = \dfrac{2\sin 4x\cos 2x + \sin 4x}{2\cos 4x\cos 2x + \cos 4x}$

$\qquad\qquad\qquad\qquad = \dfrac{\sin 4x(2\cos 2x + 1)}{\cos 4x(2\cos 2x + 1)}$

$\qquad\qquad\qquad\qquad = \dfrac{\sin 4x}{\cos 4x}$

$\qquad\qquad\qquad\qquad = \tan 4x$

95. $\cos^2 x - \sin^2 x = \cos x \cdot \cos x - \sin x \cdot \sin x$

$\qquad\qquad\qquad = \dfrac{1}{2}[\cos(x+x) + \cos(x-x)] - \dfrac{1}{2}[\cos(x-x) - \cos(x+x)]$

$\qquad\qquad\qquad = \dfrac{1}{2}\cos 2x + \dfrac{1}{2}\cos 0 - \dfrac{1}{2}\cos 0 + \dfrac{1}{2}\cos 2x$

$\qquad\qquad\qquad = \cos 2x$

Copyright © Houghton Mifflin Company. All rights reserved.

97. Let $k = \sqrt{a^2 + b^2}$, $\tan\alpha = \dfrac{a}{b}$

$$a\sin x + b\cos x = \frac{\sqrt{a^2 + b^2}}{\sqrt{a^2 + b^2}}(a\sin x + b\cos x)$$

$$= \sqrt{a^2 + b^2}\left(\frac{a}{\sqrt{a^2 + b^2}}\sin x + \frac{b}{\sqrt{a^2 + b^2}}\cos x\right)$$

$$= k(\sin\alpha \sin x + \cos\alpha \cos x) \text{ because } \sin\alpha = \frac{a}{\sqrt{a^2 + b^2}} \text{ and } \cos\alpha = \frac{b}{\sqrt{a^2 + b^2}}$$

$$= k(\cos x \cos\alpha + \sin x \sin\alpha) = k\cos(x - \alpha)$$

• •

99. A one-to-one function is a function for which each range value (y-value) is paired with one and only one domain value (x-value).

100. If every horizontal line intersects the graph of a function at most once, then the function is a one-to-one function.

101.
$$f[g(x)] = f\left[\frac{1}{2}x - 2\right]$$
$$= 2\left(\frac{1}{2}x - 2\right) + 4$$
$$= x - 4 + 4$$
$$= x$$

102. $f[f^{-1}(x)] = x$

103. The graph of f^{-1} is the reflection of the graph of f across the line given by $y = x$.

104. No, it does not pass the horizontal line test.

Section 3.5

1.
$y = \sin^{-1}1$

$\sin y = 1$ with $-\dfrac{\pi}{2} \le y \le \dfrac{\pi}{2}$

$y = \dfrac{\pi}{2}$

3.
$y = \cos^{-1}\left(-\dfrac{\sqrt{3}}{2}\right)$

$\cos y = -\dfrac{\sqrt{3}}{2} \quad 0 \le y \le \pi$

$y = \dfrac{5\pi}{6}$

5.
$y = \tan^{-1}(1)$

$\tan y = -1 \quad -\dfrac{\pi}{2} < y < \dfrac{\pi}{2}$

$y = -\dfrac{\pi}{4}$

7.
$y = \cot^{-1}\dfrac{\sqrt{3}}{3}$

$\cot y = \dfrac{\sqrt{3}}{3} \quad 0 < y < \pi$

$y = \dfrac{\pi}{3}$

9.
$y = \sec^{-1}2$

$\sec y = 2 \quad 0 \le y \le \pi$

$y = \dfrac{\pi}{3}$

11.
$y = \csc^{-1}\left(-\sqrt{2}\right)$

$\csc y = -\sqrt{2} \quad -\dfrac{\pi}{2} \le y \le \dfrac{\pi}{2}$

$y = -\dfrac{\pi}{4}$

13.
$y = \sin^{-1}\left(-\dfrac{\sqrt{3}}{2}\right)$

$\sin y = -\dfrac{\sqrt{3}}{2} \quad -\dfrac{\pi}{2} \le y \le \dfrac{\pi}{2}$

$y = -\dfrac{\pi}{3}$

15.
$y = \cos^{-1}\left(-\dfrac{1}{2}\right)$

$\cos y = -\dfrac{1}{2} \quad 0 \le y \le \pi$

$y = \dfrac{2\pi}{3}$

17.
$y = \tan^{-1}\dfrac{\sqrt{3}}{3}$

$\tan y = \dfrac{\sqrt{3}}{3} \quad -\dfrac{\pi}{2} < y < \dfrac{\pi}{2}$

$y = \dfrac{\pi}{6}$

Copyright © Houghton Mifflin Company. All rights reserved.

19.

$y = \cot^{-1}\sqrt{3}$

$\cot y = \sqrt{3} \qquad 0 < y < \pi$

$y = \dfrac{\pi}{6}$

21.

$y = \cos\left(\cos^{-1}\dfrac{1}{2}\right)$

$y = \cos\dfrac{\pi}{3}$

$y = \dfrac{1}{2}$

23.

$y = \tan(\tan^{-1}2)$

$y = 2$

25.

$y = \sin\left(\tan^{-1}\dfrac{3}{4}\right)$

$y = \dfrac{3}{5}$

27.

$y = \tan\left(\sin^{-1}\dfrac{\sqrt{2}}{2}\right)$

$y = 1$

29.

$y = \cos(\sec^{-1}2)$

$y = \dfrac{1}{2}$

31.

$y = \sin^{-1}\left(\sin\dfrac{\pi}{6}\right)$

$= \sin^{-1}\dfrac{1}{2}$

$y = \dfrac{\pi}{6}$

33.

$y = \cos^{-1}\left(\sin\dfrac{\pi}{4}\right)$

$= \cos^{-1}\dfrac{\sqrt{2}}{2}$

$y = \dfrac{\pi}{4}$

35.

$y = \sin^{-1}\left(\tan\dfrac{\pi}{3}\right)$

$= \sin^{-1}\sqrt{3}$

y is not defined.

37.

$y = \tan^{-1}\left(\sin\dfrac{\pi}{6}\right)$

$= \tan^{-1}\dfrac{1}{2}$

$y \approx 0.4636$

39.

$y = \sin^{-1}\left(\cos\left[-\dfrac{2\pi}{3}\right]\right)$

$= \sin^{-1}\left(-\dfrac{1}{2}\right)$

$y = -\dfrac{\pi}{6}$

41.

Let $\theta = \sin^{-1}\dfrac{1}{2}$ and find $y = \tan\theta$.

Then $\sin\theta = \dfrac{1}{2}$ and $-\dfrac{\pi}{2} \le \theta \le \dfrac{\pi}{2}$.

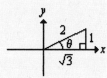

Thus $\tan\theta = \dfrac{1}{\sqrt{3}} = \dfrac{\sqrt{3}}{3}$.

$y = \dfrac{\sqrt{3}}{3}$

43.

Let $\theta = \sin^{-1}\dfrac{1}{4}$ and find $y = \sec\theta$.

Then $\sin\theta = \dfrac{1}{4}$, and $-\dfrac{\pi}{2} \le \theta \le \dfrac{\pi}{2}$.

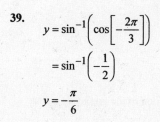

Thus $\sec\theta = \dfrac{4}{\sqrt{15}} = \dfrac{4\sqrt{15}}{15}$.

$y = \dfrac{4\sqrt{15}}{15}$

45.

Let $\theta = \sin^{-1}\dfrac{7}{25}$ and find $y = \cos\theta$.

Then $\sin\theta = \dfrac{7}{25}$, and $-\dfrac{\pi}{2} \le \theta \le \dfrac{\pi}{2}$.

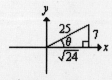

Thus $\cos\theta = \dfrac{24}{25}$.

$y = \dfrac{24}{25}$

47.

Let $\theta = \tan^{-1}\dfrac{12}{5}$ and find $y = \sec\theta$.

Then $\tan\theta = \dfrac{12}{5}$, and $-\dfrac{\pi}{2} \le \theta \le \dfrac{\pi}{2}$.

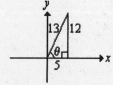

Thus $\sec\theta = \dfrac{13}{5}$.

$y = \dfrac{13}{5}$

Copyright © Houghton Mifflin Company. All rights reserved.

49.

Let $\alpha = \sin^{-1}\dfrac{\sqrt{2}}{2}$, $\alpha = \dfrac{\pi}{4}$, $\sin\alpha = \dfrac{\sqrt{2}}{2}$, $\cos\alpha = \dfrac{\sqrt{2}}{2}$.

$$y = \cos\left(2\sin^{-1}\dfrac{\sqrt{2}}{2}\right)$$
$$= \cos 2\alpha$$
$$= \cos^2\alpha - \sin^2\alpha$$
$$= \left(\dfrac{\sqrt{2}}{2}\right)^2 - \left(\dfrac{\sqrt{2}}{2}\right)^2$$
$$= 0$$

51.

Let $\alpha = \sin^{-1}\dfrac{4}{5}$, $\sin\alpha = \dfrac{4}{5}$, $\cos\alpha = \sqrt{1 - \left(\dfrac{4}{5}\right)^2} = \dfrac{3}{5}$.

$$y = \sin\left(2\sin^{-1}\dfrac{4}{5}\right)$$
$$= \sin 2\alpha = 2\sin\alpha\cos\alpha$$
$$= 2\left(\dfrac{4}{5}\right)\left(\dfrac{3}{5}\right) = \dfrac{24}{25}$$

53.

$$y = \sin\left(\sin^{-1}\dfrac{2}{3} + \cos^{-1}\dfrac{1}{2}\right)$$

Let $\alpha = \sin^{-1}\dfrac{2}{3}$, $\sin\alpha = \dfrac{2}{3}$, $\cos\alpha = \sqrt{1 - \left(\dfrac{2}{3}\right)^2} = \dfrac{\sqrt{5}}{3}$.

$\beta = \cos^{-1}\dfrac{1}{2}$, $\cos\beta = \dfrac{1}{2}$, $\sin\beta = \sqrt{1 - \left(\dfrac{1}{2}\right)^2} = \dfrac{\sqrt{3}}{2}$.

$$y = \sin(\alpha + \beta)$$
$$= \sin\alpha\cos\beta + \cos\alpha\sin\beta$$
$$= \dfrac{2}{3}\left(\dfrac{1}{2}\right) + \dfrac{\sqrt{5}}{3}\left(\dfrac{\sqrt{3}}{2}\right)$$
$$= \dfrac{1}{3} + \dfrac{\sqrt{15}}{6} = \dfrac{2 + \sqrt{15}}{6}$$

55.

$$y = \tan\left(\cos^{-1}\dfrac{1}{2} - \sin^{-1}\dfrac{3}{4}\right)$$

Let $\alpha = \cos^{-1}\dfrac{1}{2}$, $\cos\alpha = \dfrac{1}{2}$, $\sin\alpha = \sqrt{1 - \left(\dfrac{1}{2}\right)^2} = \dfrac{\sqrt{3}}{2}$, $\tan\alpha = \dfrac{\frac{\sqrt{3}}{2}}{\frac{1}{2}} = \sqrt{3}$.

$\beta = \sin^{-1}\dfrac{3}{4}$, $\sin\beta = \dfrac{3}{4}$, $\cos\beta = \sqrt{1 - \left(\dfrac{3}{4}\right)^2} = \dfrac{\sqrt{7}}{4}$, $\tan\beta = \dfrac{\frac{3}{4}}{\frac{\sqrt{7}}{4}} = \dfrac{3}{\sqrt{7}} = \dfrac{3\sqrt{7}}{7}$.

$$y = \tan(\alpha - \beta)$$
$$= \dfrac{\tan\alpha - \tan\beta}{1 + \tan\alpha\tan\beta}$$
$$= \dfrac{\sqrt{3} - \frac{3\sqrt{7}}{7}}{1 + \sqrt{3}\cdot\frac{3\sqrt{7}}{7}} = \dfrac{\sqrt{3} - \frac{3\sqrt{7}}{7}}{1 + \sqrt{3}\cdot\frac{3\sqrt{7}}{7}}\cdot\dfrac{7}{7} = \dfrac{7\sqrt{3} - 3\sqrt{7}}{7 + 3\sqrt{21}} = \dfrac{7\sqrt{3} - 3\sqrt{7}}{7 + 3\sqrt{21}}\cdot\dfrac{7 - 3\sqrt{21}}{7 - 3\sqrt{21}} = \dfrac{112\sqrt{3} - 84\sqrt{7}}{-140} = \dfrac{3\sqrt{7} - 4\sqrt{3}}{5} = \dfrac{1}{5}\left(3\sqrt{7} - 4\sqrt{3}\right)$$

57.

$$\sin^{-1}x = \cos^{-1}\dfrac{5}{13}$$
$$\sin(\sin^{-1}x) = \sin\left(\cos^{-1}\dfrac{5}{13}\right)$$
$$x = \dfrac{12}{13}$$

59.

$$\sin^{-1}(x-1) = \dfrac{\pi}{2}$$
$$(x-1) = \sin\dfrac{\pi}{2}$$
$$(x-1) = 1$$
$$x = 2$$

61.

$$\tan^{-1}\left(x + \dfrac{\sqrt{2}}{2}\right) = \dfrac{\pi}{4}$$
$$\left(x + \dfrac{\sqrt{2}}{2}\right) = \tan\dfrac{\pi}{4}$$
$$x = 1 - \dfrac{\sqrt{2}}{2}$$
$$= \dfrac{2 - \sqrt{2}}{2}$$

Copyright © Houghton Mifflin Company. All rights reserved.

63.
$$\sin^{-1}\frac{3}{5} + \cos^{-1}x = \frac{\pi}{4}$$
$$\cos^{-1}x = \frac{\pi}{4} - \sin^{-1}\frac{3}{5}$$
$$x = \cos\left(\frac{\pi}{4} - \sin^{-1}\frac{3}{5}\right)$$

Let $\alpha = \sin^{-1}\frac{3}{5}$, $\sin\alpha = \frac{3}{5}$, $\cos\alpha = \frac{4}{5}$.

$$x = \cos\left(\frac{\pi}{4} - \alpha\right)$$
$$x = \cos\frac{\pi}{4}\cos\alpha + \sin\frac{\pi}{4}\sin\alpha$$
$$x = \frac{\sqrt{2}}{2}\cdot\frac{4}{5} + \frac{\sqrt{2}}{2}\cdot\frac{3}{5}$$
$$= \frac{4\sqrt{2}}{10} + \frac{3\sqrt{2}}{10} = \frac{7\sqrt{2}}{10}$$

65.
$$\sin^{-1}\frac{\sqrt{2}}{2} + \cos^{-1}x = \frac{2\pi}{3}$$
$$\cos^{-1}x = \frac{2\pi}{3} - \sin^{-1}\frac{\sqrt{2}}{2}$$
$$x = \cos\left(\frac{2\pi}{3} - \sin^{-1}\frac{\sqrt{2}}{2}\right)$$

Let $\alpha = \sin^{-1}\frac{\sqrt{2}}{2}$, $\sin\alpha = \frac{\sqrt{2}}{2}$, $\cos\alpha = \frac{\sqrt{2}}{2}$.

$$x = \cos\left(\frac{2\pi}{3} - \alpha\right)$$
$$x = \cos\frac{2\pi}{3}\cos\alpha + \sin\frac{2\pi}{3}\sin\alpha$$
$$x = -\frac{1}{2}\cdot\frac{\sqrt{2}}{2} + \frac{\sqrt{3}}{2}\cdot\frac{\sqrt{2}}{2}$$
$$= -\frac{\sqrt{2}}{4} + \frac{\sqrt{6}}{4} = \frac{\sqrt{6}-\sqrt{2}}{4} \approx 0.2588$$

Note: Since $\sin\alpha = \frac{\sqrt{2}}{2}$ and $\cos\alpha = \frac{\sqrt{2}}{2}$, then $\alpha = \frac{\pi}{4}$.

Thus, $\cos\left(\frac{2\pi}{3} - \alpha\right) = \cos\left(\frac{2\pi}{3} - \frac{\pi}{4}\right) = \cos\left(\frac{5\pi}{12}\right) \approx 0.2588$.

67.
$$\cos(\sin^{-1}x) = \sqrt{1-x^2}$$

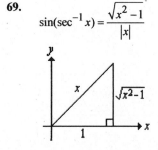

69.
$$\sin(\sec^{-1}x) = \frac{\sqrt{x^2-1}}{|x|}$$

71.
Let $\alpha = \sin^{-1}x$, $\sin\alpha = x$, $\cos\alpha = \sqrt{1-x^2}$.

Let $\beta = \sin^{-1}(-x)$, $\sin\beta = -x$, $\cos\beta = \sqrt{1-x^2}$.

$$\sin^{-1}x + \sin^{-1}(-x) = \alpha + \beta$$
$$= \sin^{-1}\left[\sin(\alpha+\beta)\right]$$
$$= \sin^{-1}(\sin\alpha\cos\beta + \cos\alpha\sin\beta)$$
$$= \sin^{-1}\left[x\sqrt{1-x^2} + \sqrt{1-x^2}(-x)\right]$$
$$= \sin^{-1}0$$
$$= 0$$

73.
Let $\alpha = \tan^{-1}x$, $\tan\alpha = x$, $\beta = \tan^{-1}\frac{1}{x}$, $\tan\beta = \frac{1}{x}$.

$$\tan^{-1}x + \tan^{-1}\frac{1}{x} = \alpha + \beta$$
$$= \tan^{-1}\left[\tan(\alpha+\beta)\right]$$
$$= \tan^{-1}\left[\frac{\tan\alpha + \tan\beta}{1-\tan\alpha\tan\beta}\right]$$
$$= \tan^{-1}\left[\frac{x+\frac{1}{x}}{1-x\cdot\frac{1}{x}}\right]$$
$$= \tan^{-1}\frac{\frac{x^2+1}{x}}{1-1}, \text{ which is undefined}$$

Thus $x = \frac{\pi}{2}$

Copyright © Houghton Mifflin Company. All rights reserved.

75. The graph of $y = \sin^{-1}(x) + 2$ (shown as a black graph) is the graph of $y = \sin^{-1} x$ (shown as a gray graph) moved two units up.

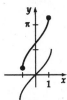

77. The graph of $y = \sin^{-1}(x+1) - 2$ (shown as a black graph) is the graph of $y = \sin^{-1} x$ (shown as a gray graph) moved one unit to the left and two units down.

79. The graph of $y = 2\cos^{-1} x$ (shown as a black graph) is the graph of $y = \cos^{-1} x$ (shown as a gray graph) stretched.

81. The graph of $y = \tan^{-1}(x+1) - 2$ (shown as a black graph) is the graph of $y = \tan^{-1} x$ (shown as a gray graph) moved one unit to the left and two units down.

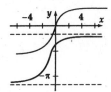

83. **a.**

$$\frac{A}{SD} = 1 - \frac{1}{2}\left[1 - \left(\frac{S}{D}\right)^2 + \frac{D}{S}\sin^{-1}\frac{S}{D}\right]$$

$$= 1 - \frac{1}{2}\left[1 - \left(\frac{1}{2}\right)^2 + 2\sin^{-1}\frac{1}{2}\right]$$

$$= 1 - \frac{1}{2}\left[\frac{3}{4} + \frac{\pi}{3}\right]$$

$$\approx 1 - \frac{1}{2}(1.79719755)$$

$$\approx 1 - 0.8985$$

$$\frac{A}{SD} \approx 0.1014$$

b.

$$\frac{A}{SD} = 1 - \frac{1}{2}\left[1 - \left(\frac{S}{D}\right)^2 + \frac{D}{S}\sin^{-1}\frac{S}{D}\right]$$

$$\approx 1 - \frac{1}{2}\left[1 - 0.390625 + 1.6\sin^{-1}0.62\right]$$

$$\approx 1 - \frac{1}{2}\left[0.609375 + 1.0802105\right]$$

$$\approx 1 - \frac{1}{2}(1.689585)$$

$$\approx 1 - 0.8448$$

$$\approx 0.1552$$

85.

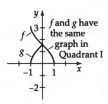

f and g have the same graph in Quadrant I

[Note: $f(x) = \cos^{-1} x$ is neither odd nor even.

$g(x) = \sin^{-1}\sqrt{1 - x^2}$ is an even function.]

No, $f(x) \neq g(x)$ on the interval $[-1, 1]$.

87.

$$y = \csc^{-1} 2x$$

$$\csc y = 2x$$

$$-\frac{\pi}{2} \leq y \leq \frac{\pi}{2}, \; y \neq 0$$

$$2x \leq -1 \text{ or } 2x \geq 1$$

$$x \leq -\frac{1}{2} \text{ or } x \geq \frac{1}{2}$$

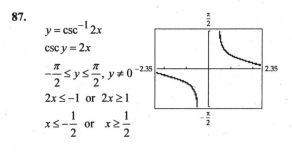

Copyright © Houghton Mifflin Company. All rights reserved.

89.
$y = \sec^{-1}(x-1)$
$\sec y = x-1$
$0 < y < \pi$
$y \neq \dfrac{\pi}{2}$
$x-1 \le -1 \qquad x-1 \ge 1$
$x \le 0 \qquad x \ge 2$

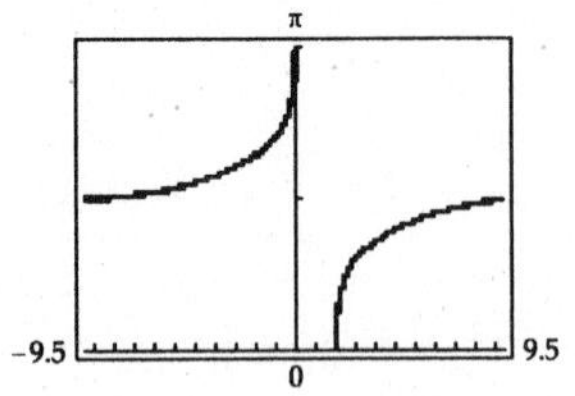

91.
$y = 2\tan^{-1} 2x$
$\dfrac{y}{2} = \tan^{-1} 2x$
$\tan y = 2x$
$-\dfrac{\pi}{2} < \dfrac{y}{2} < \dfrac{\pi}{2}$
$-\pi < y < \pi$
$-\infty < 2x < \infty$

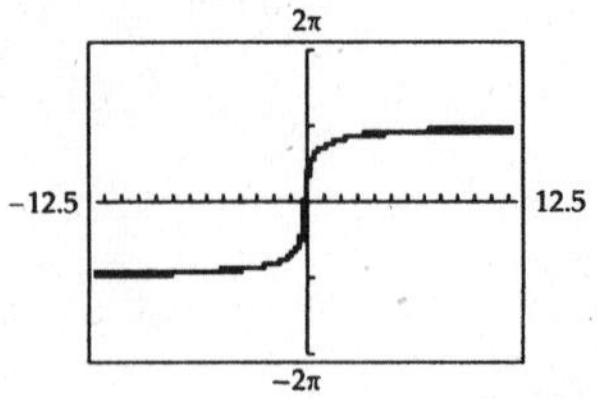

93.
$y = \cot^{-1}\dfrac{x}{3}$
$\cot y = \dfrac{\pi}{3}$
$0 < y < \pi$
$-\infty < \dfrac{x}{3} < \infty$

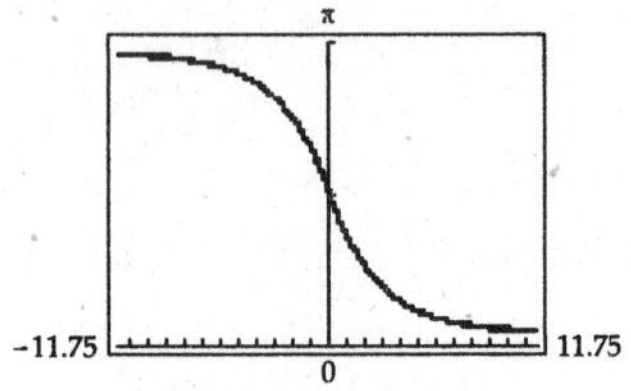

Connecting Concepts

95.
Let $\alpha = \sin^{-1} x$
$\sin\alpha = x$

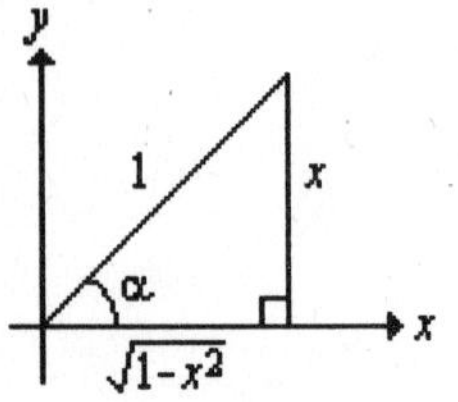

$\cos\left(\sin^{-1} x\right) = \cos\alpha$
$= \dfrac{\sqrt{1-x^2}}{1}$
$= \sqrt{1-x^2}$

97.
Let $\alpha = \csc^{-1} x$
$\csc\alpha = x$

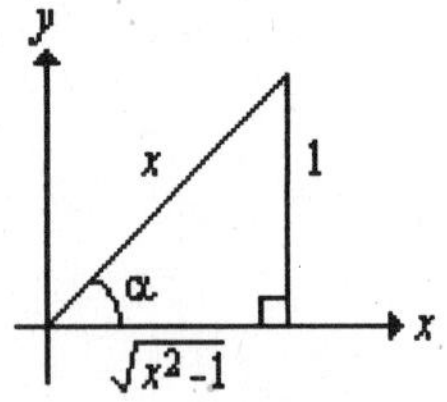

$\tan(\csc^{-1} x) = \tan\alpha$
$= \dfrac{1}{\sqrt{x^2-1}}$
$= \dfrac{\sqrt{x^2-1}}{x^2-1}$

99.
$5x = \tan^{-1} 3y$
$\tan 5x = 3y$
$y = \dfrac{1}{3}\tan 5x$

101.
$x - \dfrac{\pi}{3} = \cos^{-1}(y-3)$
$\cos\left(x - \dfrac{\pi}{3}\right) = y-3$
$y = 3 + \cos\left(x - \dfrac{\pi}{3}\right)$

Prepare for Section 3.6

103. $x = \dfrac{5 \pm \sqrt{(-5)^2 - 4(3)(-4)}}{2(3)} = \dfrac{5 \pm \sqrt{73}}{6}$

104. $\sin^2 x + \cos^2 x = 1$
$\sin^2 x = 1 - \cos^2 x$

105. $\dfrac{\pi}{2} + 2(1)\pi = \dfrac{5}{2}\pi$
$\dfrac{\pi}{2} + 2(2)\pi = \dfrac{9}{2}\pi$
$\dfrac{\pi}{2} + 2(3)\pi = \dfrac{13}{2}\pi$

106. $x^2 - \dfrac{\sqrt{3}}{2}x + x - \dfrac{\sqrt{3}}{2} = x\left(x - \dfrac{\sqrt{3}}{2}\right) + 1\left(x - \dfrac{\sqrt{3}}{2}\right)$
$= (x+1)\left(x - \dfrac{\sqrt{3}}{2}\right)$

107.

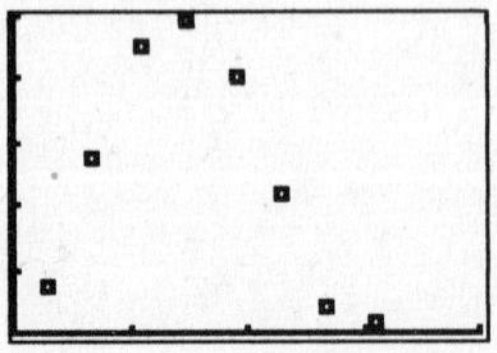

108. $2x^2 - 2x = 0$
$2x(x-1) = 0$
$x = 0 \quad x-1 = 0$
$x = 1$
The solutions are 0, 1.

Copyright © Houghton Mifflin Company. All rights reserved.

Section 3.6

1.
$$\sec x - \sqrt{2} = 0$$
$$\sec x = \sqrt{2}$$
$$x = \frac{\pi}{4}, \frac{7\pi}{4}$$

3.
$$\tan x - \sqrt{3} = 0$$
$$\tan x = \sqrt{3}$$
$$x = \frac{\pi}{3}, \frac{4\pi}{3}$$

5.
$$2\sin x \cos x = \sqrt{2}\cos x$$
$$2\sin x \cos x - \sqrt{2}\cos x = 0$$
$$\cos x(2\sin x - \sqrt{2}) = 0$$
$$\cos x = 0 \qquad 2\sin x - \sqrt{2} = 0$$
$$\sin x = \frac{\sqrt{2}}{2}$$
$$x = \frac{\pi}{2}, \frac{3\pi}{2} \qquad x = \frac{\pi}{4}, \frac{3\pi}{4}$$
The solutions are $\frac{\pi}{4}, \frac{\pi}{2}, \frac{3\pi}{4}, \frac{3\pi}{2}$.

7.
$$\sin^2 x - 1 = 0$$
$$\sin^2 x = 1$$
$$\sin x = \pm\sqrt{1}$$
$$\sin x = \pm 1$$
$$x = \frac{\pi}{2}, \frac{3\pi}{2}$$

9.
$$4\sin x \cos x - 2\sqrt{3}\sin x - 2\sqrt{2}\cos x + \sqrt{6} = 0$$
$$2\sin x(2\cos x - \sqrt{3}) - \sqrt{2}(2\cos x - \sqrt{3}) = 0$$
$$(2\cos x - \sqrt{3})(2\sin x - \sqrt{2}) = 0$$
$$2\cos x - \sqrt{3} = 0 \qquad 2\sin x - \sqrt{2} = 0$$
$$\cos x = \frac{\sqrt{3}}{2} \qquad \sin x = \frac{\sqrt{2}}{2}$$
$$x = \frac{\pi}{6}, \frac{11\pi}{6} \qquad x = \frac{\pi}{4}, \frac{3\pi}{4}$$
The solutions are $\frac{\pi}{6}, \frac{\pi}{4}, \frac{3\pi}{4}, \frac{11\pi}{6}$.

11.
$$\csc x - \sqrt{2} = 0$$
$$\csc x = \sqrt{2}$$
$$x = \frac{\pi}{4}, \frac{3\pi}{4}$$

13.
$$2\sin^2 x + 1 = 3\sin x$$
$$2\sin^2 x - 3\sin x + 1 = 0$$
$$(2\sin x - 1)(\sin x - 1) = 0$$
$$2\sin x - 1 = 0 \qquad \sin x - 1 = 0$$
$$\sin x = \frac{1}{2} \qquad \sin x = 1$$
$$x = \frac{\pi}{6}, \frac{5\pi}{6} \qquad x = \frac{\pi}{2}$$
The solutions are $\frac{\pi}{6}, \frac{\pi}{2}, \frac{5\pi}{6}$.

15.
$$4\cos^2 x - 3 = 0$$
$$\cos^2 x = \frac{3}{4}$$
$$\cos x = \pm\frac{\sqrt{3}}{2}$$
$$x = \frac{\pi}{6}, \frac{5\pi}{6}, \frac{7\pi}{6}, \frac{11\pi}{6}$$

Copyright © Houghton Mifflin Company. All rights reserved.

17.
$$2\sin^3 x = \sin x$$
$$2\sin^3 x - \sin x = 0$$
$$\sin x(2\sin^2 x - 1) = 0$$

$$\sin x = 0 \qquad\qquad 2\sin^2 x = 1$$
$$x = 0,\ \pi \qquad\qquad \sin x = \pm\frac{\sqrt{2}}{2}$$
$$x = \frac{\pi}{4},\ \frac{3\pi}{4},\ \frac{5\pi}{4},\ \frac{7\pi}{4}$$

The solutions are $0,\ \dfrac{\pi}{4},\ \dfrac{3\pi}{4},\ \pi,\ \dfrac{5\pi}{4},\ \dfrac{7\pi}{4}$.

19.
$$4\sin^2 x + 2\sqrt{3}\sin x - \sqrt{3} = 2\sin x$$
$$4\sin^2 x + 2\sqrt{3}\sin x - 2\sin x - \sqrt{3} = 0$$
$$2\sin x(2\sin x + \sqrt{3}) - (2\sin x + \sqrt{3}) = 0$$
$$(2\sin x + \sqrt{3})(2\sin x - 1) = 0$$

$$2\sin x + \sqrt{3} = 0 \qquad\qquad 2\sin x - 1 = 0$$
$$\sin x = -\frac{\sqrt{3}}{2} \qquad\qquad \sin x = \frac{1}{2}$$
$$x = \frac{4\pi}{3},\ \frac{5\pi}{3} \qquad\qquad x = \frac{\pi}{6},\ \frac{5\pi}{6}$$

The solutions are $\dfrac{\pi}{6},\ \dfrac{5\pi}{6},\ \dfrac{4\pi}{3},\ \dfrac{5\pi}{3}$.

21.
$$\sin^4 x = \sin^2 x$$
$$\sin^4 x - \sin^2 x = 0$$
$$\sin^2 x(\sin^2 x - 1) = 0$$

$$\sin^2 x = 0 \qquad\qquad \sin^2 x - 1 = 0$$
$$\sin x = 0 \qquad\qquad \sin x = \pm 1$$
$$x = 0,\ \pi \qquad\qquad x = \frac{\pi}{2},\ \frac{3\pi}{2}$$

The solutions are $0,\ \dfrac{\pi}{2},\ \pi,\ \dfrac{3\pi}{2}$.

23.
$$\cos x - 0.75 = 0$$
$$\cos x = 0.75$$
$$x \approx 41.4°,\ 318.6°$$

25.
$$3\sin x - 5 = 0$$
$$3\sin x = 5$$
$$\sin x = \frac{5}{3}$$
no solution

27.
$$3\sec x - 8 = 0$$
$$3\sec x = 8$$
$$\sec x = \frac{8}{3}$$
$$\frac{1}{\cos x} = \frac{8}{3}$$
$$\cos x = \frac{3}{8}$$
$$x \approx 68.0°,\ 292.0°$$

29.
$$\cos x + 3 = 0$$
$$\cos x = -3$$
no solution

31.
$$3 - 5\sin x = 4\sin x + 1$$
$$-9\sin x = -2$$
$$\sin x = \frac{2}{9}$$
$$x \approx 12.8°,\ 167.2°$$

33.
$$\frac{1}{2}\sin x + \frac{2}{3} = \frac{3}{4}\sin x + \frac{3}{5}$$
$$-\frac{1}{4}\sin x = \frac{3}{5} - \frac{2}{3}$$
$$-\frac{1}{4}\sin x = -\frac{1}{15}$$
$$\sin x = \frac{4}{15}$$
$$x \approx 15.5°,\ 164.5°$$

35.
$$3\tan^2 x - 2\tan x = 0$$
$$\tan x(3\tan x - 2) = 0$$

$$\tan x = 0 \qquad\qquad 3\tan x - 2 = 0$$
$$x = 0,\ 180° \qquad\qquad \tan x = \frac{2}{3}$$
$$x \approx 33.7°,\ 213.7°$$

The solutions are $0°,\ 33.7°,\ 180°,\ 213.7°$.

37.
$$3\cos x + \sec x = 0$$
$$3\cos x + \frac{1}{\cos x} = 0$$
$$3\cos^2 x + 1 = 0$$
$$\cos^2 x = -\frac{1}{3}$$
no solution

Copyright © Houghton Mifflin Company. All rights reserved.

39.
$$\tan^2 x = 3\sec^2 x - 2$$
$$\tan^2 x = 3(1 + \tan^2 x) - 2$$
$$\tan^2 x = 3 + 3\tan^2 x - 2$$
$$-2\tan^2 x = 1$$
$$\tan^2 x = -\frac{1}{2}$$
no solution

41.
$$2\sin^2 x = 1 - \cos x$$
$$2(1 - \cos^2 x) = 1 - \cos x$$
$$2 - 2\cos^2 x = 1 - \cos x$$
$$0 = 2\cos^2 x - \cos x - 1$$
$$0 = (2\cos x + 1)(\cos x - 1)$$

$$2\cos x + 1 = 0 \qquad \cos x - 1 = 0$$
$$\cos x = -\frac{1}{2} \qquad \cos x = 1$$
$$\qquad\qquad\qquad x = 0°$$
$$x = 120°, \ 240°$$

The solutions are 0°, 120°, 240°.

43.
$$3\cos^2 x + 5\cos x - 2 = 0$$
$$\cos x = \frac{-5 \pm \sqrt{5^2 - 4(3)(-2)}}{2 \cdot 3}$$
$$= \frac{-5 \pm \sqrt{49}}{6} = \frac{-5 \pm 7}{6}$$
$$\cos x = \frac{1}{3} \qquad \cos x = -2$$
$$x \approx 70.5°, \ 289.5° \qquad \text{no solution}$$
The solutions are 70.5°, 289.5°.

45.
$$2\tan^2 x - \tan x - 10 = 0$$
$$(\tan x + 2)(2\tan x - 5) = 0$$
$$\tan x + 2 = 0 \qquad 2\tan x - 5 = 0$$
$$\tan x = -2 \qquad \tan x = \frac{5}{2}$$
$$x \approx 116.6°, \ 296.6° \quad x \approx 68.2°, \ 248.2°$$
The solutions are 68.2°, 116.6°, 248.2°, 296.6°.

47.
$$3\sin x \cos x - \cos x = 0$$
$$\cos x(3\sin x - 1) = 0$$
$$\cos x = 0 \qquad\qquad 3\sin x - 1 = 0$$
$$x = 90°, \ 270° \qquad \sin x = \frac{1}{3}$$
$$\qquad\qquad\qquad\qquad x \approx 19.5°, \ 160.5°$$
The solutions are 19.5°, 90°, 160.5°, 270°.

49.
$$2\sin x \cos x - \sin x - 2\cos x + 1 = 0$$
$$\sin x(2\cos x - 1) - (2\cos x - 1) = 0$$
$$(2\cos x - 1)(\sin x - 1) = 0$$
$$2\cos x - 1 = 0 \qquad \sin x - 1 = 0$$
$$\cos x = \frac{1}{2} \qquad\qquad \sin x = 1$$
$$x = 60°, \ 300° \qquad\qquad x = 90°$$
The solutions are 60°, 90°, 300°.

51.
$$2\sin x - \cos x = 1$$
$$2\sin x - 1 = \cos x$$
$$(2\sin x - 1)^2 = (\cos x)^2$$
$$4\sin^2 x - 4\sin x + 1 = \cos^2 x$$
$$4\sin^2 x - 4\sin x + 1 = 1 - \sin^2 x$$
$$5\sin^2 x - 4\sin x = 0$$
$$\sin x(5\sin x - 4) = 0$$

$$\sin x = 0 \qquad\qquad 5\sin x - 4 = 0$$
$$x = 180° \qquad\qquad \sin x = \frac{4}{5}$$
$$\qquad\qquad\qquad x \approx 53.1° \text{ or } 126.9°$$

126.9° does not check.
The solutions are 53.1°, 180°.

53.
$$2\sin x - 3\cos x = 1$$
$$2\sin x = 3\cos x + 1$$
$$(2\sin x)^2 = (3\cos x + 1)^2$$
$$4\sin^2 x = 9\cos^2 x + 6\cos x + 1$$
$$4(1 - \cos^2 x) = 9\cos^2 x + 6\cos x + 1$$
$$0 = 13\cos^2 x + 6\cos x - 3$$

$$\cos x = \frac{-6 \pm \sqrt{6^2 - 4(13)(-3)}}{2(13)}$$
$$= \frac{-6 \pm \sqrt{192}}{26}$$
$$\cos x \approx 0.3022 \qquad\qquad \cos x \approx -0.7637$$
$$x \approx 72.4° \text{ or } 287.6° \qquad x \approx 139.8° \text{ or } 220.2°$$

287.6° and 139.8° do not check.
The solutions are 72.4°, 220.2°.

Copyright © Houghton Mifflin Company. All rights reserved.

55. $3\sin^2 x - \sin x - 1 = 0$

$$\sin x = \frac{1 \pm \sqrt{(-1)^2 - 4(3)(-1)}}{2(3)}$$

$$= \frac{1 \pm \sqrt{13}}{6}$$

$\sin x = 0.7676$ $\sin x = -0.4343$

$x = 50.1°, \ 129.9°$ $x = 205.7°, \ 334.3°$

The solutions are $50.1°, \ 129.9°, \ 205.7°, \ 334.3°$.

57. $2\cos x - 1 + 3\sec x = 0$

$$2\cos x - 1 + \frac{3}{\cos x} = 0$$

$$2\cos^2 x - \cos x + 3 = 0$$

$$\cos x = \frac{1 \pm \sqrt{(-1)^2 - 4(2)(3)}}{2(2)}$$

$$= \frac{1 \pm \sqrt{-23}}{4}$$

no solution

59. $\cos^2 x - 3\sin x + 2\sin^2 x = 0$

$1 - \sin^2 x - 3\sin x + 2\sin^2 x = 0$

$\sin^2 x - 3\sin x + 1 = 0$

$$\sin x = \frac{3 \pm \sqrt{(-3)^2 - 4(1)(1)}}{2(1)}$$

$$= \frac{3 \pm \sqrt{5}}{2}$$

$\sin x = 2.6180$ $\sin x = 0.3820$

no solution $x = 22.5°, \ 157.5°$

The solutions are $22.5°, \ 157.5°$.

61. $\tan 2x - 1 = 0$

$\tan 2x = 1$

$$2x = \frac{\pi}{4} + k\pi$$

$$x = \frac{\pi}{8} + \frac{k\pi}{2}, \text{ where } k \text{ is an integer}$$

63. $\sin 5x = 1$

$$5x = \frac{\pi}{2} + 2k\pi$$

$$x = \frac{\pi}{10} + \frac{2}{5}k\pi, \text{ where } k \text{ is an integer}$$

65. $\sin 2x - \sin x = 0$

$2\sin x \cos x - \sin x = 0$

$\sin x(2\cos x - 1) = 0$

$\sin x = 0$ $2\cos x - 1 = 0$

$\quad x = 0 + 2k\pi$ $\cos x = \dfrac{1}{2}$

$\quad\quad$ or

$\quad x = \pi + 2k\pi$ $x = \dfrac{\pi}{3} + 2k\pi$

$\quad\quad\quad\quad\quad\quad\quad\quad\quad\quad$ or

$\quad\quad\quad\quad\quad\quad\quad\quad\quad\quad x = \dfrac{5\pi}{3} + 2k\pi$

The solutions are $0 + 2k\pi, \ \dfrac{\pi}{3} + 2k\pi, \ \pi + 2k\pi, \ \dfrac{5\pi}{3} + 2k\pi$

where k is an integer.

67. $\sin\left(2x + \dfrac{\pi}{6}\right) = -\dfrac{1}{2}$

$2x + \dfrac{\pi}{6} = \dfrac{7\pi}{6} + 2k\pi$ or $2x + \dfrac{\pi}{6} = \dfrac{11\pi}{6} + 2k\pi$

$2x = \pi + 2k\pi$ $2x = \dfrac{5\pi}{3} + 2k\pi$

$x = \dfrac{\pi}{2} + k\pi$ $x = \dfrac{5\pi}{6} + k\pi$

The solutions are $\dfrac{\pi}{2} + k\pi, \ \dfrac{5\pi}{6} + k\pi$ where k is an integer.

69. $\sin^2 \dfrac{x}{2} + \cos x = 1$

$$\left(\pm\sqrt{\frac{1 - \cos x}{2}}\right)^2 + \cos x = 1$$

$$\frac{1 - \cos x}{2} + \cos x = 1$$

$1 - \cos x + 2\cos x = 2$

$\cos x = 1$

$x = 0 + 2k\pi$ where k is an integer

Copyright © Houghton Mifflin Company. All rights reserved.

71.

$$\cos 2x = 1 - 3\sin x$$

$$1 - 2\sin^2 x = 1 - 3\sin x$$

$$0 = 2\sin^2 x - 3\sin x$$
$$0 = \sin x(2\sin x - 3)$$

$$\sin x = 0 \qquad 2\sin x - 3 = 0$$

$$x = 0,\ \pi \qquad \sin x = \frac{3}{2}$$
$$\qquad\qquad\quad \text{no solution.}$$

The solutions are $0,\ \pi$.

73.

$$\sin 4x - \sin 2x = 0$$
$$2\sin 2x \cos 2x - \sin 2x = 0$$
$$\sin 2x(2\cos 2x - 1) = 0$$

$$\sin 2x = 0 \qquad\qquad 2\cos 2x - 1 = 0$$
$$2x = 0 + 2k\pi \qquad \cos 2x = \frac{1}{2}$$
$$\text{or}$$
$$2x = \pi + 2k\pi \qquad 2x = \frac{\pi}{3} + 2k\pi$$
$$\text{or}$$
$$2x = \frac{5\pi}{3} + 2k\pi$$

$$x = 0 + k\pi,\ \frac{\pi}{2} + k\pi,\ \frac{\pi}{6} + k\pi,\ \frac{5\pi}{6} + k\pi$$

The solutions are $0,\ \dfrac{\pi}{6},\ \dfrac{\pi}{2},\ \dfrac{5\pi}{6},\ \pi,\ \dfrac{7\pi}{6},\ \dfrac{3\pi}{2},\ \dfrac{11\pi}{6}$.

75.

$$\tan \frac{\pi}{2} = \sin x$$
$$\frac{1 - \cos x}{\sin x} = \sin x$$
$$1 - \cos x = \sin^2 x$$
$$1 - \cos x = 1 - \cos^2 x$$
$$\cos^2 x - \cos x = 0$$
$$\cos x(\cos x - 1) = 0$$
$$\cos x = 0 \qquad\qquad \cos x = 1$$
$$x = \frac{\pi}{2},\ \frac{3\pi}{2} \qquad\qquad x = 0$$

The solutions are $0,\ \dfrac{\pi}{2},\ \dfrac{3\pi}{2}$.

77.

$$\sin 2x \cos x + \cos 2x \sin x = 0$$
$$\sin(2x + x) = 0$$
$$\sin 3x = 0$$
$$3x = 0 + 2k\pi \quad \text{or} \quad 3x = \pi + 2k\pi$$
$$x = 0 + \frac{2}{3}k\pi \quad \text{or} \quad x = \frac{\pi}{3} + \frac{2}{3}k\pi$$

The solutions are $0,\ \dfrac{\pi}{3},\ \dfrac{2}{3}\pi,\ \pi,\ \dfrac{4}{3}\pi,\ \dfrac{5}{3}\pi$.

79.

$$\sin x \cos 2x - \cos x \sin 2x = \frac{\sqrt{3}}{2}$$
$$\sin(x - 2x) = \frac{\sqrt{3}}{2}$$
$$\sin(-x) = \frac{\sqrt{3}}{2}$$
$$\sin x = -\frac{\sqrt{3}}{2}$$
$$x = \frac{4\pi}{3},\frac{5\pi}{3}$$

81.

$$\sin 3x - \sin x = 0$$
$$2\cos\frac{3x + x}{2}\sin\frac{3x - 2}{2} = 0$$
$$2\cos 2x \sin x = 0$$
$$2(1 - 2\sin^2 x)\sin x = 0$$

$$\sin x = 0 \qquad\qquad 1 - 2\sin^2 x = 0$$
$$x = 0,\ \pi \qquad\qquad \sin^2 x = \frac{1}{2}$$
$$\qquad\qquad\qquad \sin x = \pm\frac{\sqrt{2}}{2}$$
$$\qquad\qquad x = \frac{\pi}{4},\ \frac{3\pi}{4},\ \frac{5\pi}{4},\ \frac{7\pi}{4}$$

The solutions are $0,\ \dfrac{\pi}{4},\ \dfrac{3\pi}{4},\ \pi,\ \dfrac{5\pi}{4},\ \dfrac{7\pi}{4}$.

83.

$$2\sin x \cos x + 2\sin x - \cos x - 1 = 0$$
$$2\sin x(\cos x + 1) - (\cos x + 1) = 0$$
$$(\cos x + 1)(2\sin x - 1) = 0$$

$$\cos x + 1 = 0 \qquad\qquad 2\sin x - 1 = 0$$
$$\cos x = -1 \qquad\qquad \sin x = \frac{1}{2}$$
$$x = \pi \qquad\qquad\qquad x = \frac{\pi}{6},\ \frac{5\pi}{6}$$

The solutions are $\dfrac{\pi}{6},\ \dfrac{5\pi}{6},\ \pi$.

Copyright © Houghton Mifflin Company. All rights reserved.

85.   0.7391   **87.**   −3.2957, 3.2957   **89.**   1.16

91.

Set your graphing utility to "degree" mode and graph $d = \dfrac{(288)^2}{16}\sin\theta\cos\theta$ and $d = 1295$ for $0° \le \theta \le 90°$, $-500 \le d \le 3000$.

Use the TRACE or INTERSECT feature of your graphing utility to determine the intersection of the two graphs.

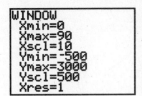 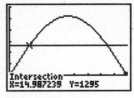 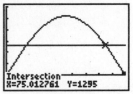

Thus, $d = 1295$ for $\theta \approx 14.99°$ and $\theta \approx 75.01°$.

The sine regression functions in Exercises 93–98 were obtained on a TI-83 calculator by using an iteration factor of 16. The use of a different iteration factor may produce a sine regression function that varies from the regression functions listed below.

93. **a.**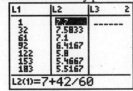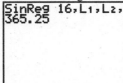

$f(x) = y \approx 1.121306016\sin(0.0159513906x + 1.836184752) + 6.625736057$

b.   Set your graphing utility to "radian" mode.

$f(71) \approx 1.121306016\sin(0.0159513906(71) + 1.836184752) + 6.625736057$

≈ 6.818600251 hours

$\approx 6 + .818600251(60) \to 6:49$

95. **a.**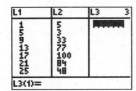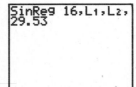

$f(x) = y \approx 49.572\sin(0.21261x - 2.1922) + 49.490$

b.   Set your graphing utility to "radian" mode.

$f(31) \approx 49.572\sin(0.21261(31) - 2.1922) + 49.490$

$\approx 2\%$

97. **a.**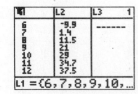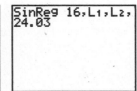

$f(x) = y \approx 35.185\sin(0.30395x - 2.1630) + 2.1515$

b.   Set your graphing utility to "radian" mode.

$f\left(9\dfrac{25}{60}\right) \approx 35.185\sin\left(0.30395\left(9\dfrac{25}{60}\right) - 2.1630\right) + 2.1515$

$\approx 24.8°$

Copyright © Houghton Mifflin Company. All rights reserved.

99. $d(t) \geq 12$ for $74 \leq t \leq 269$, so Mexico City will have at least 12 hours of daylight about $269 - 74 = 195$ days of the year.

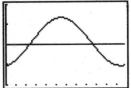

Xmin = 0, Xmax = 365, Xscl = 31,
Ymin = 10, Ymax = 14, Yscl = 1

101. a.

$\sin\theta = \dfrac{y}{3}$, so $y = 3\sin\theta$

$\cos\theta = \dfrac{x}{3}$, so $x = 3\cos\theta$

$A = 2\left(\dfrac{1}{2}bh\right) + 3y$

$\quad = (3\sin\theta)(3\cos\theta) + 3(3\sin\theta)$

$\quad = 9\sin\theta\cos\theta + 9\sin\theta$

$\quad = 9\sin\theta(\cos\theta + 1)$

b. Make sure your calculator is in degree mode.

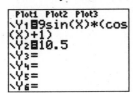   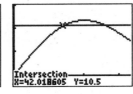   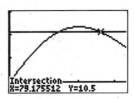

The solutions are $42°$ and $79°$.

c. Make sure your calculator is in degree mode.

   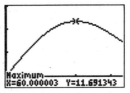

The value of θ that would maximize the area is $60°$.

Copyright © Houghton Mifflin Company. All rights reserved.

••

103. $\sqrt{3}\sin x + \cos x = \sqrt{3}$

$a = \sqrt{3},\ b = 1,\ k = \sqrt{(\sqrt{3})^2 + 1} = 2,\ \alpha$ is in first quadrant

$\tan\beta = \dfrac{1}{\sqrt{3}}$

$\beta = \dfrac{\pi}{6}$

$\alpha = \dfrac{\pi}{6}$

$2\sin\left(x + \dfrac{\pi}{6}\right) = \sqrt{3}$

$\sin\left(x + \dfrac{\pi}{6}\right) = \dfrac{\sqrt{3}}{2}$

$x + \dfrac{\pi}{6} = \dfrac{\pi}{3}\qquad\qquad x + \dfrac{\pi}{6} = \dfrac{2\pi}{3}$

$\qquad x = \dfrac{\pi}{6}\qquad\qquad\qquad x = \dfrac{\pi}{2}$

105. $-\sin x + \sqrt{3}\cos x = \sqrt{3}$

$k = \sqrt{(-1)^2 + \left(\sqrt{3}\right)^2} = 2\qquad \tan\beta = \left|\dfrac{\sqrt{3}}{-1}\right| = \sqrt{3}$

$\qquad\qquad\qquad\qquad\qquad\qquad \beta = \dfrac{\pi}{3}$

$\qquad\qquad\qquad\qquad\qquad\qquad \alpha = \dfrac{2\pi}{3}$ second quadrant

$2\sin\left(x + \dfrac{2\pi}{3}\right) = \sqrt{3}$

$\sin\left(x + \dfrac{2\pi}{3}\right) = \dfrac{\sqrt{3}}{2}$

$x + \dfrac{2\pi}{3} = \dfrac{\pi}{3}\qquad\qquad x + \dfrac{2\pi}{3} = \dfrac{2\pi}{3}$

$x = -\dfrac{\pi}{3}\qquad\qquad\qquad x = 0$

$-\dfrac{\pi}{3} + 2\pi = \dfrac{5\pi}{3}$

The solutions are 0 and $\dfrac{5\pi}{3}$.

107. $\qquad\cos 5x - \cos 3x = 0$

$-2\sin\dfrac{5x + 3x}{2}\sin\dfrac{5x - 3x}{2} = 0$

$\qquad -2\sin 4x\sin x = 0$

$\sin 4x = 0\qquad\qquad\qquad\qquad\qquad \sin x = 0$

$4x = 0 + 2k\pi\qquad 4x = \pi + 2k\pi\qquad\qquad x = 0,\ \pi$

$x = 0 + \dfrac{1}{2}k\pi\qquad x = \dfrac{\pi}{4} + \dfrac{1}{2}k\pi$

$x = 0,\ \dfrac{\pi}{2},\ \pi,\ \dfrac{3\pi}{2}\qquad x = \dfrac{\pi}{4},\ \dfrac{3\pi}{4},\ \dfrac{5\pi}{4},\ \dfrac{7\pi}{4}$

The solutions are $0,\ \dfrac{\pi}{4},\ \dfrac{\pi}{2},\ \dfrac{3\pi}{4},\ \pi,\ \dfrac{5\pi}{4},\ \dfrac{3\pi}{2},\ \dfrac{7\pi}{4}$.

109. $\qquad\sin 3x + \sin x = 0$

$2\sin\dfrac{3x + x}{2}\cos\dfrac{3x - x}{2} = 0$

$\qquad 2\sin 2x\cos x = 0$

$\sin 2x = 0\qquad\qquad\qquad\qquad \cos x = 0$

$2x = 0 + 2k\pi\qquad 2x = \pi + 2k\pi\qquad x = \dfrac{\pi}{2},\ \dfrac{3\pi}{2}$

$x = 0 + k\pi\qquad\qquad x = \dfrac{\pi}{2} + k\pi$

$x = 0,\ \pi\qquad\qquad\qquad x = \dfrac{\pi}{2},\ \dfrac{3\pi}{2}$

The solutions are $0,\ \dfrac{\pi}{2},\ \pi,\ \dfrac{3\pi}{2}$.

111. $\cos 4x + \cos 2x = 0$

$2\cos\dfrac{4x + x}{2}\cos\dfrac{4x - x}{2} = 0$

$2\cos 3x\cos x = 0$

$\cos 3x = 0\qquad\qquad\qquad\qquad \cos x = 0$

$3x = \dfrac{\pi}{2} + 2k\pi\qquad 3x = \dfrac{3\pi}{2} + 2k\pi\qquad x = \dfrac{\pi}{2},\ \dfrac{3\pi}{2}$

$x = \dfrac{\pi}{6} + \dfrac{2}{3}k\pi\qquad x = \dfrac{\pi}{2} + \dfrac{2}{3}k\pi$

$x = \dfrac{\pi}{6},\ \dfrac{5\pi}{6},\ \dfrac{3\pi}{2}\qquad x = \dfrac{\pi}{2},\ \dfrac{7\pi}{6},\ \dfrac{11\pi}{6}$

The solutions are $\dfrac{\pi}{6},\ \dfrac{\pi}{2},\ \dfrac{5\pi}{6},\ \dfrac{7\pi}{6},\ \dfrac{3\pi}{2},\ \dfrac{11\pi}{6}$.

Copyright © Houghton Mifflin Company. All rights reserved.

113. $\theta = 20°$

$$x = \sqrt{(4 + 18\cot 20°)^2 + 100} - (4 + 18\cot 20°)$$

$x \approx 0.93$ ft.

$\theta = 30°$

$$x = \sqrt{(4 + 18\cot 30°)^2 + 100} - (4 + 18\cot 30°)$$

$x \approx 1.39$ ft.

....   ███ **Chapter 3 True/False Exercises** ███

1.
False; if $\alpha = 45°$ and $\beta = 60°$, then $\dfrac{\tan\alpha}{\tan\beta} = \dfrac{\tan 45°}{\tan 60°} = \dfrac{1}{\sqrt{3}/2} = \dfrac{2\sqrt{3}}{3} \neq \dfrac{45°}{60°}$

2.
False; if $y = 0$, then $\dfrac{\sin x}{\cos y} = \dfrac{\sin x}{\cos 0} = \dfrac{\sin x}{1} = \sin x$, but $\tan\dfrac{x}{y} = \tan\dfrac{x}{0}$ which is undefined.

3.
False; if $x = 1$, then $\sin^{-1} x = \sin^{-1} 1 = \dfrac{\pi}{2}$ but $\csc x^{-1} = \csc 1^{-1} = \csc 1 = \dfrac{1}{\sin 1} \approx 1.188$.

4.
False; if $\alpha = \dfrac{\pi}{2}$, then $\sin 2\alpha = \sin\pi = 0$ but $2\sin\dfrac{\pi}{2} = 2$.

5.
False; if $\alpha = \beta = \dfrac{\pi}{4}$, we get $\sin\dfrac{\pi}{2} = 1 \neq \dfrac{\sqrt{2}}{2} + \dfrac{\sqrt{2}}{2} = \sqrt{2}$.

6.
False; $\sin x = 0$ has an infinite number of solutions $x = k\pi$, but $\sin x = 0$ is not an identity.

7.
False; $\tan 45° = \tan 225°$ but $45° \neq 225°$.

8.
False; $\cos^{-1}[\cos(3\pi/2)] = \cos^{-1}(0) = \pi/2 \neq 3\pi/2$.

9.
False; $\cos\left(\cos^{-1} 2\right) \neq 2$ because $\cos^{-1} 2$ is undefined.

10.
False; if $\alpha = 1$, then we get $\dfrac{\pi}{2} \neq \dfrac{1}{\csc 1} \approx 0.8415$.

11.
False; if $\theta = 30°$, then $\cos 30° = \dfrac{\sqrt{3}}{2} \neq \dfrac{1}{2} = \sin 150°$.

12.
False; since $\sin^2\theta \geq 0$ for all θ, but $\sin\theta^2$ can be < 0.

....   ███ **Chapter Review** ███

1. $\cos(45° + 30°) = \cos 45°\cos 30° - \sin 45°\sin 30°$ [3.2]

$$= \dfrac{\sqrt{2}}{2} \cdot \dfrac{\sqrt{3}}{2} - \dfrac{\sqrt{2}}{2} \cdot \dfrac{1}{2}$$

$$= \dfrac{\sqrt{6}}{4} - \dfrac{\sqrt{2}}{4}$$

$$= \dfrac{\sqrt{6} - \sqrt{2}}{4}$$

2. $\tan(210° - 45°) = \dfrac{\tan 210° - \tan 45°}{1 + \tan 210°\tan 45°}$ [3.2]

$$= \dfrac{\dfrac{1}{\sqrt{3}} - 1}{1 + \dfrac{1}{\sqrt{3}} \cdot 1} = \dfrac{\dfrac{1}{\sqrt{3}} - 1}{1 + \dfrac{1}{\sqrt{3}}} \cdot \dfrac{\sqrt{3}}{\sqrt{3}}$$

$$= \dfrac{1 - \sqrt{3}}{\sqrt{3} + 1} \cdot \dfrac{1 - \sqrt{3}}{1 - \sqrt{3}} = \dfrac{1 - 2\sqrt{3} + 3}{1 - 3}$$

$$= \dfrac{4 - 2\sqrt{3}}{-2} = \sqrt{3} - 2$$

Copyright © Houghton Mifflin Company. All rights reserved.

3.
$$\sin\left(\frac{2\pi}{3}+\frac{\pi}{4}\right)=\sin\frac{2\pi}{3}\cos\frac{\pi}{4}+\cos\frac{2\pi}{3}\sin\frac{\pi}{4}\quad[3.2]$$
$$=\frac{\sqrt{3}}{2}\cdot\frac{\sqrt{2}}{2}-\frac{1}{2}\cdot\frac{\sqrt{2}}{2}$$
$$=\frac{\sqrt{6}}{4}-\frac{\sqrt{2}}{4}$$
$$=\frac{\sqrt{6}-\sqrt{2}}{4}$$

4.
$$\sec\left(\frac{4\pi}{3}-\frac{\pi}{4}\right)=\frac{1}{\cos\left(\frac{4\pi}{3}-\frac{\pi}{4}\right)}\quad[3.2]$$
$$=\frac{1}{\cos\frac{4\pi}{3}\cos\frac{\pi}{4}+\sin\frac{4\pi}{3}\sin\frac{\pi}{4}}$$
$$=\frac{1}{-\frac{1}{2}\cdot\frac{\sqrt{2}}{2}+\left(-\frac{\sqrt{3}}{2}\right)\cdot\frac{\sqrt{2}}{2}}\cdot\frac{-4}{-4}$$
$$=\frac{-4}{\sqrt{2}+\sqrt{6}}\cdot\frac{\sqrt{2}-\sqrt{6}}{\sqrt{2}-\sqrt{6}}$$
$$=\frac{-4\left(\sqrt{2}-\sqrt{6}\right)}{2-6}=\sqrt{2}-\sqrt{6}$$

5.
$$\sin(60°-135°)=\sin60°\cos135°-\cos60°\sin135°\quad[3.2]$$
$$=\frac{\sqrt{3}}{2}\left(-\frac{\sqrt{2}}{2}\right)-\frac{1}{2}\cdot\frac{\sqrt{2}}{2}$$
$$=-\frac{\sqrt{6}}{4}-\frac{\sqrt{2}}{4}=-\frac{\sqrt{6}+\sqrt{2}}{4}$$

6.
$$\cos\left(\frac{5\pi}{3}-\frac{7\pi}{4}\right)=\cos\frac{5\pi}{3}\cos\frac{7\pi}{4}+\sin\frac{5\pi}{3}\sin\frac{7\pi}{4}\quad[3.2]$$
$$=\frac{1}{2}\cdot\frac{\sqrt{2}}{2}+\left(-\frac{\sqrt{3}}{2}\right)\left(-\frac{\sqrt{2}}{2}\right)$$
$$=\frac{\sqrt{2}}{4}+\frac{\sqrt{6}}{4}=\frac{\sqrt{2}+\sqrt{6}}{4}$$

7.
$$\sin22.5°=\sin\frac{45°}{2}\quad[3.3]$$
$$=\sqrt{\frac{1-\cos45°}{2}}$$
$$=\sqrt{\frac{1-\frac{\sqrt{2}}{2}}{2}}$$
$$=\sqrt{\frac{2-\sqrt{2}}{4}}$$
$$=\frac{\sqrt{2-\sqrt{2}}}{2}$$

8.
$$\cos105°=\cos\frac{210°}{2}\quad[3.3]$$
$$=-\sqrt{\frac{1+\cos210°}{2}}$$
$$=-\sqrt{\frac{1+\left(-\frac{\sqrt{3}}{2}\right)}{2}}$$
$$=-\sqrt{\frac{2-\sqrt{3}}{4}}$$
$$=-\frac{\sqrt{2-\sqrt{3}}}{2}$$

9.
$$\tan67.5°=\tan\frac{135°}{2}\quad[3.3]$$
$$=\frac{1-\cos135°}{\sin135°}$$
$$=\frac{1-\left(-\frac{1}{\sqrt{2}}\right)}{\frac{1}{\sqrt{2}}}$$
$$=\frac{1+\frac{1}{\sqrt{2}}}{\frac{1}{\sqrt{2}}}$$
$$=\sqrt{2}+1$$

10.
$$\sin112.5°=\sin\frac{225°}{2}\quad[3.3]$$
$$=\sqrt{\frac{1-\cos225°}{2}}$$
$$=\sqrt{\frac{1-\left(-\frac{\sqrt{2}}{2}\right)}{2}}$$
$$=\sqrt{\frac{2+\sqrt{2}}{4}}$$
$$=\frac{\sqrt{2+\sqrt{2}}}{2}$$

Copyright © Houghton Mifflin Company. All rights reserved.

11.
$\sin\alpha = \dfrac{1}{2}, \cos\alpha = \dfrac{\sqrt{3}}{2}$, quadrant I, $\tan\alpha = \dfrac{\sqrt{3}}{3}$ [3.2/3.3]

$\cos\beta = \dfrac{1}{2}, \sin\beta = -\dfrac{\sqrt{3}}{2}$, quadrant IV, $\tan\beta = -\sqrt{3}$

a. $\cos(\alpha - \beta) = \cos\alpha\cos\beta + \sin\alpha\sin\beta$

$$= \frac{\sqrt{3}}{2}\cdot\frac{1}{2} + \frac{1}{2}\left(-\frac{\sqrt{3}}{2}\right)$$

$$= \frac{\sqrt{3}}{4} - \frac{\sqrt{3}}{4} = 0$$

b. $\tan 2\alpha = \dfrac{2\tan\alpha}{1 - \tan^2\alpha}$

$$= \frac{2\left(\frac{\sqrt{3}}{3}\right)}{1 - \left(\frac{\sqrt{3}}{3}\right)^2} = \frac{\frac{2\sqrt{3}}{3}}{1 - \frac{1}{3}}\cdot\frac{3}{3}$$

$$= \frac{2\sqrt{3}}{3 - 1} = \sqrt{3}$$

c. $\sin\dfrac{\beta}{2} = \sqrt{\dfrac{1 - \cos\beta}{2}}$

$$= \sqrt{\frac{1 - \frac{1}{2}}{2}}$$

$$= \sqrt{\frac{1}{4}}$$

$$= \frac{1}{2}$$

12.
$\sin\alpha = \dfrac{\sqrt{3}}{2}, \cos\alpha = -\dfrac{1}{2}$, quadrant II   [3.2/3.3]

$\cos\beta = -\dfrac{1}{2}, \sin\beta = -\dfrac{\sqrt{3}}{2}$, quadrant III

a. $\sin(\alpha + \beta) = \sin\alpha\cos\beta + \cos\alpha\sin\beta$

$$= \frac{\sqrt{3}}{2}\cdot\left(-\frac{1}{2}\right) + \left(-\frac{1}{2}\right)\left(-\frac{\sqrt{3}}{2}\right)$$

$$= -\frac{\sqrt{3}}{4} + \frac{\sqrt{3}}{4} = 0$$

b. $\sec 2\beta = \dfrac{1}{\cos 2\beta} = \dfrac{1}{\cos^2\beta - \sin^2\beta}$

$$= \frac{1}{\left(-\frac{1}{2}\right)^2 - \left(-\frac{\sqrt{3}}{2}\right)^2}$$

$$= \frac{1}{\frac{1}{4} - \frac{3}{4}} = \frac{1}{-\frac{2}{4}} = -2$$

c. $\cos\dfrac{\alpha}{2} = \sqrt{\dfrac{1 + \cos\alpha}{2}}$

$$= \sqrt{\frac{1 + \left(-\frac{1}{2}\right)}{2}} = \sqrt{\frac{\frac{1}{2}}{2}}$$

$$= \sqrt{\frac{1}{4}}$$

$$= \frac{1}{2}$$

Copyright © Houghton Mifflin Company. All rights reserved.

13.
$\sin\alpha = -\dfrac{1}{2}$, $\cos\alpha = \dfrac{\sqrt{3}}{2}$, quadrant IV, $\tan\alpha = -\dfrac{\sqrt{3}}{3}$ [3.2/3.3]

$\cos\beta = -\dfrac{\sqrt{3}}{2}$, $\sin\beta = -\dfrac{1}{2}$, quadrant III

a. $\sin(\alpha - \beta) = \sin\alpha\cos\beta - \cos\alpha\sin\beta$

$= -\dfrac{1}{2}\left(-\dfrac{\sqrt{3}}{2}\right) - \dfrac{\sqrt{3}}{2}\left(-\dfrac{1}{2}\right)$

$= \dfrac{\sqrt{3}}{4} + \dfrac{\sqrt{3}}{4} = \dfrac{2\sqrt{3}}{4}$

$= \dfrac{\sqrt{3}}{2}$

b. $\tan 2\alpha = \dfrac{2\tan\alpha}{1 - \tan^2\alpha}$

$= \dfrac{2\left(-\dfrac{\sqrt{3}}{3}\right)}{1 - \left(-\dfrac{\sqrt{3}}{3}\right)^2}$

$= \dfrac{-\dfrac{2\sqrt{3}}{3}}{1 - \dfrac{1}{3}} \cdot \dfrac{3}{3}$

$= \dfrac{-2\sqrt{3}}{3 - 1}$

$= -\sqrt{3}$

c. $\cos\dfrac{\beta}{2} = -\sqrt{\dfrac{1 + \cos\beta}{2}}$

$= -\sqrt{\dfrac{1 + \left(-\dfrac{\sqrt{3}}{2}\right)}{2}}$

$= -\sqrt{\dfrac{2 - \sqrt{3}}{4}}$

$= -\dfrac{\sqrt{2 - \sqrt{3}}}{2}$

14.
$\sin\alpha = \dfrac{\sqrt{2}}{2}$, $\cos\alpha = \dfrac{\sqrt{2}}{2}$, quadrant I   [3.2/3.3]

$\cos\beta = \dfrac{\sqrt{3}}{2}$, $\sin\beta = -\dfrac{1}{2}$, quadrant IV, $\tan\beta = -\dfrac{\sqrt{3}}{3}$

a. $\cos(\alpha - \beta) = \cos\alpha\cos\beta + \sin\alpha\sin\beta$

$= \dfrac{\sqrt{2}}{2} \cdot \dfrac{\sqrt{3}}{2} + \dfrac{\sqrt{2}}{2}\left(-\dfrac{1}{2}\right)$

$= \dfrac{\sqrt{6}}{4} - \dfrac{\sqrt{2}}{4} = \dfrac{\sqrt{6} - \sqrt{2}}{4}$

b. $\tan 2\beta = \dfrac{2\tan\beta}{1 - \tan^2\beta}$

$= \dfrac{2\left(-\dfrac{\sqrt{3}}{3}\right)}{1 - \left(-\dfrac{\sqrt{3}}{3}\right)^2}$

$= \dfrac{-2\left(\dfrac{\sqrt{3}}{3}\right)}{1 - \dfrac{1}{3}} \cdot \dfrac{3}{3}$

$= \dfrac{2\sqrt{3}}{3 - 1}$

$= -\sqrt{3}$

c. $\sin 2\alpha = 2\sin\alpha\cos\alpha$

$= 2\dfrac{\sqrt{2}}{2} \cdot \dfrac{\sqrt{2}}{2}$

$= 1$

15. $2\sin 3x\cos 3x = \sin 2(3x)$ [3.3]

$= \sin 6x$

16. $\dfrac{\tan 2x + \tan x}{1 - \tan 2x\tan x} = \tan(2x + x)$ [3.2]

$= \tan 3x$

17. $\sin 4x\cos x - \cos 4x\sin x = \sin(4x - x)$ [3.2]

$= \sin 3x$

18. $\cos^2 2\theta - \sin^2 2\theta = \cos 2(2\theta)$ [3.3]

$= \cos 4\theta$

19. $\dfrac{\sin 2\theta}{\cos 2\theta} = \tan 2\theta$ [3.1]

20. $\dfrac{1 - \cos 2\theta}{\sin 2\theta} = \dfrac{1 - (\cos^2\theta - \sin^2\theta)}{2\sin\theta\cos\theta}$ [3.3]

$= \dfrac{\sin^2\theta + \cos^2\theta - \cos^2\theta + \sin^2\theta}{2\sin\theta\cos\theta}$

$= \dfrac{2\sin^2\theta}{2\sin\theta\cos\theta}$

$= \dfrac{\sin\theta}{\cos\theta}$

$= \tan\theta$

Copyright © Houghton Mifflin Company. All rights reserved.

21.
$$\cos 2\theta - \cos 4\theta = -2\sin\frac{2\theta + 4\theta}{2}\sin\frac{2\theta - 4\theta}{2} \quad [3.4]$$
$$= -2\sin 3\theta \sin(-\theta)$$
$$= 2\sin 3\theta \sin\theta$$

22.
$$\sin 3\theta - \sin 5\theta = 2\cos\frac{3\theta + 5\theta}{2}\sin\frac{3\theta - 5\theta}{2} \quad [3.4]$$
$$= 2\cos 4\theta \sin(-\theta)$$
$$= -2\cos 4\theta \sin\theta$$

23.
$$\sin 6\theta + \sin 2\theta = 2\sin\frac{6\theta + 2\theta}{2}\cos\frac{6\theta - 2\theta}{2} \quad [3.4]$$
$$= 2\sin 4\theta \cos 2\theta$$

24.
$$\sin 5\theta - \sin\theta = 2\cos\frac{5\theta + \theta}{2}\sin\frac{5\theta - \theta}{2} \quad [3.4]$$
$$= 2\cos 3\theta \sin 2\theta$$

25.
$$\frac{1}{\sin x - 1} + \frac{1}{\sin x + 1} = \frac{(\sin x + 1) + (\sin x - 1)}{(\sin x - 1)(\sin x + 1)}$$
$$= \frac{2\sin x}{\sin^2 x - 1}$$
$$= \frac{2\sin x}{-\cos^2 x}$$
$$= -2\tan x \sec x$$

26.
$$\frac{\sin x}{1 - \cos x} = \frac{\sin x(1 + \cos x)}{(1 - \cos x)(1 + \cos x)}$$
$$= \frac{\sin x + \sin x \cos x}{1 - \cos^2 x}$$
$$= \frac{\sin x + \sin x \cos x}{\sin^2 x}$$
$$= \frac{\sin x}{\sin^2 x} + \frac{\sin x \cos x}{\sin^2 x}$$
$$= \csc x + \cot x, \ 0 < x < \frac{\pi}{2}$$

27.
$$\frac{1 + \sin x}{\cos^2 x} = \frac{1}{\cos^2 x} + \frac{\sin x}{\cos^2 x}$$
$$= \sec^2 x + \tan x \sec x$$
$$= \tan^2 x + 1 + \tan x \sec x$$

28.
$$\frac{\cos^2 2x - \sin^2 2x}{\cos 2x + \sin 2x} = \frac{(\cos 2x - \sin 2x)(\cos 2x + \sin 2x)}{\cos 2x + \sin 2x}$$
$$= \cos 2x - \sin 2x$$
$$= \cos^2 x - \sin^2 x - \sin 2x$$

29.
$$\frac{1}{\cos x} - \cos x = \frac{1 - \cos^2 x}{\cos x}$$
$$= \frac{\sin^2 x}{\cos x}$$
$$= \tan x \sin x$$

30.
$$\sin(270° - \theta) - \cos(270° - \theta) = \sin 270°\cos\theta - \cos 270°\sin\theta - \cos 270°\cos\theta - \sin 270°\sin\theta$$
$$= (-1)\cos\theta - 0 - 0 - (-1)\sin\theta$$
$$= -\cos\theta + \sin\theta$$
$$= \sin\theta - \cos\theta$$

31.
$$\sin\left(\frac{\pi}{4} - \alpha\right) = \sin\frac{\pi}{4}\cos\alpha - \cos\frac{\pi}{4}\sin\alpha$$
$$= \frac{\sqrt{2}}{2}\cos\alpha - \frac{\sqrt{2}}{2}\sin\alpha$$
$$= \frac{\sqrt{2}}{2}(\cos\alpha - \sin\alpha)$$

32.
$$\sin(180° - \alpha + \beta) = \sin[180° - (\alpha - \beta)]$$
$$= \sin 180°\cos(\alpha - \beta) - \cos 180°\sin(\alpha - \beta)$$
$$= 0[\cos(\alpha - \beta)] - (-1)[\sin\alpha\cos\beta - \cos\alpha\sin\beta]$$
$$= \sin\alpha\cos\beta - \cos\alpha\sin\beta$$

33.
$$\frac{\sin 4x - \sin 2x}{\cos 4x - \cos 2x} = \frac{2\cos\frac{4x + 2x}{2}\sin\frac{4x - 2x}{2}}{-2\sin\frac{4x + 2x}{2}\sin\frac{4x - 2x}{2}}$$
$$= -\frac{\cos 3x \sin x}{\sin 3x \sin x}$$
$$= -\cot 3x$$

34.
$$2\sin x \sin 3x = \cos(x - 3x) - \cos(x + 3x)$$
$$= \cos 2x - \cos 4x$$
$$= \cos 2x - (2\cos^2 2x - 1)$$
$$= 1 + \cos 2x - 2\cos^2 2x$$
$$= (1 - \cos 2x)(1 + 2\cos 2x)$$

Copyright © Houghton Mifflin Company. All rights reserved.

35.
$$\sin x - \cos 2x = \sin x - (1 - 2\sin^2 x)$$
$$= 2\sin^2 x + \sin x - 1$$
$$= (2\sin x - 1)(\sin x + 1)$$

36.
$$\cos 4x = 1 - 2\sin^2 2x$$
$$= 1 - 2(2\sin x \cos x)^2$$
$$= 1 - 2(4\sin^2 x \cos^2 x)$$
$$= 1 - 8\sin^2 x \cos^2 x$$
$$= 1 - 8\sin^2 x(1 - \sin^2 x)$$
$$= 1 - 8\sin^2 x + 8\sin^4 x$$

37.
$$\tan 4x = \frac{2\tan 2x}{1 - \tan^2 2x}$$
$$= \frac{2\left(\dfrac{2\tan x}{1 - \tan^2 x}\right)}{1 - \left(\dfrac{2\tan x}{1 - \tan^2 x}\right)^2}$$
$$= \frac{\dfrac{4\tan x}{1 - \tan^2 x}}{\dfrac{(1 - \tan^2 x)^2 - (2\tan x)^2}{(1 - \tan^2 x)^2}} \cdot \frac{(1 - \tan^2 x)^2}{(1 - \tan^2 x)^2}$$
$$= \frac{4\tan x(1 - \tan^2 x)}{(1 - \tan^2 x)^2 - 4\tan^2 x}$$
$$= \frac{4\tan x - 4\tan^3 x}{1 - 2\tan^2 x + \tan^4 x - 4\tan^2 x}$$
$$= \frac{4\tan x - 4\tan^3 x}{1 - 6\tan^2 x + \tan^4 x}$$

38.
$$\frac{\sin 2x - \sin x}{\cos 2x + \cos x} = \frac{2\cos\frac{2x + x}{2}\sin\frac{2x - x}{2}}{2\cos\frac{2x + x}{2}\cos\frac{2x - x}{2}}$$
$$= \frac{\sin\frac{\pi}{2}}{\cos\frac{\pi}{2}}$$
$$= \tan\frac{\pi}{2}$$
$$= \frac{1 - \cos x}{\sin x}$$

39.
$$2\sin 3x \cos 3x - 2\sin x \cos x = \sin 6x - \sin 2x$$
$$= 2\cos\frac{6x + 2x}{2}\sin\frac{6x - 2x}{2}$$
$$= 2\cos 4x \sin 2x$$

40.
$$2\sin x \sin 2x = 2\sin x(2\sin x \cos x)$$
$$= 4\cos x \sin^2 x$$

41.
$$\cos(x + y)\cos(x - y) = \frac{1}{2}\Big[\cos(x + y + x - y) + \cos(x + y - x + y)\Big]$$
$$= \frac{1}{2}(\cos 2x + \cos 2y)$$
$$= \frac{1}{2}\Big[2\cos^2 x - 1 + 2\cos^2 y - 1\Big]$$
$$= \cos^2 x + \cos^2 y - 1$$

42.
$$\cos(x + y)\sin(x - y) = \frac{1}{2}\Big[\sin(x + y + x - y) - \sin(x + y - x + y)\Big]$$
$$= \frac{1}{2}(\sin 2x - \sin 2y)$$
$$= \frac{1}{2}\Big[2\sin x \cos x - 2\sin y \cos y\Big]$$
$$= \sin x \cos x - \sin y \cos y$$

43.
$$y = \sec\left(\sin^{-1}\frac{12}{13}\right), \quad \alpha = \sin^{-1}\frac{12}{13}, \quad \sin\alpha = \frac{12}{13}, \quad \cos\alpha = \frac{5}{13}, \quad \sec\alpha = \frac{13}{5} \quad [3.5]$$
$$y = \sec\alpha = \frac{13}{5}$$

Copyright © Houghton Mifflin Company. All rights reserved.

44.

$$y = \cos\left(\sin^{-1}\frac{3}{5}\right) \qquad \text{[3.5]}$$

$$\alpha = \sin^{-1}\frac{3}{5}, \quad \sin\alpha = \frac{3}{5}, \quad y = \cos\alpha = \frac{4}{5}$$

45.

$$\alpha = \sin^{-1}\left(-\frac{3}{5}\right) \qquad\qquad \beta = \cos^{-1}\frac{5}{13} \quad \text{[3.5]}$$

$$\sin\alpha = -\frac{3}{5} \qquad\qquad \cos\beta = \frac{5}{13}$$

$$\cos\alpha = \frac{4}{5} \qquad\qquad \sin\beta = \frac{12}{13}$$

$$y = \cos\left[\sin^{-1}\left(-\frac{3}{5}\right) + \cos^{-1}\frac{5}{13}\right]$$

$$= \cos(\alpha + \beta)$$

$$= \cos\alpha\cos\beta - \sin\alpha\sin\beta$$

$$= \frac{4}{5}\cdot\frac{5}{13} - \left(-\frac{3}{5}\right)\cdot\frac{12}{13}$$

$$= \frac{20}{65} + \frac{36}{65}$$

$$= \frac{56}{65}$$

46.

$$y = \cos\left(2\sin^{-1}\frac{3}{5}\right) \qquad\qquad \alpha = \sin^{-1}\frac{3}{5} \quad \text{[3.5]}$$

$$y = \cos 2\alpha \qquad\qquad \sin\alpha = \frac{3}{5}$$

$$= \cos^2\alpha - \sin^2\alpha$$

$$= \left(\frac{4}{5}\right)^2 - \left(\frac{3}{5}\right)^2 \qquad\qquad \cos\alpha = \frac{4}{5}$$

$$= \frac{16}{25} - \frac{9}{25}$$

$$= \frac{7}{25}$$

47.

$$2\sin^{-1}(x-1) = \frac{\pi}{3} \qquad \text{[3.6]}$$

$$\sin^{-1}(x-1) = \frac{\pi}{6}$$

$$x - 1 = \sin\frac{\pi}{6}$$

$$x - 1 = \frac{1}{2}$$

$$x = \frac{3}{2}$$

48.

$$\sin^{-1}x + \cos^{-1}\frac{4}{5} = \frac{\pi}{2} \qquad\qquad \alpha = \cos^{-1}\frac{4}{5} \quad \text{[3.6]}$$

$$\sin^{-1}x + \alpha = \frac{\pi}{2} \qquad\qquad \cos\alpha = \frac{4}{5}$$

$$\sin^{-1}x = \frac{\pi}{2} - \alpha \qquad\qquad \sin\alpha = \frac{3}{5}$$

$$x = \sin\left(\frac{\pi}{2} - \alpha\right)$$

$$= \sin\frac{\pi}{2}\cos\alpha - \cos\frac{\pi}{2}\sin\alpha$$

$$= 1\cdot\frac{4}{5} - 0\cdot\frac{3}{5}$$

$$= \frac{4}{5}$$

49.

$$4\sin^2 x + 2\sqrt{3}\sin x - 2\sin x - \sqrt{3} = 0 \quad \text{[3.6]}$$

$$2\sin x\left(2\sin x + \sqrt{3}\right) - \left(2\sin x + \sqrt{3}\right) = 0$$

$$\left(2\sin x + \sqrt{3}\right)(2\sin x - 1) = 0$$

$$2\sin x + \sqrt{3} = 0 \qquad\qquad 2\sin x - 1 = 0$$

$$\sin x = -\frac{\sqrt{3}}{2} \qquad\qquad \sin x = \frac{1}{2}$$

$$x = 240°, 300° \qquad\qquad x = 30°, 150°$$

The solutions are $30°, 150°, 240°, 300°$.

50.

$$2\sin x\cos x - \sqrt{2}\cos x - 2\sin x + \sqrt{2} = 0 \quad \text{[3.6]}$$

$$\cos x(2\sin x - \sqrt{2}) - (2\sin x - \sqrt{2}) = 0$$

$$(2\sin x - \sqrt{2})(\cos x - 1) = 0$$

$$2\sin x - \sqrt{2} = 0 \qquad\qquad \cos x - 1 = 0$$

$$\sin x = \frac{\sqrt{2}}{2} \qquad\qquad \cos x = 1$$

$$x = 45°, 135° \qquad\qquad x = 0°$$

The solutions are $0°, 45°, 135°$.

51.

$$3\cos^2 x + \sin x = 1 \qquad\qquad \text{[3.6]}$$

$$3(1 - \sin^2 x) + \sin x = 1$$

$$0 = 3\sin^2 x - \sin x - 2$$

$$0 = (3\sin x + 2)(\sin x - 1)$$

$$3\sin x + 2 = 0 \qquad\qquad \sin x - 1 = 0$$

$$\sin x = -\frac{2}{3} \qquad\qquad \sin x = 1$$

$$x = 3.8713 \text{ or } 5.553 \qquad\qquad x = \frac{\pi}{2}$$

The solutions are $\frac{\pi}{2} + 2k\pi, \ 3.8713 + 2k\pi, \ 5.553 + 2k\pi$

where k is an integer.

Copyright © Houghton Mifflin Company. All rights reserved.

52.
$$\tan^2 x - 2\tan x - 3 = 0 \quad [3.6]$$
$$(\tan x + 1)(\tan x - 3) = 0$$

$$\tan x = -1 \qquad\qquad \tan x = 3$$

$$x = -\frac{\pi}{4} + k\pi \qquad\qquad x = 1.2490 + k\pi$$

The solutions are $-\dfrac{\pi}{4} + k\pi$, $1.2490 + k\pi$ where k is an integer.

53.
$$\sin 3x \cos x - cox3x\sin x = \frac{1}{2} \quad [3.6]$$

$$\sin(3x - x) = \frac{1}{2}$$

$$\sin 2x = \frac{1}{2}$$

$$2x = \frac{\pi}{6} + 2k\pi \qquad\qquad 2x = \frac{5\pi}{6} + 2k\pi$$

$$x = \frac{\pi}{12} + k\pi \qquad\qquad x = \frac{5\pi}{12} + k\pi$$

The solutions are $\dfrac{\pi}{12}, \dfrac{5\pi}{12}, \dfrac{13\pi}{12}, \dfrac{17\pi}{12}$.

54.
$$\cos\left(2x - \frac{\pi}{3}\right) = -\frac{\sqrt{3}}{2} \quad [3.6]$$

$$2x - \frac{\pi}{3} = \frac{5\pi}{6} + 2k\pi \qquad\qquad 2x - \frac{\pi}{3} = \frac{7\pi}{6} + 2k\pi$$

$$2x = \frac{7\pi}{6} + 2k\pi \qquad\qquad 2x = \frac{3\pi}{2} + 2k\pi$$

$$x = \frac{7\pi}{12} + k\pi \qquad\qquad x = \frac{3\pi}{4} + k\pi$$

The solutions are $\dfrac{7\pi}{12}, \dfrac{19\pi}{12}, \dfrac{3\pi}{4}, \dfrac{7\pi}{4}$.

55. $f(x) = \sqrt{3}\sin x + \cos x \quad [3.4]$

$$f(x) = 2\sin\left(x + \frac{\pi}{6}\right)$$

amplitude $= 2$

phase shift $= -\dfrac{\pi}{6}$

56. $f(x) = -2\sin x - 2\cos x \quad [3.4]$

$$f(x) = 2\sqrt{2}\sin\left(x + \frac{5\pi}{4}\right)$$

amplitude $= 2\sqrt{2}$

phase shift $= -\dfrac{5\pi}{4}$

57. $f(x) = -\sin x - \sqrt{3}\cos x \quad [3.4]$

$$f(x) = 2\sin\left(x + \frac{4\pi}{3}\right)$$

amplitude $= 2$

phase shift $= -\dfrac{4\pi}{3}$

58.
$$f(x) = \frac{\sqrt{3}}{2}\sin x - \frac{1}{2}\cos x \quad [3.4]$$

$$f(x) = \sin\left(x + \frac{11\pi}{6}\right) \text{ or } \sin\left(x - \frac{\pi}{6}\right)$$

amplitude $= 1$

phase shift $= \dfrac{\pi}{6}$

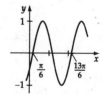

59. $f(x) = 2\cos^{-1} x$

60. $f(x) = \sin^{-1}(x - 1)$

61. $f(x) = \sin^{-1}\dfrac{x}{2}$

62. $f(x) = \sec^{-1} 2x$

Copyright © Houghton Mifflin Company. All rights reserved.

63. a.

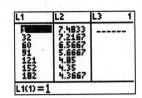 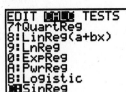 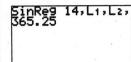 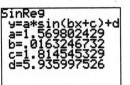

$$y \approx 1.5698\sin(0.01632x + 1.8145) + 5.9360$$

b. $y \approx 1.5698\sin(0.01632(104) + 1.8145) + 5.9360$ [3.6]
$$\approx 5.368$$
$$= 5:22$$

1.
$$1 + \sin^2 x \sec^2 x = 1 + \sin^2 x \frac{1}{\cos^2 x}$$
$$= 1 + \tan^2 x$$
$$= \sec^2 x$$

2.
$$\frac{1}{\sec x - \tan x} - \frac{1}{\sec x + \tan x} = \frac{\sec x + \tan x - \sec x + \tan x}{\sec^2 x - \tan^2 x}$$
$$= \frac{2\tan x}{1}$$
$$= 2\tan x$$

3.
$$\cos^3 x + \cos x \sin^2 x = \cos x(\cos^2 x + \sin^2 x)$$
$$= \cos x$$

4.
$$\csc x - \cot x = \frac{1}{\sin x} - \frac{\cos x}{\sin x}$$
$$= \frac{1 - \cos x}{\sin x}$$

5.
$$\sin 195° = \sin(150° + 45°) \qquad [3.2]$$
$$= \sin 150° \cos 45° + \cos 150° \sin 45°$$
$$= \frac{1}{2}\left(\frac{\sqrt{2}}{2}\right) + \left(-\frac{\sqrt{3}}{2}\right)\left(\frac{\sqrt{2}}{2}\right)$$
$$= \frac{\sqrt{2} - \sqrt{6}}{4}$$

6.
$$\sin\alpha = -\frac{3}{5}, \cos\alpha = -\frac{4}{5}, \cos\beta = -\frac{\sqrt{2}}{2}, \sin\beta \frac{\sqrt{2}}{2} \quad [3.2]$$
$$\sin(\alpha + \beta) = \sin\alpha\cos\beta + \cos\alpha\sin\beta$$
$$= \left(-\frac{3}{5}\right)\left(-\frac{\sqrt{2}}{2}\right) + \left(-\frac{4}{5}\right)\left(\frac{\sqrt{2}}{2}\right)$$
$$= \frac{3\sqrt{2} - 4\sqrt{2}}{10}$$
$$= -\frac{\sqrt{2}}{10}$$

7.
$$\sin\left(\theta - \frac{3\pi}{2}\right) = \sin\theta\cos\frac{3\pi}{2} - \cos\theta\sin\frac{3\pi}{2}$$
$$= \sin\theta(0) - \cos\theta(-1)$$
$$= \cos\theta$$

8.
$$\cos 6x \sin 3x + \sin 6x \cos 3x = \sin(6x + 3x) \quad [3.2]$$
$$= \sin 9x$$

9.
$$\sin\theta = \frac{4}{5}, \cos\theta = -\frac{3}{5} \quad [3.3]$$
$$\cos 2\theta = 2\cos^2\theta - 1$$
$$= 2\left(-\frac{3}{5}\right)^2 - 1$$
$$= \frac{18}{25} - 1 = -\frac{7}{25}$$

10.
$$\tan\frac{\theta}{2} + \frac{\cos\theta}{\sin\theta} = \frac{1 - \cos\theta}{\sin\theta} + \frac{\cos\theta}{\sin\theta}$$
$$= \frac{1 - \cos\theta + \cos\theta}{\sin\theta}$$
$$= \csc\theta$$

Copyright © Houghton Mifflin Company. All rights reserved.

11.
$$\sin^2 2x + 4\cos^4 x = (2\sin x \cos x)^2 + 4\cos^4 x$$
$$= 4\sin^2 x \cos^2 x + 4\cos^4 x$$
$$= 4\cos^2 x(\sin^2 x + \cos^2 x)$$
$$= 4\cos^2 x$$

12.
$$\sin 15° \cos 75° = \frac{1}{2}\big[\sin(15° + 75°) + \sin(15° - 75°)\big] \quad [3.4]$$
$$= \frac{1}{2}(\sin 90° - \sin 60°)$$
$$= \frac{1}{2}\left(1 - \frac{\sqrt{3}}{2}\right)$$
$$= \frac{1}{2} - \frac{\sqrt{3}}{4}$$
$$= \frac{2 - \sqrt{3}}{4}$$

13.
$$y = -\frac{\sqrt{3}}{2}\sin x + \frac{1}{2}\cos x \quad [3.4]$$
$$a = -\frac{\sqrt{3}}{2}, \; b = \frac{1}{2}$$
$$k = \sqrt{\left(-\frac{\sqrt{3}}{2}\right)^2 + \left(\frac{1}{2}\right)^2} = 1$$
$$\sin \beta = \left|\frac{\frac{1}{2}}{1}\right| = \frac{1}{2}$$
$$\beta = \frac{\pi}{6}$$
$$\alpha = \pi - \frac{\pi}{6}$$
$$= \frac{5\pi}{6}$$
$$y = \sin\left(x + \frac{5\pi}{6}\right)$$

14.
$$\theta = \cos^{-1}(0.7644) \quad [3.5]$$
$$\theta = 0.701$$

15.
$$\sin\left(\cos^{-1}\frac{12}{13}\right) \quad [3.5]$$
Let $\theta = \cos^{-1}\frac{12}{13}$ and find $\sin\theta$.
Then $\cos\theta = \frac{12}{13}$ and $0 \le \theta \le \pi$.
$$\sin\theta = \frac{5}{13}$$
$$\sin\left(\cos^{-1}\frac{12}{13}\right) = \frac{5}{13}$$

16. The graph of $y = \sin^{-1}(x + 2)$ is the graph of $y = \sin^{-1} x$ moved two units to the left.

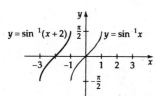

17.
$$3\sin x - 2 = 0 \quad\quad [3.6]$$
$$\sin x = \frac{2}{3}$$
$$x = 41.8°, \; 138.2°$$

18.
$$\sin x \cos x - \frac{\sqrt{3}}{2}\sin x = 0 \quad [3.6]$$
$$\sin x\left(\cos x - \frac{\sqrt{3}}{2}\right) = 0$$

$$\sin x = 0 \qquad\qquad \cos x - \frac{\sqrt{3}}{2} = 0$$
$$x = 0, \; \pi \qquad\qquad \cos x = \frac{\sqrt{3}}{2}$$
$$\cos x = \frac{\sqrt{3}}{2}$$
$$x = \frac{\pi}{6}, \; \frac{11\pi}{6}$$

The solutions are $0, \; \frac{\pi}{6}, \; \pi, \; \frac{11\pi}{6}$.

19.
$$\sin 2x + \sin x - 2\cos x - 1 = 0$$
$$2\sin x \cos x + \sin x - 2\cos x - 1 = 0$$
$$\sin x(2\cos x + 1) - (2\cos x + 1) = 0$$
$$(2\cos x + 1)(\sin x - 1) = 0$$

$$\cos x = -\frac{1}{2} \qquad\qquad \sin x = 1$$
$$x = \frac{2\pi}{3}, \; \frac{4\pi}{3} \qquad\qquad x = \frac{\pi}{2}$$

The solutions are $\frac{\pi}{2}, \; \frac{2\pi}{3}, \; \frac{4\pi}{3}$.

Copyright © Houghton Mifflin Company. All rights reserved.

20. **a.**

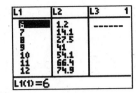 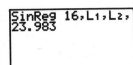

$$f(x) = y \approx 38.965\sin(0.40089x + 2.9143) + 33.197$$

b. $f\left(10\frac{40}{60}\right) \approx 38.965\sin\left(0.40089\left(10\frac{40}{60}\right) + 2.9143\right) + 33.197$ [3.6]

$$\approx 63.9°$$

....

1. $-2x + 1 < 7$
$-2x < 6$
$x > -3$

2. Shift the graph of $y = f(x)$ horizontally 1 unit to the left and up 2 units.

3. Reflect the graph of $y = f(x)$ across the x-axis.

4. $f(-x) = -x - \sin(-x)$
$= -x + \sin x$
$= -(x - \sin x)$
$= -f(x)$
odd function

5. $f(x) = \dfrac{5x}{x-1}$

$y = \dfrac{5x}{x-1}$

$x = \dfrac{5y}{y-1}$

$x(y-1) = 5y$

$xy - 5y = x$

$y(x-5) = x$

$y = \dfrac{x}{x-5}$

$f^{-1}(x) = \dfrac{x}{x-5}$

6. $240° = 240°\left(\dfrac{\pi}{180°}\right) = \dfrac{4\pi}{3}$ [2.1]

7. $\dfrac{5\pi}{3} = \dfrac{5\pi}{3}\left(\dfrac{180°}{\pi}\right) = 300°$ [2.1]

8. $\sin\dfrac{\pi}{3} = \dfrac{\sqrt{3}}{2}$

9. $\csc 60° = \dfrac{1}{\sin 60°} = \dfrac{2}{\sqrt{3}} = \dfrac{2\sqrt{3}}{3}$

10. $\sin\theta = \dfrac{\text{opp}}{\text{hyp}} = \dfrac{2}{3}$

adjacent side $= \sqrt{3^2 - 2^2}$
$= \sqrt{9-4}$
$= \sqrt{5}$

$\tan\theta = \dfrac{\text{opp}}{\text{adj}} = \dfrac{2}{\sqrt{5}} = \dfrac{2\sqrt{5}}{5}$

11. $\cot\theta > 0$ in quadrant III
Positive [2.3]

12. $\theta = 310°$ [2.3]
Since $270° < \theta < 360°$,
$\theta = \theta' = 360°$
$\theta' = 50°$

13. $\theta = \dfrac{5\pi}{3}$ [2.3]
Since $\dfrac{3\pi}{2} < \theta < 2\pi$,
$\theta = \theta' = 2\pi$
$\theta' = \dfrac{\pi}{3}$

Copyright © Houghton Mifflin Company. All rights reserved.

14. $t = \dfrac{\pi}{3}$ [2.4]

$y = \sin t = \sin \dfrac{\pi}{3} = \dfrac{\sqrt{3}}{2}$

$x = \cos t = \cos \dfrac{\pi}{3} = \dfrac{1}{2}$

The point on the unit circle

corresponding to $t = \dfrac{\pi}{3}$ is $\left(\dfrac{1}{2}, \dfrac{\sqrt{3}}{2} \right)$.

15. $y = 0.43 \cos \left(2x - \dfrac{\pi}{6} \right)$ [2.7]

amplitude: 0.43

$0 \le 2x - \dfrac{\pi}{6} \le 2\pi$

$\dfrac{\pi}{6} \le 2x \le \dfrac{13\pi}{6}$

$\dfrac{\pi}{12} \le x \le \dfrac{13\pi}{12}$

period $= \pi$,   phase shift $= \dfrac{\pi}{12}$

16. $y = \sin^{-1} \dfrac{1}{2}$ [3.5]

$\sin y = \dfrac{1}{2}$ $-\dfrac{\pi}{2} \le y \le \dfrac{\pi}{2}$

$y = \dfrac{\pi}{6}$

17. $\cos^{-1}(-0.8) = 2.498$ [3.5]

18. Domain: $[-1, 1]$.   [3.5]

19. Range: $\left(-\dfrac{\pi}{2}, \dfrac{\pi}{2} \right)$ [3.5]

20. $2\cos^2 x - 1 = -\sin x$

$1 - 2\sin^2 x = -\sin x$

$0 = 2\sin^2 x - \sin x - 1$

$0 = (2\sin x + 1)(\sin x - 1)$

$2\sin x + 1 = 0$ $\sin x - 1 = 0$

$\sin x = -\dfrac{1}{2}$ $\sin x = 1$

$x = \dfrac{7\pi}{6}, \dfrac{11\pi}{6}$ $x = \dfrac{\pi}{2}$

The solutions are $\dfrac{\pi}{2}, \dfrac{7\pi}{6}, \dfrac{11\pi}{6}$.

Copyright © Houghton Mifflin Company. All rights reserved.

Chapter 4
Applications of Trigonometry

1.

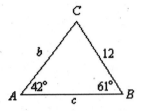

$C = 180° - 42° - 61°$

$C = 77°$

$$\frac{b}{\sin B} = \frac{a}{\sin A}$$

$$\frac{b}{\sin 61°} = \frac{12}{\sin 42°}$$

$$b = \frac{12 \sin 61°}{\sin 42°}$$

$$b \approx 16$$

$$\frac{c}{\sin C} = \frac{a}{\sin A}$$

$$\frac{c}{\sin 77°} = \frac{12}{\sin 42°}$$

$$c = \frac{12 \sin 77°}{\sin 42°}$$

$$c \approx 17$$

3.

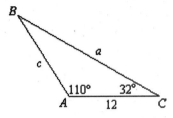

$B = 180° - 110° - 32°$

$B = 38°$

$$\frac{b}{\sin B} = \frac{a}{\sin A}$$

$$\frac{12}{\sin 38°} = \frac{a}{\sin 110°}$$

$$a = \frac{12 \sin 110°}{\sin 38°}$$

$$a \approx 18$$

$$\frac{c}{\sin C} = \frac{b}{\sin B}$$

$$\frac{c}{\sin 32°} = \frac{12}{\sin 38°}$$

$$c = \frac{12 \sin 32°}{\sin 38°}$$

$$c \approx 10$$

5.

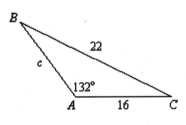

$$\frac{a}{\sin A} = \frac{b}{\sin B}$$

$$\frac{22}{\sin 132°} = \frac{16}{\sin B}$$

$$\sin B = \frac{16 \sin 132°}{22}$$

$$\sin B \approx 0.5405$$

$$B \approx 33°$$

$C \approx 180° - 33° - 132°$

$C \approx 15°$

$$\frac{a}{\sin A} = \frac{c}{\sin C}$$

$$\frac{22}{\sin 132°} \approx \frac{c}{\sin 15°}$$

$$c \approx \frac{22 \sin 15°}{\sin 132°}$$

$$c \approx 7.7$$

7.

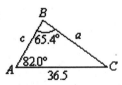

$C = 180° - 65.4° - 82.0°$

$C = 32.6°$

$$\frac{c}{\sin C} = \frac{b}{\sin B}$$

$$\frac{c}{\sin 32.6°} = \frac{36.5}{\sin 65.4°}$$

$$c = \frac{36.5 \sin 32.6°}{\sin 65.4°}$$

$$c \approx 21.6$$

$$\frac{a}{\sin A} = \frac{b}{\sin B}$$

$$\frac{a}{\sin 82.0°} = \frac{36.5}{\sin 65.4°}$$

$$a = \frac{36.5 \sin 82.0°}{\sin 65.4°}$$

$$a \approx 39.8$$

9.

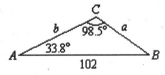

$B = 180° - 98.5° - 33.8°$

$B = 47.7°$

$$\frac{a}{\sin A} = \frac{c}{\sin C}$$

$$\frac{a}{\sin 33.8°} = \frac{102}{\sin 98.5°}$$

$$a = \frac{102 \sin 33.8°}{\sin 98.5°}$$

$$a \approx 57.4$$

$$\frac{b}{\sin B} = \frac{c}{\sin C}$$

$$\frac{b}{\sin 47.7°} = \frac{102}{\sin 98.5°}$$

$$b = \frac{102 \sin 47.7°}{\sin 98.5°}$$

$$b \approx 76.3$$

Copyright © Houghton Mifflin Company. All rights reserved.

11.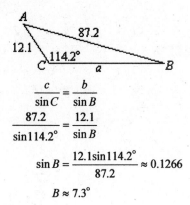

$$\frac{c}{\sin C} = \frac{b}{\sin B}$$

$$\frac{87.2}{\sin 114.2^\circ} = \frac{12.1}{\sin B}$$

$$\sin B = \frac{12.1 \sin 114.2^\circ}{87.2} \approx 0.1266$$

$$B \approx 7.3^\circ$$

$$A \approx 180^\circ - 114.2^\circ - 7.3^\circ$$

$$A \approx 58.5^\circ$$

$$\frac{c}{\sin C} = \frac{a}{\sin A}$$

$$\frac{87.2}{\sin 114.2^\circ} \approx \frac{a}{\sin 58.5^\circ}$$

$$a = \frac{87.2 \sin 58.5^\circ}{\sin 114.2^\circ} \approx 81.5$$

13.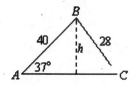

$$\sin 37^\circ = \frac{h}{40}$$

$$h = 40 \sin 37^\circ$$

$$h \approx 24$$

Since $h < 28$, two triangles exist.

$$\frac{c}{\sin C} = \frac{a}{\sin A}$$

$$\frac{40}{\sin C} = \frac{28}{\sin 37^\circ}$$

$$\sin C = \frac{40 \sin 37^\circ}{28} = 0.8597$$

$$C \approx 59^\circ \text{ or } 121^\circ$$

$C = 59^\circ$	$C = 121^\circ$
$B = 180^\circ - 37^\circ - 59^\circ = 84^\circ$	$B = 180^\circ - 121^\circ - 37^\circ = 22^\circ$
$\dfrac{b}{\sin 84^\circ} = \dfrac{28}{\sin 37^\circ}$	$\dfrac{b}{\sin 22^\circ} = \dfrac{28}{\sin 37^\circ}$
$b = \dfrac{28 \sin 84^\circ}{\sin 37^\circ} = 46$	$b = \dfrac{28 \sin 22^\circ}{\sin 37^\circ} \approx 17$

15.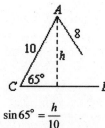

$$\sin 65^\circ = \frac{h}{10}$$

$$h = 10 \sin 65^\circ$$

$$h \approx 9.06$$

Since $h > 8$, no triangle is formed.

17.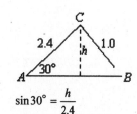

$$\sin 30^\circ = \frac{h}{2.4}$$

$$h = 2.4 \sin 30^\circ$$

$$h \approx 1.2$$

Since $h > 1$, no triangle is formed.

Copyright © Houghton Mifflin Company. All rights reserved.

19.

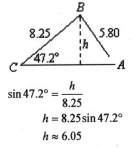

$$\sin 14.8° = \frac{h}{6.35}$$

$$h = 6.35 \sin 14.8°$$

$$h \approx 1.62$$

Since $h < 4.80,$ two solutions exist.

$$\frac{c}{\sin C} = \frac{a}{\sin A}$$

$$\frac{6.35}{\sin C} = \frac{4.80}{\sin 14.8°}$$

$$\sin C = \frac{6.35 \sin 14.8°}{4.80}$$

$$\sin C = 0.3379$$

$$C \approx 19.8° \text{ or } 160.2°$$

$C = 19.8°$

$B = 180° - 19.8° - 14.8°$

$\quad = 145.4°$

$$\frac{b}{\sin 145.4°} = \frac{4.80}{\sin 14.8°}$$

$$b = \frac{4.80 \sin 145.4°}{\sin 14.8°}$$

$$b \approx 10.7$$

$C = 160.2°$

$B = 180° - 160.2° - 14.8°$

$\quad = 5.0°$

$$\frac{b}{\sin 5.0°} = \frac{4.80}{\sin 14.8°}$$

$$b = \frac{4.80 \sin 5.0°}{\sin 14.8°}$$

$$b \approx 1.64$$

21.

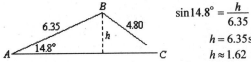

$$\sin 47.2° = \frac{h}{8.25}$$

$$h = 8.25 \sin 47.2°$$

$$h \approx 6.05$$

Since $h > 5.80,$ no triangle is formed.

23.

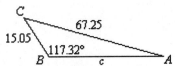

Since $b > a$, one triangle exists.

$$\frac{a}{\sin A} = \frac{b}{\sin B}$$

$$\frac{15.05}{\sin A} = \frac{67.25}{\sin 117.32°}$$

$$\sin A = \frac{15.04 \sin 117.32°}{67.25} \approx 0.1988$$

$$A = 11.47°$$

$$C = 180° - 11.47° - 117.32°$$

$$C = 51.21°$$

$$\frac{c}{\sin C} = \frac{b}{\sin B}$$

$$\frac{c}{\sin 51.21°} = \frac{67.25}{\sin 117.32°}$$

$$c = \frac{67.25 \sin 51.21°}{\sin 117.32°}$$

$$c \approx 59.00$$

Copyright © Houghton Mifflin Company. All rights reserved.

25.

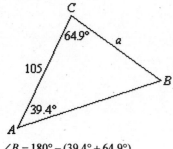

$\angle B = 180° - (39.4° + 64.9°)$

$\angle B = 75.7°$

$$\frac{a}{\sin A} = \frac{b}{\sin B}$$

$$\frac{a}{\sin 39.4°} = \frac{105}{\sin 75.7°}$$

$$a = \frac{105\sin 39.4°}{\sin 75.7°}$$

$$a \approx 68.8 \text{ miles}$$

27. $a = 155$ yd, $c = 165$ yd, $A = 42.0°$

$$\frac{c}{\sin C} = \frac{a}{\sin A}$$

$$\frac{165}{\sin C} = \frac{155}{\sin 42.0°}$$

$$\sin C = \frac{155}{165\sin 42.0°}$$

$$C = \sin^{-1}\left(\frac{155}{165\sin 42.0°}\right) = 45.4°$$

$B = 180° - 42.0° - 45.4° = 92.6°$

$$\frac{b}{\sin B} = \frac{a}{\sin A}$$

$$\frac{b}{\sin 92.6°} = \frac{155}{\sin 42.0°}$$

$$b = \frac{155\sin 92.6°}{165\sin 42.0°} = 231 \text{ yd}$$

29.

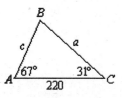

$B = 180° - 67° - 31°$

$B = 82°$

$$\frac{c}{\sin C} = \frac{b}{\sin B}$$

$$\frac{c}{\sin 31°} = \frac{220}{\sin 82°}$$

$$c = \frac{220\sin 31°}{\sin 82°}$$

$$c \approx 110 \text{ feet}$$

31.

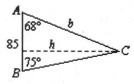

$C = 180° - 68° - 75°$

$C = 37°$

$$\frac{b}{\sin B} = \frac{c}{\sin C}$$

$$\frac{b}{\sin 75°} = \frac{85}{\sin 37°}$$

$$b = \frac{85\sin 75°}{\sin 37°}$$

$$b \approx 136.4267957$$

$$\sin 68° = \frac{h}{b}$$

$$h = b\sin 68°$$

$$h \approx 130 \text{ yards}$$

33.

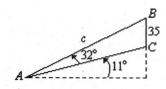

$A = 32° - 11°$

$A = 21°$

$B = 180° - 90° - 32°$

$B = 58°$

$C = 180° - 58° - 21°$

$C = 101°$

$$\frac{c}{\sin C} = \frac{a}{\sin A}$$

$$\frac{c}{\sin 101°} = \frac{35}{\sin 21°}$$

$$c = \frac{35\sin 101°}{\sin 21°}$$

$$c \approx 96 \text{ feet}$$

Copyright © Houghton Mifflin Company. All rights reserved.

35.

$A = 5°$
$B = 180° - 90° - 75°$
$B = 15°$
$C = 180° - 15° - 5°$
$C = 160°$

$\dfrac{b}{\sin B} = \dfrac{a}{\sin A}$

$\dfrac{b}{\sin 15°} = \dfrac{12}{\sin 5°}$

$b = \dfrac{12\sin 15°}{\sin 5°}$

$\sin 70° = \dfrac{h}{b}$

$h = b\sin 70°$

$h = \left(\dfrac{12\sin 15°}{\sin 5°}\right)\sin 70°$

$h \approx 33$ feet

37.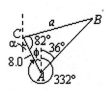

$\phi = 360° - 332°$
$\phi = 28°$
$\alpha = 28°$

$C = 82° - 28°$
$C = 54°$
$A = 28° + 36°$
$A = 64°$
$B = 180° - 64° - 54°$
$B = 62°$

$\dfrac{a}{\sin A} = \dfrac{b}{\sin B}$

$\dfrac{a}{\sin 64°} = \dfrac{8.0}{\sin 62°}$

$a = \dfrac{8.0\sin 64°}{\sin 62°}$

$a \approx 8.1$ miles

39.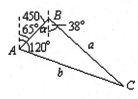

$A = 120° - 65°$
$A = 55°$
$\alpha = 65°$
$B = 38° + 65°$
$B = 103°$
$C = 180° - 103° - 55°$
$C = 22°$

$\dfrac{b}{\sin B} = \dfrac{c}{\sin C}$

$\dfrac{b}{\sin 103°} = \dfrac{450}{\sin 22°}$

$b = \dfrac{450\sin 103°}{\sin 22°}$

$b \approx 1200$ miles

••

Connecting Concepts

41.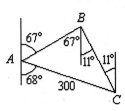

$A = 180° - 67° - 68°$
$A = 45°$
$B = 67° + 11°$
$B = 78°$
$C = 180° - 45° - 78°$
$C = 57°$

$\dfrac{c}{\sin C} = \dfrac{b}{\sin B}$

$\dfrac{c}{\sin 57°} = \dfrac{300}{\sin 78°}$

$c = \dfrac{300\sin 57°}{\sin 78°}$

$c \approx 260$ meters

43.

$\dfrac{a}{\sin A} = \dfrac{b}{\sin B}$

$\dfrac{a}{b} = \dfrac{\sin A}{\sin B}$

$\dfrac{a}{b} + 1 = \dfrac{\sin A}{\sin B} + 1$

$\dfrac{a + b}{b} = \dfrac{\sin A + \sin B}{\sin B}$

45.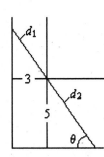

$\dfrac{3}{d_1} = \cos\theta,\ \dfrac{5}{d_2} = \sin\theta$

$d_1 = \dfrac{3}{\cos\theta},\ d_2 = \dfrac{5}{\sin\theta}$

$L = d_1 + d_2$

$L(\theta) = \dfrac{3}{\cos\theta} + \dfrac{5}{\sin\theta}$

The graph of L is shown.
The minimum value of L is approximately 11.19 m.

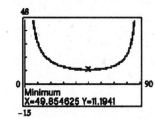

Copyright © Houghton Mifflin Company. All rights reserved.

46. $\sqrt{(10.0)^2 + (15.0)^2 - 2(10.0)(15.0)\cos 110.0^\circ} \approx 20.7$

47. $A = \frac{1}{2}bh = \frac{1}{2}(6)(8.5) = 22.5 \text{ in.}^2$

48.
$$c^2 = a^2 + b^2 - 2ab\cos C$$
$$c^2 - a^2 - b^2 = -2ab\cos C$$
$$\cos C = \frac{c^2 - a^2 - b^2}{-2ab}$$
$$C = \cos^{-1}\left(\frac{c^2 - a^2 - b^2}{-2ab}\right)$$

49.
$$P = 6 + 9 + 10 = 25$$
$$\text{semiperimeter} = \frac{1}{2}(25) = 12.5 \text{ m}$$

50.
$$s = \frac{a + b + c}{2} = \frac{3 + 4 + 5}{2} = 6$$
$$\sqrt{6(6-3)(6-4)(6-5)} = \sqrt{6(3)(2)(1)} = 6$$

51. $c^2 = a^2 + b^2$

Section 4.2

1.
$$c^2 = a^2 + b^2 - 2ab\cos C$$
$$c^2 = 12^2 + 18^2 - 2(12)(18)\cos 44^\circ$$
$$c^2 = 468 - 432\cos 44^\circ$$
$$c = \sqrt{468 - 432\cos 44^\circ}$$
$$c \approx 13$$

3.
$$b^2 = a^2 + c^2 - 2ac\cos B$$
$$b^2 = 120^2 + 180^2 - 2(120)(180)\cos 56^\circ$$
$$b^2 = 46{,}800 - 43{,}200\cos 56^\circ$$
$$b = \sqrt{46{,}800 - 43{,}200\cos 56^\circ}$$
$$b \approx 150$$

5.
$$a^2 = b^2 + c^2 - 2bc\cos A$$
$$a^2 = 60^2 + 84^2 - 2(60)(84)\cos 13^\circ$$
$$a^2 = 10{,}656 - 10{,}080\cos 13^\circ$$
$$a = \sqrt{10{,}656 - 10{,}080\cos 13^\circ}$$
$$a \approx 29$$

7.
$$c^2 = a^2 + b^2 - 2ab\cos C$$
$$c^2 = 9.0^2 + 7.0^2 - 2(9.0)(7.0)\cos 72^\circ$$
$$c^2 = 130 - 126\cos 72^\circ$$
$$c = \sqrt{130 - 126\cos 72^\circ}$$
$$c \approx 9.5$$

9.
$$c^2 = a^2 + b^2 - 2ab\cos C$$
$$c^2 = 4.6^2 + 7.2^2 - 2(4.6)(7.2)\cos 124^\circ$$
$$c^2 = 73 - 66.24\cos 124^\circ$$
$$c = \sqrt{73 - 66.24\cos 124^\circ}$$
$$c \approx 10$$

11.
$$b^2 = a^2 + c^2 - 2ac\cos B$$
$$b^2 = 25.9^2 + 33.4^2 - 2(25.9)(33.4)\cos 84.0^\circ$$
$$b^2 = 1786.37 - 1730.12\cos 84.0^\circ$$
$$b = \sqrt{1786.37 - 1730\cos 84.0^\circ}$$
$$b \approx 40.1$$

13.
$$b^2 = a^2 + c^2 - 2ac\cos B$$
$$b^2 = 122^2 + 55.9^2 - 2(122)(55.9)\cos 44.2^\circ$$
$$b^2 = 18{,}008.81 - 13{,}639.6\cos 44.2^\circ$$
$$b = \sqrt{18{,}008.81 - 13{,}639.6\cos 44.2^\circ}$$
$$b \approx 90.7$$

15.
$$\cos A = \frac{b^2 + c^2 - a^2}{2bc}$$
$$\cos A = \frac{32^2 + 40^2 - 25^2}{2(32)(40)}$$
$$\cos A = \frac{1999}{2560}$$
$$A = \cos^{-1}\left(\frac{1999}{2560}\right) \approx 39^\circ$$

Copyright © Houghton Mifflin Company. All rights reserved.

17.
$$\cos C = \frac{a^2 + b^2 - c^2}{2ab}$$
$$\cos C = \frac{8.0^2 + 9.0^2 - 12^2}{2(8.0)(9.0)}$$
$$\cos C = \frac{1}{144}$$
$$C = \cos^{-1}\left(\frac{1}{144}\right) \approx 90°$$

19.
$$\cos B = \frac{a^2 + c^2 - b^2}{2ac}$$
$$\cos B = \frac{80^2 + 124^2 - 92^2}{2(80)(124)}$$
$$\cos B = \frac{13,312}{19,840}$$
$$B = \cos^{-1}\left(\frac{13312}{19840}\right) \approx 47.9°$$

21.
$$\cos C = \frac{a^2 + b^2 - c^2}{2ab}$$
$$\cos C = \frac{1025^2 + 625^2 - 1420^2}{2(1025)(625)}$$
$$\cos C = \frac{-575,150}{1,281,250}$$
$$C = \cos^{-1}\left(\frac{-575,150}{1,281,250}\right) \approx 116.67°$$

23.
$$\cos B = \frac{a^2 + c^2 - b^2}{2ac}$$
$$\cos B = \frac{32.5^2 + 29.6^2 - 40.1^2}{2(32.5)(29.6)}$$
$$\cos B = \frac{324.4}{1924}$$
$$B = \cos^{-1}\left(\frac{324.4}{1924}\right)$$
$$B \approx 80.3°$$

25.
$$K = \frac{1}{2}bc \sin A$$
$$K = \frac{1}{2}(12)(24)\sin 105°$$
$$K \approx 140 \text{ square units}$$

27.
$$C = 180° - 42° - 76°$$
$$C = 62°$$

$$K = \frac{c^2 \sin A \sin B}{2 \sin C}$$
$$K = \frac{12^2 \sin 42° \sin 76°}{2 \sin 62°}$$
$$K \approx 53 \text{ square units}$$

29.
$$s = \frac{1}{2}(a + b + c)$$
$$s = \frac{1}{2}(16 + 12 + 14)$$
$$s = 21$$

$$K = \sqrt{s(s-a)(s-b)(s-c)}$$
$$K = \sqrt{21(21-16)(21-12)(21-14)}$$
$$K = \sqrt{21(5)(9)(7)}$$
$$K \approx 81 \text{ square units}$$

31.
$$\frac{a}{\sin A} = \frac{b}{\sin B}$$
$$\frac{22.4}{\sin A} = \frac{26.9}{\sin 54.3°}$$
$$\sin A = \frac{22.4 \sin 54.3°}{26.9}$$
$$\sin A \approx 0.6762$$
$$A \approx 42.5°$$
$$C = 180° - 42.5° - 54.3°$$
$$C = 83.2°$$

$$K = \frac{1}{2}ab \sin C$$
$$K = \frac{1}{2}(22.4)(26.9)\sin 83.2°$$
$$K \approx 299 \text{ square units}$$

33.
$$C = 180° - 116° - 34°$$
$$C = 30°$$

$$K = \frac{c^2 \sin A \sin B}{2 \sin C}$$
$$K = \frac{8.5^2 \sin 116° \sin 34°}{2 \sin 30°}$$
$$K \approx 36 \text{ square units}$$

Copyright © Houghton Mifflin Company. All rights reserved.

35.

$$s = \frac{1}{2}(a + b + c)$$

$$s = \frac{1}{2}(3.6 + 4.2 + 4.8)$$

$$s = 6.3$$

$$K = \sqrt{s(s-a)(s-b)(s-c)}$$

$$K = \sqrt{6.3(6.3 - 3.6)(6.3 - 4.2)(6.3 - 4.8)}$$

$$K = \sqrt{6.3(2.7)(2.1)(1.5)}$$

$K \approx 7.3$ square units

37.

$\alpha = 32°$

$\beta = 72°$

$B = 72° + 32°$

$B = 104°$

$$b^2 = a^2 + c^2 - 2ac\cos 104°$$

$$b^2 = 320^2 + 560^2 - 2(320)(560)\cos 104°$$

$$b^2 = 416,000 - 358,400\cos 104°$$

$$b = \sqrt{416,000 - 358,400\cos 104°}$$

$b \approx 710$ miles

39.

$$a^2 = b^2 + c^2 - 2bc\cos A$$

$$a^2 = 26^2 + 90^2 - 2(26)(90)\cos 45°$$

$$a^2 = 8776 - 4680\cos 45°$$

$$a = \sqrt{8776 - 4680\cos 45°}$$

$a \approx 74$ feet

41.

Let a = the length of the diagonal on the front of the box.
Let b = the length of the diagonal on the right side of the box.
Let c = the length of the diagonal on the top of the box.

$$a^2 = (4.75)^2 + (6.50)^2 = 64.8125$$

$$a = \sqrt{64.8125}$$

$$b^2 = (3.25)^2 + (4.75)^2 = 33.125$$

$$b = \sqrt{33.125}$$

$$c^2 = (6.50)^2 + (3.25)^2 = 52.8125$$

$$\theta = C$$

$$\cos C = \frac{a^2 + b^2 - c^2}{2ab}$$

$$\cos\theta = \frac{64.8125 + 33.125 - 52.8125}{2\sqrt{64.8125}\sqrt{33.125}}$$

$$\cos\theta = \frac{45.125}{2\sqrt{64.8125}\sqrt{33.125}}$$

$$\theta = \cos^{-1}\left(\frac{45.125}{2\sqrt{64.8125}\sqrt{33.125}}\right)$$

$\theta \approx 60.9°$

Copyright © Houghton Mifflin Company. All rights reserved.

43.

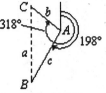

$b = (18 \text{ mph})(10 \text{ hours}) = 180 \text{ miles}$

$c = (22 \text{ mph})(10 \text{ hours}) = 220 \text{ miles}$

$A = 318° - 198°$

$A = 120°$

$a^2 = b^2 + c^2 - 2bc \cos A$

$a^2 = 180^2 + 220^2 - 2(180)(220)\cos 120°$

$a^2 = 120,400$

$a \approx 350 \text{ miles}$

45.

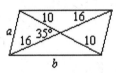

$A = \dfrac{360°}{6}$

$A = 60°$

$a^2 = 40^2 + 40^2 - 2(40)(40)\cos 60°$

$a^2 = 1600$

$a = 40 \text{ cm}$

47.

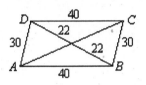

$a^2 = 16^2 + 10^2 - 2(16)(10)\cos 35°$

$a^2 = 356 - 320 \cos 35°$

$a = \sqrt{356 - 320 \cos 35°}$

$a \approx 9.7 \text{ inches}$

$b^2 = 10^2 + 16^2 - 2(10)(16)\cos 145°$

$b^2 = 356 - 320 \cos 145°$

$b = \sqrt{356 - 320 \cos 145°}$

$b \approx 25 \text{ inches}$

49.

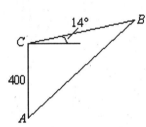

$\cos A = \dfrac{30^2 + 40^2 - 44^2}{2(30)(40)}$

$\cos A = 0.235$

$A = \cos^{-1}(0.235)$

$B = 180° - \cos^{-1}(0.235)$

$(AC)^2 = 40^2 + 30^2 - 2(40)(30)\cos(180° - \cos^{-1}(0.235))$

$AC = \sqrt{40^2 + 30^2 - 2(40)(30)\cos(180° - \cos^{-1}(0.235))}$

$\approx 55 \text{ centimeters}$

51.

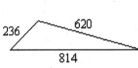

$C = 90° + 14°$

$C = 104°$

$a = \dfrac{180(5280)}{3600} \cdot 10$

$a = 2640 \text{ feet}$

$c^2 = 2640^2 + 400^2 - 2(2640)(400)(\cos 104°)$

$c^2 \approx 7,640,539$

$c \approx 2800 \text{ feet}$

53.

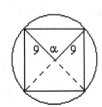

$s = \dfrac{1}{2}(a + b + c)$

$s = \dfrac{1}{2}(236 + 620 + 814)$

$s = 835$

$K = \sqrt{s(s-a)(s-b)(s-c)}$

$K = \sqrt{835(835 - 236)(835 - 620)(835 - 814)}$

$K = \sqrt{835(599)(215)(21)}$

$K \approx \sqrt{2,258,240,000}$

$K \approx 47,500 \text{ square meters}$

55.

$\alpha = 90°$

$K = 4\left[\dfrac{1}{2}(9)(9)\sin 90°\right]$

$K = 162 \text{ in}^2$

Copyright © Houghton Mifflin Company. All rights reserved.

57.

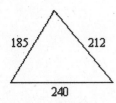

$s = \frac{1}{2}(185 + 212 + 240)$

$s = 318.5$

$K = \sqrt{318.5(318.5 - 185)(318.5 - 212)(318.5 - 240)}$

$K \approx 18,854 \text{ ft}^2$

$\text{cost} = 2.20(18,854)$

$\text{cost} \approx \$41,000$

59.

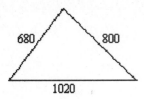

$s = \frac{1}{2}(680 + 800 + 1020)$

$s = 1250$

$K = \sqrt{1250(1250 - 680)(1250 - 800)(1250 - 1020)}$

$K \approx 271,558 \text{ ft}^2$

$\text{Acres} = \dfrac{271,558}{43,560}$

$\text{Acres} \approx 6.23$

61.   For ABC,

$\dfrac{13.0 - 16.1}{10.4} \approx -0.2981$

$\dfrac{\sin\left(\dfrac{53.5 - 86.5}{2}\right)}{\cos\left(\dfrac{40.0}{2}\right)} \approx -0.3022$

Triangle ABC has correct dimensions.

For DEF,

$\dfrac{17.2 - 21.3}{22.8} \approx -0.1798$

$\dfrac{\sin\left(\dfrac{52.1 - 59.9}{2}\right)}{\cos\left(\dfrac{68.0}{2}\right)} \approx -0.0820$

Triangle DEF has an incorrect dimension.

Connecting Concepts

63. a.

$A = (\text{side})^2$

$= 8^2$

$= 64 \text{ square units}$

$s = \dfrac{8 + 8 + 8 + 8}{2}$

$= \dfrac{32}{2}$

$= 16$

$K = \sqrt{(16 - 8)(16 - 8)(16 - 8)(16 - 8)}$

$= \sqrt{8 \cdot 8 \cdot 8 \cdot 8}$

$= 8 \cdot 8$

$= 64 \text{ square units}$

b.

$A = LW$

$= 11(5)$

$= 55 \text{ square units}$

$s = \dfrac{11 + 5 + 11 + 5}{2}$

$= \dfrac{32}{2}$

$= 16$

$K = \sqrt{(16 - 11)(16 - 5)(16 - 11)(16 - 5)}$

$= \sqrt{5 \cdot 11 \cdot 5 \cdot 11}$

$= 5 \cdot 11$

$= 55 \text{ square units}$

65.

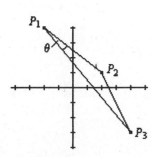

$d(P_1, P_2) = \sqrt{[2 - (-2)]^2 + (1 - 4)^2} = 5$

$d(P_1, P_3) = \sqrt{(-2 - 4)^2 + (4 - (-3))^2} = \sqrt{85}$

$d(P_2, P_3) = \sqrt{(2 - 4)^2 + (1 - (-3))^2} = 2\sqrt{5}$

$\cos\theta = \dfrac{5^2 + \left(\sqrt{85}\right)^2 - \left(2\sqrt{5}\right)^2}{2 \cdot 5 \cdot \sqrt{85}}$

$\cos\theta \approx 0.9762$

$\theta \approx 12.5°$

67.

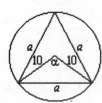

$\alpha = \dfrac{360}{3}$

$= 120°$

$a^2 = 10^2 + 10^2 - 2(10)(10)\cos 120°$

$a^2 = 300$

$a = \sqrt{300}$

$\text{perimeter} = 3a$

$= 3\sqrt{300}$

$\approx 52.0 \text{ cm}$

69.

$\cos A = \dfrac{b^2 + c^2 - a^2}{2bc} = \dfrac{b^2 + 2bc + c^2 - a^2 - 2bc}{2bc} = \dfrac{(b + c)^2 - a^2}{2bc} - \dfrac{2bc}{2bc} = \dfrac{(b + c - a)(b + c + a)}{2bc} - 1$

Copyright © Houghton Mifflin Company. All rights reserved.

71.

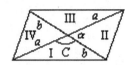

$$\alpha + C = 180°$$
$$\alpha = 180° - C$$
$$\sin\alpha = \sin C$$

$$K_I = K_{III} = \frac{1}{2}ab\sin C$$

$$K_{II} = K_{IV} = \frac{1}{2}ab\sin\alpha = \frac{1}{2}ab\sin C$$

$$K = K_I + K_{II} + K_{III} + K_{IV}$$

$$= \frac{1}{2}ab\sin C + \frac{1}{2}ab\sin C + \frac{1}{2}ab\sin C + \frac{1}{2}ab\sin C$$

$$K = 2ab\sin C$$

73.　$V = 18K$, where K = area of triangular base

$$V = \frac{1}{2}(4)(4)(\sin 72°)(18)$$

$$V \approx 140 \text{ in}^3$$

· ·

Prepare for Section 4.3

75.
$$\sqrt{\left(\frac{3}{5}\right)^2 + \left(-\frac{4}{5}\right)^2} = \sqrt{\frac{9}{25} + \frac{16}{25}} = \sqrt{\frac{25}{25}} = 1$$

76.　$10\cos 228° \approx -6.691$

77.
$$\tan\alpha = \left|\frac{-\sqrt{3}}{3}\right|$$
$$\tan\alpha = \frac{\sqrt{3}}{3}$$
$$\alpha = \tan^{-1}\left(\frac{\sqrt{3}}{3}\right) = 30°$$

78.
$$\cos\alpha = \frac{-17}{\sqrt{338}}$$
$$\alpha = \cos^{-1}\left(\frac{-17}{\sqrt{338}}\right) \approx 157.6°$$

79.
$$\frac{1}{\sqrt{5}} \cdot \frac{\sqrt{5}}{\sqrt{5}} = \frac{\sqrt{5}}{5}$$

80.
$$\frac{28}{\sqrt{68}} = \frac{28}{2\sqrt{17}} = \frac{14}{\sqrt{17}} \cdot \frac{\sqrt{17}}{\sqrt{17}} = \frac{14\sqrt{17}}{17}$$

Section 4.3

1.　$a = 4 - (-3) = 7$
$b = -1 - 0 = -1$

A vector equivalent to $\mathbf{P_1P_2}$ is $\mathbf{v} = \langle 7, -1 \rangle$.

3.　$a = -3 - 4 = -7$
$b = -3 - 2 = -5$

A vector equivalent to $\mathbf{P_1P_2}$ is $\mathbf{v} = \langle -7, -5 \rangle$.

5.　$a = 2 - 2 = 0$
$b = 3 - (-5) = 8$

A vector equivalent to $\mathbf{P_1P_2}$ is $\mathbf{v} = \langle 0, 8 \rangle$.

7.
$$\|\mathbf{v}\| = \sqrt{(-3)^2 + 4^2} \qquad \alpha = \tan^{-1}\left|\frac{4}{-3}\right| = \tan^{-1}\frac{4}{3}$$
$$\|\mathbf{v}\| = \sqrt{9 + 16} \qquad\qquad \alpha \approx 53.1°$$
$$\|\mathbf{v}\| = 5 \qquad\qquad\qquad \theta = 180° - \alpha$$
$$\theta \approx 180° - 53.1°$$
$$\theta \approx 126.9°$$

$$\mathbf{u} = \left\langle \frac{-3}{5}, \frac{4}{5} \right\rangle$$

A unit vector in the direction of $\mathbf{v}$ is $\mathbf{u} = \left\langle -\frac{3}{5}, \frac{4}{5} \right\rangle$.

Copyright © Houghton Mifflin Company. All rights reserved.

9.

$$\|\mathbf{v}\| = \sqrt{20^2 + (-40)^2} \qquad \alpha = \tan^{-1}\left|\frac{-40}{20}\right| = \tan^{-1} 2$$

$$\|\mathbf{v}\| = \sqrt{400 + 1600}$$

$$\|\mathbf{v}\| = \sqrt{2000} = 20\sqrt{5} \qquad \alpha \approx 63.4°$$

$$\approx 44.7 \qquad\qquad \theta = 360° - \alpha$$

$$\theta \approx 360° - 63.4°$$

$$\theta \approx 296.6°$$

$$\mathbf{u} = \left\langle \frac{20}{20\sqrt{5}}, \frac{-40}{20\sqrt{5}} \right\rangle = \left\langle \frac{\sqrt{5}}{5}, \frac{-2\sqrt{5}}{5} \right\rangle$$

A unit vector in the direction of $\mathbf{v}$ is $\mathbf{u} = \left\langle \frac{\sqrt{5}}{5}, -\frac{2\sqrt{5}}{5} \right\rangle$.

11.

$$\|\mathbf{v}\| = \sqrt{2^2 + (-4)^2} \quad \alpha = \tan^{-1}\left|\frac{-4}{2}\right| = \tan^{-1} 2$$

$$\|\mathbf{v}\| = \sqrt{4 + 16}$$

$$\|\mathbf{v}\| = \sqrt{20} = 2\sqrt{5} \qquad \alpha \approx 63.4°$$

$$\approx 4.5 \qquad\qquad \theta = 360° - \alpha$$

$$\theta \approx 360° - 63.4°$$

$$\approx 296.6°$$

$$\mathbf{u} = \left\langle \frac{2}{2\sqrt{5}}, \frac{-4}{2\sqrt{5}} \right\rangle = \left\langle \frac{\sqrt{5}}{5}, -\frac{2\sqrt{5}}{5} \right\rangle$$

A unit vector in the direction of $\mathbf{v}$ is $\mathbf{u} = \left\langle \frac{\sqrt{5}}{5}, -\frac{2\sqrt{5}}{5} \right\rangle$.

13.

$$\|\mathbf{v}\| = \sqrt{42^2 + (-18)^2} \qquad \alpha = \tan^{-1}\left|\frac{-18}{42}\right| = \tan^{-1}\frac{3}{7}$$

$$\|\mathbf{v}\| = \sqrt{1764 + 324}$$

$$\|\mathbf{v}\| = \sqrt{2088} \qquad\qquad \alpha \approx 23.2°$$

$$= 6\sqrt{58} \qquad\qquad \theta = 360° - \alpha$$

$$\approx 45.7 \qquad\qquad \theta \approx 360° - 23.2°$$

$$\theta \approx 336.8°$$

$$\mathbf{u} = \left\langle \frac{42}{6\sqrt{58}}, \frac{-18}{6\sqrt{58}} \right\rangle = \left\langle \frac{7\sqrt{58}}{58}, -\frac{3\sqrt{58}}{58} \right\rangle$$

A unit vector in the direction of $\mathbf{v}$ is $\mathbf{u} = \left\langle \frac{7\sqrt{58}}{58}, -\frac{3\sqrt{58}}{58} \right\rangle$.

15. $3\mathbf{u} = 3\langle -2, 4 \rangle = \langle -6, 12 \rangle$

17.

$$2\mathbf{u} - \mathbf{v} = 2\langle -2, 4 \rangle - \langle -3, -2 \rangle$$

$$= \langle -4, 8 \rangle - \langle -3, -2 \rangle$$

$$= \langle -1, 10 \rangle$$

19.

$$\frac{2}{3}\mathbf{u} + \frac{1}{6}\mathbf{v} = \frac{2}{3}\langle -2, 4 \rangle + \frac{1}{6}\langle -3, -2 \rangle$$

$$= \left\langle -\frac{4}{3}, \frac{8}{3} \right\rangle + \left\langle -\frac{1}{2}, -\frac{1}{3} \right\rangle$$

$$= \left\langle -\frac{11}{6}, \frac{7}{3} \right\rangle$$

21. $\|\mathbf{u}\| = \sqrt{(-2)^2 + 4^2} = \sqrt{20} = 2\sqrt{5}$

23.

$$3\mathbf{u} - 4\mathbf{v} = 3\langle -2, 4 \rangle - 4\langle -3, -2 \rangle$$

$$= \langle -6, 12 \rangle - \langle -12, -8 \rangle$$

$$= \langle 6, 20 \rangle$$

$$\|3\mathbf{u} - 4\mathbf{v}\| = \sqrt{6^2 + 20^2} = \sqrt{436} = 2\sqrt{109}$$

25.

$$4\mathbf{v} = 4(-2\mathbf{i} + 3\mathbf{j})$$

$$= -8\mathbf{i} + 12\mathbf{j}$$

27.

$$6\mathbf{u} + 2\mathbf{v} = 6(3\mathbf{i} - 2\mathbf{j}) + 2(-2\mathbf{i} + 3\mathbf{j})$$

$$= (18\mathbf{i} - 12\mathbf{j}) + (-4\mathbf{i} + 6\mathbf{j})$$

$$= (18 - 4)\mathbf{i} + (-12 + 6)\mathbf{j}$$

$$= 14\mathbf{i} - 6\mathbf{j}$$

29.

$$\frac{2}{3}\mathbf{v} + \frac{3}{4}\mathbf{u} = \frac{2}{3}(-2\mathbf{i} + 3\mathbf{j}) + \frac{3}{4}(3\mathbf{i} - 2\mathbf{j})$$

$$= \left(-\frac{4}{3}\mathbf{i} + 2\mathbf{j} \right) + \left(\frac{9}{4}\mathbf{i} - \frac{3}{2}\mathbf{j} \right)$$

$$= \left(-\frac{4}{3} + \frac{9}{4} \right)\mathbf{i} + \left(2 - \frac{3}{2} \right)\mathbf{j}$$

$$= \frac{11}{12}\mathbf{i} + \frac{1}{2}\mathbf{j}$$

31.

$$\mathbf{u} - 2\mathbf{v} = (3\mathbf{i} - 2\mathbf{j}) - 2(-2\mathbf{i} + 3\mathbf{j})$$

$$= (3\mathbf{i} - 2\mathbf{j}) - (-4\mathbf{i} + 6\mathbf{j})$$

$$= (3 + 4)\mathbf{i} + (-2 - 6)\mathbf{j}$$

$$= 7\mathbf{i} - 8\mathbf{j}$$

$$\|\mathbf{u} - 2\mathbf{v}\| = \sqrt{7^2 + (-8)^2} = \sqrt{113}$$

33.

$$a_1 = 5\cos 27° \approx 4.5$$

$$a_2 = 5\sin 27° \approx 2.3$$

$$\mathbf{v} = a_1\mathbf{i} + a_2\mathbf{j} \approx 4.5\mathbf{i} + 2.3\mathbf{j}$$

Copyright © Houghton Mifflin Company. All rights reserved.

35.
$$a_1 = 4\cos\frac{\pi}{4} \approx 2.8$$

$$a_2 = 4\sin\frac{\pi}{4} \approx 2.8$$

$$\mathbf{v} = a_1\mathbf{i} + a_2\mathbf{j} \approx 2.8\mathbf{i} + 2.8\mathbf{j}$$

37.   heading = 124° ⇒   wind from the west ⇒
    direction angle = −34°   direction angle = 0°

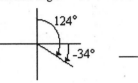  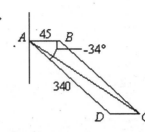

$$\mathbf{AB} = 45\mathbf{i}$$
$$\mathbf{AD} = 340\cos(-34°)\mathbf{i} + 340\sin(-34°)\mathbf{j}$$
$$\mathbf{AD} \approx 281.9\mathbf{i} - 190.1\mathbf{j}$$
$$\mathbf{AC} = \mathbf{AB} + \mathbf{AD}$$
$$\mathbf{AC} = 45\mathbf{i} + 281.9\mathbf{i} - 190.1\mathbf{j}$$
$$\mathbf{AC} \approx 327\mathbf{i} - 190\mathbf{j}$$
$$\|\mathbf{AC}\| = \sqrt{327^2 + (-190)^2}$$
$$\|\mathbf{AC}\| \approx 380 \text{ mph}$$

The ground speed of the plane is approximately 380 mph.

39.   heading = 96° ⇒   heading = 37° ⇒
    direction angle = −6°   direction angle = 53°

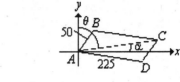

$$\mathbf{AB} = 50\cos 53°\mathbf{i} + 50\sin 53°\mathbf{j}$$
$$\approx 30.1\mathbf{i} + 39.9\mathbf{j}$$
$$\mathbf{AD} = 225\cos(-6°)\mathbf{i} + 225\sin(-6°)\mathbf{j}$$
$$\approx 223.8\mathbf{i} - 23.5\mathbf{j}$$
$$\mathbf{AC} = \mathbf{AB} + \mathbf{AD}$$
$$\approx 30.1\mathbf{i} + 39.9\mathbf{j} + 223.8\mathbf{i} - 23.5\mathbf{j}$$
$$\approx 253.9\mathbf{i} + 16.4\mathbf{j}$$
$$\|\mathbf{AC}\| = \sqrt{(253.9)^2 + (16.4)^2} \approx 250$$

$$\alpha = \tan^{-1}\left|\frac{16.4}{253.9}\right|$$
$$= \tan^{-1}\frac{16.4}{253.9}$$
$$\alpha \approx 4°$$
$$\theta = 90° - \alpha$$
$$\theta \approx 90° - 4°$$
$$\theta \approx 86°$$

The ground speed of the plane is about 250 mph at a heading of approximately 86°.

41.

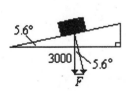

$$\sin 5.6° = \frac{F}{3000}$$
$$F = 3000\sin 5.6°$$
$$F \approx 293 \text{ lb}$$

43.

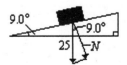

$$\cos 9.0° = \frac{N}{25}$$
$$N = 25\cos 9.0°$$
$$N \approx 24.7 \text{ lb}$$

45. $\mathbf{v} \cdot \mathbf{w} = \langle 3, -2 \rangle \cdot \langle 1, 3 \rangle$
    $= 3(1) + (-2)3$
    $= 3 - 6$
    $= -3$

47. $\mathbf{v} \cdot \mathbf{w} = \langle 4, 1 \rangle \cdot \langle -1, 4 \rangle$
    $= 4(-1) + 1(4)$
    $= -4 + 4$
    $= 0$

49. $\mathbf{v} \cdot \mathbf{w} = (\mathbf{i} + 2\mathbf{j}) \cdot (-\mathbf{i} + \mathbf{j})$
    $= 1(-1) + 2(1)$
    $= -1 + 2$
    $= 1$

Copyright © Houghton Mifflin Company. All rights reserved.

51.
$$\mathbf{v} \cdot \mathbf{w} = (6\mathbf{i} - 4\mathbf{j}) \cdot (-2\mathbf{i} - 3\mathbf{j})$$
$$= 6(-2) + (-4)(-3)$$
$$= -12 + 12$$
$$= 0$$

53.
$$\cos\theta = \frac{\mathbf{v} \cdot \mathbf{w}}{\|\mathbf{v}\| \, \|\mathbf{w}\|}$$
$$\cos\theta = \frac{\langle 2, -1 \rangle \cdot \langle 3, 4 \rangle}{\sqrt{2^2 + (-1)^2} \sqrt{3^2 + 4^2}}$$
$$\cos\theta = \frac{2(3) + (-1)4}{\sqrt{5}\sqrt{25}}$$
$$\cos\theta = \frac{2}{5\sqrt{5}} \approx 0.1789$$
$$\theta \approx 79.7°$$

55.
$$\cos\theta = \frac{\mathbf{v} \cdot \mathbf{w}}{\|\mathbf{v}\| \, \|\mathbf{w}\|}$$
$$\cos\theta = \frac{\langle 0, 3 \rangle \cdot \langle 2, 2 \rangle}{\sqrt{0^2 + 3^2} \sqrt{2^2 + 2^2}}$$
$$\cos\theta = \frac{0(2) + 3(2)}{\sqrt{9}\sqrt{8}}$$
$$\cos\theta = \frac{6}{6\sqrt{2}} \approx 0.7071$$
$$\theta = 45°$$

57.
$$\cos\theta = \frac{\mathbf{v} \cdot \mathbf{w}}{\|\mathbf{v}\| \, \|\mathbf{w}\|}$$
$$\cos\theta = \frac{(5\mathbf{i} - 2\mathbf{j}) \cdot (2\mathbf{i} + 5\mathbf{j})}{\sqrt{5^2 + (-2)^2} \sqrt{2^2 + 5^2}}$$
$$\cos\theta = \frac{5(2) + (-2)(5)}{\sqrt{29}\sqrt{29}}$$
$$\cos\theta = \frac{0}{\sqrt{29}\sqrt{29}} = 0$$
$$\theta = 90°$$

Thus, the vectors are orthogonal.

59.
$$\cos\theta = \frac{\mathbf{v} \cdot \mathbf{w}}{\|\mathbf{v}\| \, \|\mathbf{w}\|}$$
$$\cos\theta = \frac{(5\mathbf{i} + 2\mathbf{j}) \cdot (-5\mathbf{i} - 2\mathbf{j})}{\sqrt{5^2 + 2^2} \sqrt{(-5)^2 + (-2)^2}}$$
$$\cos\theta = \frac{5(-5) + 2(-2)}{\sqrt{29}\sqrt{29}}$$
$$\cos\theta = \frac{-29}{\sqrt{29}\sqrt{29}} = -1$$
$$\theta = 180°$$

61.
$$\text{proj}_{\mathbf{w}}\mathbf{v} = \frac{\mathbf{v} \cdot \mathbf{w}}{\|\mathbf{w}\|}$$
$$\text{proj}_{\mathbf{w}}\mathbf{v} = \frac{\langle 6, 7 \rangle \cdot \langle 3, 4 \rangle}{\sqrt{3^2 + 4^2}} = \frac{18 + 28}{\sqrt{25}} = \frac{46}{5}$$

63.
$$\text{proj}_{\mathbf{w}}\mathbf{v} = \frac{\mathbf{v} \cdot \mathbf{w}}{\|\mathbf{w}\|}$$
$$\text{proj}_{\mathbf{w}}\mathbf{v} = \frac{\langle -3, 4 \rangle \cdot \langle 2, 5 \rangle}{\sqrt{2^2 + 5^2}} = \frac{-6 + 20}{\sqrt{29}} = \frac{14}{\sqrt{29}} = \frac{14\sqrt{29}}{29} \approx 2.6$$

65.
$$\text{proj}_{\mathbf{w}}\mathbf{v} = \frac{\mathbf{v} \cdot \mathbf{w}}{\|\mathbf{w}\|}$$
$$\text{proj}_{\mathbf{w}}\mathbf{v} = \frac{(2\mathbf{i} + \mathbf{j}) \cdot (6\mathbf{i} + 3\mathbf{j})}{\sqrt{6^2 + 3^2}} = \frac{12 + 3}{\sqrt{45}} = \frac{5}{\sqrt{5}} = \sqrt{5} \approx 2.2$$

67.
$$\text{proj}_{\mathbf{w}}\mathbf{v} = \frac{\mathbf{v} \cdot \mathbf{w}}{\|\mathbf{w}\|}$$
$$\text{proj}_{\mathbf{w}}\mathbf{v} = \frac{(3\mathbf{i} - 4\mathbf{j}) \cdot (3\mathbf{i} - 4\mathbf{j})}{\sqrt{(-6)^2 + 12^2}} = \frac{-18 - 48}{\sqrt{180}} = -\frac{11}{\sqrt{5}}$$
$$= -\frac{11\sqrt{5}}{5} \approx -4.9$$

69.
$$W = \mathbf{F} \cdot \mathbf{s}$$
$$W = \|\mathbf{F}\| \, \|\mathbf{s}\| \cos\alpha$$
$$W = (75)(15)(\cos 32°)$$
$$W \approx 954 \text{ foot-pounds}$$

71.

$$W = \mathbf{F} \cdot \mathbf{s}$$
$$W = \|\mathbf{F}\| \, \|\mathbf{s}\| \cos\alpha$$
$$W = (75)(12)(\cos 30°)$$
$$W \approx 779 \text{ foot-pounds}$$

Copyright © Houghton Mifflin Company. All rights reserved.

73.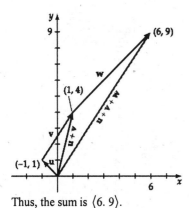

Thus, the sum is $\langle 6.\,9 \rangle$.

Connecting Concepts

75.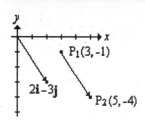

The vector from $P_1(3, -1)$ to $P_2(5, -4)$ is equivalent to $2\mathbf{i} - 3\mathbf{j}$.

77.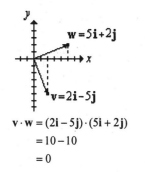

$$\mathbf{v} \cdot \mathbf{w} = (2\mathbf{i} - 5\mathbf{j}) \cdot (5\mathbf{i} + 2\mathbf{j})$$
$$= 10 - 10$$
$$= 0$$

The two vectors are perpendicular.

79. $\mathbf{v} = \langle -2,\, 7 \rangle$

$$\langle -2,\, 7 \rangle \cdot \langle a,\, b \rangle = 0$$
$$-2a + 7b = 0$$
$$a = \frac{7}{2}b$$
$$\text{Let } b = 2$$
$$a = 7$$

Thus, $\mathbf{u} = \langle 7,\, 2 \rangle$ is one example.

81.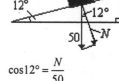

$$\cos 12° = \frac{N}{50}$$
$$N = 50\cos 12°$$
$$F\mu = 0.13N$$
$$= 0.13(50\cos 12°)$$
$$F\mu = 6.4 \text{ pounds}$$

83. Let $\mathbf{u} = c\mathbf{i} + b\mathbf{j}$, $\mathbf{v} = c\mathbf{i} + d\mathbf{j}$, and $\mathbf{w} = e\mathbf{i} + f\mathbf{j}$.

$$(\mathbf{u} \cdot \mathbf{v}) \cdot \mathbf{w} = \left[(a\mathbf{i} + b\mathbf{j}) \cdot (c\mathbf{i} + d\mathbf{j}) \right] \cdot (e\mathbf{i} + f\mathbf{j}).$$
$$= (ac + bd) \cdot (e\mathbf{i} + f\mathbf{j})$$

$ac + bd$ is a scalar quantity. The product of a scalar and a vector is not defined. Therefore, no, $(\mathbf{u} \cdot \mathbf{v}) \cdot \mathbf{w}$ does not equal $\mathbf{u} \cdot (\mathbf{v} \cdot \mathbf{w})$.

85. Let $\mathbf{v} = \langle a,b \rangle$ and $\mathbf{w} = \langle d,e \rangle$

$$c\mathbf{v} = \langle ca, ab \rangle$$

$$c(\mathbf{v} \cdot \mathbf{w}) = c\langle a,b \rangle \cdot \langle d,e \rangle = c(ad + be) = cad + cbe$$
$$(c\mathbf{v} \cdot \mathbf{w}) = \langle ca, ab \rangle \cdot \langle d,e \rangle = cad + cbe$$

Therefore, $c(\mathbf{v} \cdot \mathbf{w}) = (c\mathbf{v}) \cdot \mathbf{w}$.

87. Neither. If the force and the distance are the same, the work will be the same.

Chapter 4 True/False Exercises

1. False, we cannot solve a triangle using Law of Cosines if we are only given two sides and the angle opposite one of the given sides.

2. True

3. True

4. False, $2\mathbf{i} \neq 2\mathbf{j}$

5. True

6. True

7. True

8. True

9. True

10. False, $\mathbf{v} \cdot \mathbf{v} = a^2 + b^2$

11. False, let $\mathbf{v} = \mathbf{i} + \mathbf{j}, \mathbf{w} = \mathbf{i} - \mathbf{j}$.
Then $\mathbf{v} \cdot \mathbf{w} = 1 - 1 = 0$

Copyright © Houghton Mifflin Company. All rights reserved.

1.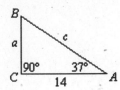

$B = 180° - 92° - 37°$
$B = 51°$

$\tan A = \dfrac{a}{4}$

$a = 14 \tan 37°$

$a \approx 11$

$\cos A = \dfrac{14}{c}$ [4.1]

$c = \dfrac{14}{\cos 37°}$

$c \approx 18$

2.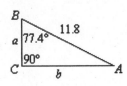

$A = 180° - 94.0° - 77.4°$
$A = 8.6°$

$\sin B = \dfrac{b}{11.8}$

$b = 11.8 \sin 77.4°$

$b \approx 11.5$

$\cos B = \dfrac{a}{11.8}$ [4.1]

$a = 11.8 \cos 77.4°$

$a \approx 2.57$

3.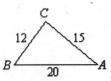

$\cos B = \dfrac{12^2 + 20^2 - 15^2}{2(12)(20)}$

$\cos B \approx 0.6646$

$B \approx 48°$

$\cos C = \dfrac{12^2 + 15^2 - 20^2}{2(12)(15)}$

$\cos C \approx -0.0861$

$C \approx 95°$

$A = 180° - 48° - 95°$ [4.2]
$A = 37°$

4.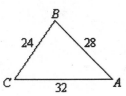

$\cos C = \dfrac{24^2 + 32^2 - 28^2}{2(24)(32)}$

$\cos C \approx 0.5313$

$C \approx 58°$

$\cos A = \dfrac{32^2 + 28^2 - 24^2}{2(32)(28)}$

$\cos A \approx 0.6875$

$A \approx 47°$

$B \approx 180° - 58° - 47°$ [4.2]
$B \approx 75°$

5.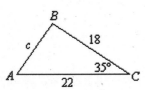

$c^2 = 22^2 + 18^2 - 2(22)(18)\cos 35°$
$c^2 \approx 159$
$c = \sqrt{159}$
$c \approx 13$

$\dfrac{18}{\sin A} = \dfrac{\sqrt{159}}{\sin 35°}$

$\sin A = \dfrac{18 \sin 35°}{\sqrt{159}}$

$\sin A \approx 0.8188$

$A \approx 55°$

$B \approx 180 - 35° - 55°$ [4.2]
$B \approx 90°$

6.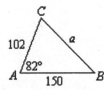

$a^2 = 102^2 + 150^2 - 2(102)(150)\cos 82°$
$a^2 \approx 28645$
$a = \sqrt{28645}$
$a \approx 169$

$\dfrac{150}{\sin C} = \dfrac{\sqrt{28645}}{\sin 82°}$

$\sin C = \dfrac{150 \sin 82°}{\sqrt{28645}}$

$\sin C \approx 0.8776$

$C \approx 61°$

$B \approx 180° - 61° - 82°$ [4.2]
$B \approx 37°$

7.

$\dfrac{10}{\sin C} = \dfrac{8}{\sin 105°}$

$\sin C = \dfrac{10 \sin 105°}{8}$

$\sin C \approx 1.207$

No triangle is formed.   [7.1]

8.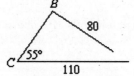

$\dfrac{110}{\sin B} = \dfrac{80}{\sin 55°}$

$\sin B = \dfrac{110 \sin 55°}{80}$

$\sin B \approx 1.1263$

No triangle is formed.   [4.1]

9.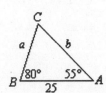

$C = 180° - 80° - 55°$
$C = 45°$

$\dfrac{25}{\sin 45°} = \dfrac{a}{\sin 55°}$

$a = \dfrac{25 \sin 55°}{\sin 45°}$

$a \approx 29$

$\dfrac{25}{\sin 45°} = \dfrac{b}{\sin 80°}$ [4.1]

$b = \dfrac{25 \sin 80°}{\sin 45°}$

$b \approx 35$

Copyright © Houghton Mifflin Company. All rights reserved.

10.

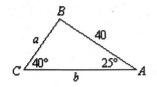

$A = 180° - 40° - 25°$

$A = 115°$

$\dfrac{a}{\sin 115°} = \dfrac{40}{\sin 40°}$

$a = \dfrac{40 \sin 115°}{\sin 40°}$

$a \approx 56$

$\dfrac{b}{\sin 25°} = \dfrac{40}{\sin 40°}$ [4.1]

$b = \dfrac{40 \sin 25°}{\sin 40°}$

$b \approx 26$

11.

$s = \dfrac{1}{2}(a + b + c)$ [4.2]

$s = \dfrac{1}{2}(24 + 30 + 36)$

$s = 45$

$K = \sqrt{s(s-a)(s-b)(s-c)}$

$K = \sqrt{45(45-24)(45-30)(45-36)}$

$K = \sqrt{127,575}$

$K \approx 360$ square units

12.

$s = \dfrac{1}{2}(a + b + c)$ [4.2]

$s = \dfrac{1}{2}(9.0 + 7.0 + 12)$

$s = 14$

$K = \sqrt{s(s-a)(s-b)(s-c)}$

$K = \sqrt{14(14-9.0)(14-7.0)(14-12)}$

$K = \sqrt{980}$

$K \approx 31$ square units

13.

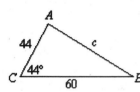

$K = \dfrac{1}{2}ab \sin C$

$K = \dfrac{1}{2}(60)(44)\sin 44°$

$K \approx 920$ square units

[4.2]

14.

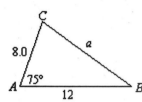

$K = \dfrac{1}{2}bc \sin A$

$K = \dfrac{1}{2}(8.0)(12)\sin 75°$

$K \approx 46$ square units

[4.2]

15.

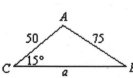

$\dfrac{50}{\sin B} = \dfrac{75}{\sin 15°}$

$\sin B = \dfrac{50 \sin 15°}{75}$

$\sin B \approx 0.1725$

$B \approx 10°$

$A \approx 180° - 10° - 15°$

$A \approx 155°$

$K \approx \dfrac{1}{2}(50)(75)\sin 155°$ [4.2]

$K \approx 790$ square units

16.

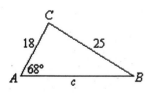

$\dfrac{18}{\sin B} = \dfrac{25}{\sin 68°}$

$\sin B = \dfrac{18 \sin 68°}{25}$

$\sin B \approx 0.6676$

$B \approx 42°$

$C \approx 180° - 42° - 68°$

$C \approx 70°$

$K \approx \dfrac{1}{2}(18)(25)\sin 70°$ [4.2]

$K \approx 210$ square units

17.

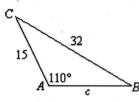

$\dfrac{15}{\sin B} = \dfrac{32}{\sin 110°}$

$\sin B = \dfrac{15 \sin 110°}{32}$

$\sin B \approx 0.4405$

$B \approx 26°$

$C \approx 180° - 110° - 26°$

$C \approx 44°$

$K \approx \dfrac{1}{2}(15)(32)\sin 44°$ [4.2]

$K \approx 170$ square units

Copyright © Houghton Mifflin Company. All rights reserved.

18.

$$\frac{18}{\sin B} = \frac{22}{\sin 45°}$$

$$\sin B \approx \frac{18 \sin 45°}{22}$$

$$\sin B \approx 0.5785$$

$$B \approx 35°$$

$A \approx 180° - 45° - 35°$

$A \approx 100°$

$K \approx \frac{1}{2}(18)(22)\sin 100°$ [4.2]

$K \approx 190$ square units

19. Let $\mathbf{P_1P_2} = a_1\mathbf{i} + a_2\mathbf{j}$.   [4.3]

$a_1 = 3 - (-2) = 5$

$a_2 = 7 - 4 = 3$

A vector equivalent to $\mathbf{P_1P_2}$ is $\mathbf{v} = \langle 5,\ 3 \rangle$.

20. Let $\mathbf{P_1P_2} = a_1\mathbf{i} + a_2\mathbf{j}$.   [4.3]

$a_1 = -3 - (-4) = 1$

$a_2 = 6 - 0 = 6$

A vector equivalent to $\mathbf{P_1P_2}$ is $\mathbf{v} = \langle 1,\ 6 \rangle$.

21.

$\|\mathbf{v}\| = \sqrt{(-4)^2 + 2^2}$

$\|\mathbf{v}\| = \sqrt{16 + 4}$

$\|\mathbf{v}\| \approx 4.5$

$\alpha \approx \tan^{-1}\left|\frac{2}{-4}\right| = \tan^{-1}\frac{1}{2}$ [4.3]

$\alpha \approx 26.6°$

$\theta \approx 180° - 26.6°$

$\theta \approx 153.4°$

22.

$\|\mathbf{v}\| = \sqrt{6^2 + (-3)^2}$

$\|\mathbf{v}\| = \sqrt{36 + 9}$

$\|\mathbf{v}\| \approx 6.7$

$\alpha \approx \tan^{-1}\left|\frac{-3}{6}\right| = \tan^{-1}\frac{1}{2}$ [4.3]

$\alpha \approx 26.6°$

$\theta \approx 360° - 26.6°$

$\theta \approx 333.4°$

23.

$\|\mathbf{u}\| = \sqrt{(-2)^2 + 3^2}$

$\|\mathbf{u}\| = \sqrt{4 + 9}$

$\|\mathbf{u}\| \approx 3.6$

$\alpha = \tan^{-1}\left|\frac{3}{-2}\right| = \tan^{-1}\frac{3}{2}$ [4.3]

$\alpha \approx 56.3°$

$\theta \approx 180° - 56.3°$

$\theta \approx 123.7°$

24.

$\|\mathbf{u}\| = \sqrt{(-4)^2 + (-7)^2}$

$\|\mathbf{u}\| = \sqrt{16 + 49}$

$\|\mathbf{u}\| \approx 8.1$

$\alpha = \tan^{-1}\left|\frac{-7}{-4}\right| = \tan^{-1}\frac{7}{4}$ [4.3]

$\alpha \approx 60.3°$

$\theta \approx 180° + 60.3°$

$\theta \approx 240.3°$

25.

$\|\mathbf{w}\| = \sqrt{(-8)^2 + 5^2}$

$\|\mathbf{w}\| = \sqrt{89}$

$\|\mathbf{u}\| = \left\langle \frac{-8}{\sqrt{89}},\ \frac{5}{\sqrt{89}} \right\rangle = \left\langle -\frac{8\sqrt{89}}{89},\ \frac{5\sqrt{89}}{89} \right\rangle$ [4.3]

A unit vector in the direction of $\|\mathbf{w}\|$ is $\|\mathbf{u}\| = \left\langle -\frac{8\sqrt{89}}{89},\ \frac{5\sqrt{89}}{89} \right\rangle$.

26.

$\|\mathbf{w}\| = \sqrt{7^2 + (-12)^2}$

$\|\mathbf{w}\| = \sqrt{193}$

$\mathbf{u} = \left\langle \frac{7}{\sqrt{193}},\ \frac{-12}{\sqrt{193}} \right\rangle = \left\langle \frac{7\sqrt{193}}{193},\ -\frac{12\sqrt{193}}{193} \right\rangle$ [4.3]

A unit vector in the direction of $\mathbf{w}$ is $\mathbf{u} = \left\langle \frac{7\sqrt{193}}{193},\ -\frac{12\sqrt{193}}{193} \right\rangle$.

27.

$\|\mathbf{v}\| = \sqrt{5^2 + 1^2}$

$\|\mathbf{v}\| = \sqrt{26}$

$\mathbf{u} = \frac{5}{\sqrt{26}}\mathbf{i} + \frac{1}{\sqrt{26}}\mathbf{j} = \frac{5\sqrt{26}}{26}\mathbf{i} + \frac{\sqrt{26}}{26}\mathbf{j}$ [4.3]

A unit vector in the direction of $\mathbf{v}$ is $\mathbf{u} = \frac{5\sqrt{26}}{26}\mathbf{i} + \frac{\sqrt{26}}{26}\mathbf{j}$.

28.

$\|\mathbf{v}\| = \sqrt{3^2 + (-5)^2}$

$\|\mathbf{v}\| = \sqrt{34}$

$\mathbf{u} = \frac{3}{\sqrt{34}}\mathbf{i} - \frac{5}{\sqrt{34}}\mathbf{j} = \frac{3\sqrt{34}}{34}\mathbf{i} - \frac{5\sqrt{34}}{34}\mathbf{j}$ [4.3]

A unit vector in the direction of $\mathbf{v}$ is $\mathbf{u} = \frac{3\sqrt{34}}{34}\mathbf{i} - \frac{5\sqrt{34}}{34}\mathbf{j}$.

Copyright © Houghton Mifflin Company. All rights reserved.

29. $\mathbf{v} - \mathbf{u} = \langle -4, -1 \rangle - \langle 3, 2 \rangle$ [4.3]

$= \langle -7, -3 \rangle$

30. $2\mathbf{u} - 3\mathbf{v} = 2\langle 3, 2 \rangle - 3\langle -4, -1 \rangle$ [4.3]

$= \langle 6, 4 \rangle - \langle -12, -3 \rangle$

$= \langle 18, 7 \rangle$

31. $-\mathbf{u} + \dfrac{1}{2}\mathbf{v} = -(10\mathbf{i} + 6\mathbf{j}) + \dfrac{1}{2}(8\mathbf{i} - 5\mathbf{j})$ [4.3]

$= (-10\mathbf{i} - 6\mathbf{j}) + \left(4\mathbf{i} - \dfrac{5}{2}\mathbf{j}\right)$

$= (-10 + 4)\mathbf{i} + \left(-6 - \dfrac{5}{2}\right)\mathbf{j}$

$= -6\mathbf{i} - \dfrac{17}{2}\mathbf{j}$

32. $\dfrac{2}{3}\mathbf{v} - \dfrac{3}{4}\mathbf{u} = \dfrac{2}{3}(8\mathbf{i} - 5\mathbf{j}) - \dfrac{3}{4}(10\mathbf{i} + 6\mathbf{j})$ [4.3]

$= \left(\dfrac{16}{3}\mathbf{i} - \dfrac{10}{3}\mathbf{j}\right) - \left(\dfrac{15}{2}\mathbf{i} + \dfrac{9}{2}\mathbf{j}\right)$

$= \left(\dfrac{16}{3} - \dfrac{15}{2}\right)\mathbf{i} + \left(-\dfrac{10}{3} - \dfrac{9}{2}\right)\mathbf{j}$

$= \dfrac{32 - 45}{6}\mathbf{i} + \dfrac{-20 = 27}{6}\mathbf{j}$

$= -\dfrac{13}{6}\mathbf{i} - \dfrac{47}{6}\mathbf{j}$

33.

$\mathbf{v} = 400\sin 204°\mathbf{i} + 400\cos 204°\mathbf{j}$

$\mathbf{v} \approx -162.7\mathbf{i} - 365.4\mathbf{j}$

$\mathbf{w} = -45\mathbf{i}$

$\mathbf{R} = \mathbf{v} + \mathbf{w}$

$\mathbf{R} \approx -162.7\mathbf{i} - 365.4\mathbf{j} - 45\mathbf{i}$

$\mathbf{R} \approx -207.7\mathbf{i} - 365.4\mathbf{j}$

$\|\mathbf{R}\| \approx \sqrt{(-207.7)^2 + (-365.4)^2}$

$\|\mathbf{R}\| \approx 420$ mph

$\alpha = \tan^{-1}\left|\dfrac{-365.4}{-207.7}\right| = \tan^{-1}\dfrac{365.4}{207.7}$

$\alpha \approx 60°$

$\theta \approx 180° + 60°$

$\theta \approx 240°$

The ground speed is approximately 420 mph at a heading of 240° [4.3]

34. $\theta = \sin^{-1}\dfrac{40}{320}$ [4.3]

$\theta \approx 7°$

35. $\mathbf{u} \cdot \mathbf{v} = \langle 3, 7 \rangle \cdot \langle -1, 3 \rangle$ [4.3]

$= (3)(-1) + (7)(3)$

$= 18$

36. $\mathbf{v} \cdot \mathbf{u} = \langle -8, 5 \rangle \cdot \langle 2, -1 \rangle$ [4.3]

$= (-8)(2) + (5)(-1)$

$= -21$

37. $\mathbf{v} \cdot \mathbf{u} = (-4\mathbf{i} - \mathbf{j}) \cdot (2\mathbf{i} + \mathbf{j})$ [4.3]

$= (-4)(2) + (-1)(1)$

$= -9$

38. $\mathbf{u} \cdot \mathbf{v} = (-3\mathbf{i} + 7\mathbf{j}) \cdot (-2\mathbf{i} + 2\mathbf{j})$ [4.3]

$= (-3)(-2) + (7)(2)$

$= 20$

39. $\cos\alpha = \dfrac{\langle 7, -4 \rangle \cdot \langle 2, 3 \rangle}{\sqrt{7^2 + (-4)^2}\sqrt{2^2 + 3^2}}$ [4.3]

$\cos\alpha = \dfrac{14 + (-12)}{\sqrt{65}\sqrt{13}}$

$\cos\alpha \approx 0.0688$

$\alpha \approx 86°$

40. $\cos\alpha = \dfrac{\langle -5, 2 \rangle \cdot \langle 2, -4 \rangle}{\sqrt{(-5)^2 + 2^2}\sqrt{2^2 + (-4)^2}}$

$\cos\alpha = \dfrac{-10 - 8}{\sqrt{29}\sqrt{20}}$

$\cos\alpha \approx -0.7474$ [4.3]

$\alpha = 138°$

41. $\cos\alpha = \dfrac{(6\mathbf{i} - 11\mathbf{j}) \cdot (2\mathbf{i} - 4\mathbf{j})}{\sqrt{6^2 + (-11)^2}\sqrt{2^2 + 4^2}}$

$\cos\alpha = \dfrac{12 - 44}{\sqrt{157}\sqrt{20}}$

$\cos\alpha \approx -0.5711$ [4.3]

$\cos\alpha \approx 125°$

42. $\cos\alpha = \dfrac{(\mathbf{i} - 5\mathbf{j}) \cdot (\mathbf{i} + 5\mathbf{j})}{\sqrt{1^2 + (-5)^2}\sqrt{1^2 + 5^2}}$

$\cos\alpha = \dfrac{1 - 25}{\sqrt{26}\sqrt{26}}$

$\cos\alpha \approx -0.9231$ [4.3]

$\alpha \approx 157°$

Copyright © Houghton Mifflin Company. All rights reserved.

43. $\text{proj}_{\mathbf{w}}\mathbf{v} = \dfrac{\mathbf{v}\cdot\mathbf{w}}{\|\mathbf{w}\|}$ [4.3]

$\text{proj}_{\mathbf{w}}\mathbf{v} = \dfrac{\langle -2,5\rangle \cdot \langle 5,4\rangle}{\sqrt{5^2 + 4^2}}$

$= \dfrac{-10+20}{\sqrt{41}}$

$= \dfrac{10}{\sqrt{41}}$

$= \dfrac{10\sqrt{41}}{41}$

44. $\text{proj}_{\mathbf{w}}\mathbf{v} = \dfrac{\mathbf{v}\cdot\mathbf{w}}{\|\mathbf{w}\|}$ [4.3]

$\text{proj}_{\mathbf{w}}\mathbf{v} = \dfrac{(4\mathbf{i}-7\mathbf{j})\cdot(-2\mathbf{i}-5\mathbf{j})}{\sqrt{(-2)^2 + (-5)^2}}$

$= \dfrac{-8+35}{\sqrt{29}}$

$= \dfrac{27}{\sqrt{29}}$

$= \dfrac{27\sqrt{29}}{29}$

45. $\mathbf{w} = \|\mathbf{F}\|\;\|\mathbf{S}\|\cos\theta$ [4.3]

$\mathbf{w} = 60\cdot 14\cos 38°$

$\mathbf{w} \approx 662$ foot-pounds

●●

Chapter Test

1.

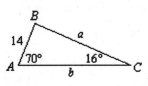

$B = 180° - 70° - 16°$
$B = 94°$

$\dfrac{c}{\sin C} = \dfrac{a}{\sin A}$

$a = \dfrac{14\sin 70°}{\sin 16°}$

$a \approx 48$

$\dfrac{c}{\sin C} = \dfrac{b}{\sin B}$ [4.1]

$b = \dfrac{14\sin 94°}{\sin 16°}$

$b \approx 51$

2.

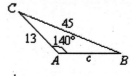

$\dfrac{b}{\sin B} = \dfrac{a}{\sin A}$

$B = \sin^{-1}\left(\dfrac{b\,\sin A}{a}\right)$ [4.1]

$B = \sin^{-1}\left(\dfrac{13\sin 140°}{45}\right)$

$B \approx 11°$

3.

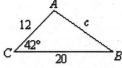

$c^2 = a^2 + b^2 - 2ab\cos C$

$c^2 = 20^2 + 12^2 - 2(20)(12)\cos 42°$ [4.2]

$c \approx 14$

4.

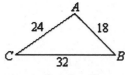

$\cos B = \dfrac{a^2 + c^2 - b^2}{2ac}$

$B = \cos^{-1}\left(\dfrac{32^2 + 18^2 - 24^2}{2(32)(18)}\right)$ [4.2]

$B \approx 48°$

5.

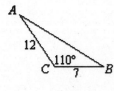

$K = \dfrac{1}{2}ab\sin C$

$K = \dfrac{1}{2}(7)(12)(\sin 110°)$

$K \approx 39$ square units
[4.2]

6.

$A = 180° - 42° - 75°$
$A = 63°$

$K = \dfrac{b^2 \sin A\,\sin C}{2\sin B}$

$K = \dfrac{12^2 \sin 63°\sin 75°}{2\sin 42°}$

$K \approx 93$ square units
[4.2]

Copyright © Houghton Mifflin Company. All rights reserved.

7.

C, 55, 17, B, 42, A

$$s = \frac{1}{2}(a+b+c)$$

$$s = \frac{1}{2}(17+55+42) = 57$$

$$K = \sqrt{s(s-a)(s-b)(s-c)}$$

$$K = \sqrt{57(57-17)(57-55)(57-42)}$$

$$K \approx 260 \text{ square units} \qquad [4.2]$$

8.

$$|\mathbf{v}| = \sqrt{(-2)^2 + (3)^2} = \sqrt{13} \quad [4.3]$$

9.

$$a_1 = 12\cos 220° \approx -9.2 \quad [4.3]$$
$$a_2 = 12\sin 220° \approx -7.7$$
$$\mathbf{v} = a_1\mathbf{i} + a_2\mathbf{j}$$
$$\mathbf{v} = -9.2\mathbf{i} - 7.7\mathbf{j}$$

10.

$$3\mathbf{u} - 5\mathbf{v} = 3(2\mathbf{i} - 3\mathbf{j}) - 5(5\mathbf{i} + 4\mathbf{j}) \quad [4.3]$$
$$= (6\mathbf{i} - 9\mathbf{j}) - (25\mathbf{i} + 20\mathbf{j})$$
$$= (6-25)\mathbf{i} + (-9-20)\mathbf{j}$$
$$= -19\mathbf{i} - 29\mathbf{j}$$

11.

$$\mathbf{u} \cdot \mathbf{v} = (-2\mathbf{i} + 3\mathbf{j}) \cdot (5\mathbf{i} + 3\mathbf{j}) \quad [4.3]$$
$$= (-2 \cdot 5) + (3 \cdot 3)$$
$$= -10 + 9$$
$$= -1$$

12.

$$\cos\theta = \frac{\mathbf{u} \cdot \mathbf{v}}{\|\mathbf{u}\| \, \|\mathbf{v}\|} = \frac{\langle 3,5 \rangle \cdot \langle -6,2 \rangle}{\sqrt{3^2 + 5^2}\sqrt{(-6)^2 + 2^2}} \qquad [4.3]$$

$$\cos\theta = \frac{-18+10}{\sqrt{34}\sqrt{40}} = \frac{-8}{\sqrt{34}\sqrt{40}}$$

$$\theta \approx 103°$$

13.

$$A = 142° - 65° = 77°$$

$$R^2 = 24^2 + 18^2 - 2(24)(18)\cos 77°$$

$$R \approx 27 \text{ miles} \qquad [4.3]$$

14.

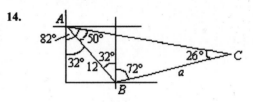

$$A = 82° - 32° = 50°$$

$$C = 180° - 50° - 32° - 72° = 26°$$

$$\frac{a}{\sin 50°} = \frac{12}{\sin 26°}$$

$$a = \frac{12\sin 50°}{\sin 26°}$$

$$a \approx 21 \text{ miles} \qquad [4.3]$$

15.

112, 165, 140

$$S = \frac{1}{2}(112+165+140) = 208.5$$

$$K = \sqrt{208.5(208.5-112)(208.5-165)(208.5-140)}$$

$$K \approx 7743$$

$$\text{cost} \approx 8.50(7743)$$

$$\text{cost} \approx \$66,000 \quad [4.2]$$

Copyright © Houghton Mifflin Company. All rights reserved.

••

1. $d = \sqrt{(-3-4)^2 + (4-(-1))^2} = \sqrt{49+25} = \sqrt{74}$ [1.2]

2. $f(x) + g(x) = \sin x + \cos x$ [1.5]

3. $(f \circ g)(x) = f[g(x)]$ [1.5]
$= f[\cos x]$
$= \sec(\cos x))$

4.
$f(x) = \frac{1}{2}x - 3$ [1.6]
$y = \frac{1}{2}x - 3$
$x = \frac{1}{2}y - 3$
$2(x+3) = y$
$f^{-1}(x) = 2x + 6$

5. Shifted 2 units to the right and 3 units up.   [1.4]

6. $\sin 27° = \dfrac{15}{a}$ [2.2]
$a = \dfrac{15}{\sin 27°} \approx 33$ cm

7. $y = 3\sin(2\pi x)$

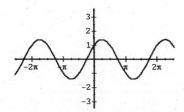

[2.5]

8. $y = \dfrac{1}{4}\tan(2x)$

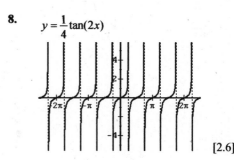

[2.6]

9. $y = 2\sin\left(\dfrac{\pi x}{2}\right) + 1$

[2.7]

10. $y = 3\sin\left(\dfrac{1}{3}x - \dfrac{\pi}{2}\right)$ [2.7]

$0 \le \dfrac{1}{3}x - \dfrac{\pi}{2} \le 2\pi$

$\dfrac{\pi}{2} \le \dfrac{1}{3}x \le \dfrac{5\pi}{2}$

$\dfrac{3\pi}{2} \le x \le \dfrac{15\pi}{2}$

amplitude $= 3$, period $= 6\pi$, phase shift $= \dfrac{3\pi}{2}$

11. $y = \sin x + \cos x$ [2.7]

Amplitude: $\sqrt{2}$, period: 2π, phase shift: $\dfrac{\pi}{4}$

12. $c^2 = a^2 + b^2 - 2ab\cos C$ [4.2]
$c^2 = (10)^2 + (12)^2 - 2(10)(12)\cos 50°$
$= 244 - 240\cos 50°$
≈ 9.473
$c \approx 9.5$ cm

13. $\dfrac{1}{\cos x} - \cos x = \dfrac{1 - \cos^2 x}{\cos x}$ [3.1]

$= \dfrac{\sin^2 x}{\cos x}$

$= \sin x \tan x$

14. $\sin^{-1}\sin\left(\dfrac{2\pi}{3}\right) = \sin^{-1}\left(\dfrac{\sqrt{3}}{2}\right) = \dfrac{\pi}{3}$ [3.5]

Copyright © Houghton Mifflin Company. All rights reserved.

15. $\tan\left(\cos^{-1}\left(\dfrac{12}{13}\right)\right)$ [3.5]

Let $\theta = \cos^{-1}\dfrac{12}{13}$ and find $y = \tan\theta$.

Then $\cos\theta = \dfrac{12}{13}$, and $0 \le \theta \le \pi$.

$\sqrt{13^2 - 12^2} = \sqrt{25} = 5$

Thus $\tan\theta = \dfrac{5}{12}$.

$y = \dfrac{5}{12}$

16. $\sin x \tan x - \dfrac{1}{2}\tan x = 0$ [3.6]

$\tan x\left(\sin x - \dfrac{1}{2}\right) = 0$

$\tan x = 0 \qquad\qquad \sin x - \dfrac{1}{2} = 0$

$\quad x = 0, \pi \qquad\qquad\qquad \sin x = \dfrac{1}{2}$

$\qquad\qquad\qquad\qquad\qquad x = \dfrac{\pi}{6}, \dfrac{5\pi}{6}$

17. $\|\mathbf{v}\| = \sqrt{4^2 + (-3)^2}$ $\alpha = \tan^{-1}\left|\dfrac{-3}{4}\right| = \tan^{-1}\dfrac{3}{4}$ [4.3]

$\|\mathbf{v}\| = \sqrt{16 + 9}$ $\alpha \approx 36.9°$

$\|\mathbf{v}\| = 5$ $\theta = 360° - \alpha$

$\qquad\qquad\qquad\qquad \theta \approx 360° - 36.9°$

$\qquad\qquad\qquad\qquad \theta \approx 323.1°$

18. $\cos\theta = \dfrac{\mathbf{v}\cdot\mathbf{w}}{\|\mathbf{v}\|\,\|\mathbf{w}\|}$ [4.3]

$\cos\theta = \dfrac{\langle 1,\,2\rangle\cdot\langle -2,\,3\rangle}{\sqrt{1^2 + (2)^2}\,\sqrt{(-2)^2 + 3^2}}$

$\cos\theta = \dfrac{1(-2) + (2)(3)}{\sqrt{5}\,\sqrt{13}}$

$\cos\theta = \dfrac{4}{\sqrt{5}\,\sqrt{13}} \approx 0.49613$

$\theta = 60.3°$

19.

$\alpha = \sin^{-1}\left|\dfrac{1}{3}\right| = \sin^{-1}\dfrac{1}{3}$

$\alpha \approx 19.5°$

heading $= \theta = 270° + \alpha$ [4.3]

$\theta \approx 270° + 19.5°$

$\theta \approx 289.5°$

20. $\mathbf{AB} = 515(\cos 36\mathbf{i} + \sin 36\mathbf{j}) \approx 416.6\mathbf{i} + 302.7\mathbf{j}$ [4.3]

$\mathbf{AD} = 150[\cos(-30°)\mathbf{i} + \sin(-30°)\mathbf{j}] \approx 129.9\mathbf{i} - 75\mathbf{j}$

$\mathbf{AC} = \mathbf{AB} + \mathbf{AD}$

$\mathbf{AC} = 416.6\mathbf{i} + 302.7\mathbf{j} + 129.9\mathbf{i} - 75\mathbf{j}$

$\mathbf{AC} \approx 546.5\mathbf{i} + 227.7\mathbf{j}$

$\|\mathbf{AC}\| = \sqrt{546.5^2 + (227.7)^2}$

$\|\mathbf{AC}\| \approx 592$ mph

$\alpha = 90° - \theta = 90° - \tan^{-1}\left(\dfrac{227.7}{546.5}\right) \approx 67.4°$

Copyright © Houghton Mifflin Company. All rights reserved.

Chapter 5
Complex Numbers

Section 5.1

1. $\sqrt{-81} = i\sqrt{81} = 9i$

3. $\sqrt{-98} = i\sqrt{98} = 7i\sqrt{2}$

5. $\sqrt{16} + \sqrt{-81} = 4 + i\sqrt{81}$
$= 4 + 9i$

7. $5 + \sqrt{-49} = 5 + i\sqrt{49} = 5 + 7i$

9. $8 - \sqrt{-18} = 8 - i\sqrt{18} = 8 - 3i\sqrt{2}$

11. $(5 + 2i) + (6 - 7i) = 5 + 2i + 6 - 7i$
$= (5 + 6) + (2i - 7i)$
$= 11 - 5i$

13. $(-2 - 4i) - (5 - 8i) = -2 - 4i - 5 + 8i$
$= (-2 - 5) + (-4i + 8i)$
$= -7 + 4i$

15. $(1 - 3i) + (7 - 2i) = 1 - 3i + 7 - 2i$
$= (1 + 7) + (-3i - 2i)$
$= 8 - 5i$

17. $(-3 - 5i) - (7 - 5i) = -3 - 5i - 7 + 5i$
$= (-3 - 7) + (-5i + 5i)$
$= -10$

19. $8i - (2 - 8i) = 8i - 2 + 8i$
$= -2 + (8i + 8i)$
$= -2 + 16i$

21. $5i \cdot 8i = 40i^2$
$= 40(-1)$
$= -40$

23. $\sqrt{-50} \cdot \sqrt{-2} = i\sqrt{50} \cdot i\sqrt{2} = 5i\sqrt{2} \cdot i\sqrt{2}$
$= 5i^2(\sqrt{2})^2 = 5(-1)(2)$
$= -10$

25. $3(2 + 5i) - 2(3 - 2i) = 6 + 15i - 6 + 4i$
$= (6 - 6) + (15i + 4i)$
$= 19i$

27. $(4 + 2i)(3 - 4i) = 4(3 - 4i) + (2i)(3 - 4i)$
$= 12 - 16i + 6i - 8i^2$
$= 12 - 16i + 6i - 8(-1)$
$= 12 - 16i + 6i + 8$
$= (12 + 8) + (-16i + 6i)$
$= 20 - 10i$

29. $(-3 - 4i)(2 + 7i) = -3(2 + 7i) - 4i(2 + 7i)$
$= -6 - 21i - 8i - 28i^2$
$= -6 - 21i - 8i - 28(-1)$
$= -6 - 21i - 8i + 28$
$= (-6 + 28) + (-21i - 8i)$
$= 22 - 29i$

31. $(4 - 5i)(4 + 5i) = 4(4 + 5i) - 5i(4 + 5i)$
$= 16 + 20i - 20i - 25i^2$
$= 16 + 20i - 20i - 25(-1)$
$= 16 + 20i - 20i + 25$
$= (16 + 25) + (20i - 20i)$
$= 41$

33. $(3 + \sqrt{-4})(2 - \sqrt{-9}) = (3 + i\sqrt{4})(2 - i\sqrt{9})$
$= (3 + 2i)(2 - 3i)$
$= 3(2 - 3i) + 2i(2 - 3i)$
$= 6 - 9i + 4i - 6i^2$
$= 6 - 9i + 4i - 6(-1)$
$= 6 - 9i + 4i + 6$
$= (6 + 6) + (-9i + 4i)$
$= 12 - 5i$

35. $(3 + 2\sqrt{-18})(2 + 2\sqrt{-50}) = (3 + 2i\sqrt{18})(2 + 2i\sqrt{50}) = [3 + 2i(3\sqrt{2})][2 + 2i(5\sqrt{2})] = (3 + 6i\sqrt{2})(2 + 10i\sqrt{2})$
$= 3(2 + 10i\sqrt{2}) + 6i\sqrt{2}(2 + 10i\sqrt{2}) = 6 + 30i\sqrt{2} + 12i\sqrt{2} + 60i^2(\sqrt{2})^2$
$= 6 + 30i\sqrt{2} + 12i\sqrt{2} + 60(-1)(2) = 6 + 30i\sqrt{2} + 12i\sqrt{2} - 120$
$= (6 - 120) + (30i\sqrt{2} + 12i\sqrt{2}) = -114 + 42i\sqrt{2}$

Copyright © Houghton Mifflin Company. All rights reserved.

37. $\dfrac{6}{i} = \dfrac{6}{i} \cdot \dfrac{i}{i} = \dfrac{6i}{i^2} = \dfrac{6i}{-1} = -6i$

39. $\dfrac{6+3i}{i} = \dfrac{6+3i}{i} \cdot \dfrac{i}{i} = \dfrac{6i+3i^2}{i^2} = \dfrac{6i+3(-1)}{-1} = \dfrac{6i-3}{-1} = 3-6i$

41. $\dfrac{1}{7+2i} = \dfrac{1}{7+2i} \cdot \dfrac{7-2i}{7-2i} = \dfrac{1(7-2i)}{(7+2i)(7-2i)} = \dfrac{7-2i}{49-4i^2} = \dfrac{7-2i}{49-4(-1)} = \dfrac{7-2i}{49+4} = \dfrac{7-2i}{53} = \dfrac{7}{53} - \dfrac{2}{53}i$

43. $\dfrac{2i}{1+i} = \dfrac{2i}{1+i} \cdot \dfrac{1-i}{1-i} = \dfrac{2i(1-i)}{(1+i)(1-i)} = \dfrac{2i-2i^2}{1-i^2} = \dfrac{2i-2(-1)}{1-(-1)} = \dfrac{2i+2}{1+1} = \dfrac{2+2i}{2} = \dfrac{2}{2} + \dfrac{2}{2}i = 1+i$

45. $\dfrac{5-i}{4+5i} = \dfrac{5-i}{4+5i} \cdot \dfrac{4-5i}{4-5i} = \dfrac{(5-i)(4-5i)}{(4+5i)(4-5i)} = \dfrac{5(4-5i)-i(4-5i)}{4(4-5i)+5i(4-5i)} = \dfrac{20-25i-4i+5i^2}{16-20i+20i-25i^2}$

$= \dfrac{20-25i-4i+5(-1)}{16-25(-1)} = \dfrac{20-25i-4i-5}{16+25} = \dfrac{(20-5)+(-25i-4i)}{16+25} = \dfrac{15-29i}{41} = \dfrac{15}{41} - \dfrac{29}{41}i$

47. $\dfrac{3+2i}{3-2i} = \dfrac{3+2i}{3-2i} \cdot \dfrac{3+2i}{3+2i} = \dfrac{(3+2i)^2}{(3-2i)(3+2i)} = \dfrac{3^2+2(3)(2i)+(2i)^2}{3^2-(2i)^2} = \dfrac{9+12i+4i^2}{9-4i^2} = \dfrac{9+12i+4(-1)}{9-4(-1)}$

$= \dfrac{9+12i-4}{9+4} = \dfrac{5+12i}{13} = \dfrac{5}{13} + \dfrac{12}{13}i$

49. $\dfrac{-7+26i}{4+3i} = \dfrac{-7+26i}{4+3i} \cdot \dfrac{4-3i}{4-3i} = \dfrac{(-7+26i)(4-3i)}{(4+3i)(4-3i)} = \dfrac{-7(4-3i)+26i(4-3i)}{4^2-(3i)^2} = \dfrac{-28+21i+104i-78i^2}{16-9i^2}$

$= \dfrac{-28+21i+104i-78(-1)}{16-9(-1)} = \dfrac{-28+21i+104i+78}{16+9} = \dfrac{50+125i}{25} = \dfrac{50}{25} + \dfrac{125}{25}i = 2+5i$

51. $(3-5i)^2 = 3^3 + 2(3)(-5i) + (-5i)^2$

$= 9 - 30i + 25i^2$

$= 9 - 30i + 25(-1)$

$= 9 - 30i - 25$

$= -16 - 30i$

53. $(1+2i)^3 = (1+2i)(1+2i)^2$

$= (1+2i)[1^2 + 2(1)(2i) + (2i)^2]$

$= (1+2i)[1 + 4i + 4i^2]$

$= (1+2i)[1 + 4i + 4(-1)]$

$= (1+2i)[1 + 4i - 4]$

$= (1+2i)(-3+4i)$

$= 1(-3+4i) + 2i(-3+4i)$

$= -3 + 4i - 6i + 8i^2$

$= -3 + 4i - 6i - 8$

$= -11 - 2i$

55. Use the Powers of i Theorem.
The remainder of $15 \div 4$ is 3.

$i^{15} = i^3 = -i$

57. Use the Powers of i Theorem.
The remainder of $40 \div 4$ is 0.

$-i^{40} = -(i^0) = -1$

59. Use the Powers of i Theorem.
The remainder of $25 \div 4$ is 1.

$\dfrac{1}{i^{25}} = \dfrac{1}{i} = \dfrac{1}{i} \cdot \dfrac{i}{i} = \dfrac{i}{i^2} = \dfrac{i}{-1} = -i$

61. Use the Powers of i Theorem.
The remainder of $34 \div 4$ is 2.

$i^{-34} = \dfrac{1}{i^{34}} = \dfrac{1}{i^2} = \dfrac{1}{-1} = -1$

Copyright © Houghton Mifflin Company. All rights reserved.

63. Use $a = 3$, $b = -3$, $c = 3$.

$$\frac{-b+\sqrt{b^2-4ac}}{2a} = \frac{-(-3)+\sqrt{(-3)^2-4(3)(3)}}{2(3)}$$

$$= \frac{3+\sqrt{9-36}}{6} = \frac{3+\sqrt{-27}}{6}$$

$$= \frac{3+i\sqrt{27}}{6} = \frac{3+3i\sqrt{3}}{6}$$

$$= \frac{3}{6}+\frac{3\sqrt{3}}{6}i = \frac{1}{2}+\frac{\sqrt{3}}{2}i$$

65. Use $a = 2$ $b = 6$, $c = 6$.

$$\frac{-b+\sqrt{b^2-4ac}}{2a} = \frac{-(6)+\sqrt{(6)^2-4(2)(6)}}{2(2)}$$

$$= \frac{-6+\sqrt{36-48}}{4} = \frac{-6+\sqrt{-12}}{4}$$

$$= \frac{-6+i\sqrt{12}}{4} = \frac{-6+2i\sqrt{3}}{4}$$

$$= \frac{-6}{4}+\frac{2i\sqrt{3}}{4} = -\frac{3}{2}+\frac{\sqrt{3}}{2}i$$

67. Use $a = 4$, $b = -4$, $c = 2$.

$$\frac{-b+\sqrt{b^2-4ac}}{2a} = \frac{-(-4)+\sqrt{(-4)^2-4(4)(2)}}{2(4)}$$

$$= \frac{4+\sqrt{16-32}}{8} = \frac{4+\sqrt{-16}}{8}$$

$$= \frac{4+i\sqrt{16}}{8} = \frac{4+4i}{8}$$

$$= \frac{4}{8}+\frac{4i}{8} = \frac{1}{2}+\frac{1}{2}i$$

●●●

Connecting Concepts

69. $x^2+16 = x^2+4^2 = (x+4i)(x-4i)$

71. $z^2+25 = z^2+5^2 = (z+5i)(z-5i)$

73. $4x^2+81 = (2x)^2+9^2 = (2x+9i)(2x-9i)$

75. If $x = 1+2i$, then $x^2-2x+5 = (1+2i)^2-2(1+2i)+5 = 1+4i+4i^2-2-4i+5 = 1+4i+4(-1)-2-4i+5$

$$= 1+4i-4-2-4i+5 = (1-4-2+5)+(4i-4i) = 0$$

77. Verify that $(-1+i\sqrt{3})^3 = 8$.

$(-1+i\sqrt{3})^3 = (-1+i\sqrt{3})(-1+i\sqrt{3})^2 = (-1+i\sqrt{3})[(-1)^2+2(-1)(i\sqrt{3})+(i\sqrt{3})^2]$

$\quad = (-1+i\sqrt{3})[1-2i\sqrt{3}+3i^2] = (-1+i\sqrt{3})[1-2i\sqrt{3}+3(-1)] = (-1+i\sqrt{3})[1-2i\sqrt{3}-3]$

$\quad = (-1+i\sqrt{3})(-2-2i\sqrt{3}) = -1(-2-2i\sqrt{3})+i\sqrt{3}(-2-2i\sqrt{3}) = 2+2i\sqrt{3}-2i\sqrt{3}-2i^2(\sqrt{3})^2$

$\quad = 2+2i\sqrt{3}-2i\sqrt{3}-2(-1)(3) = 2+2i\sqrt{3}-2i\sqrt{3}+6 = (2+6)+(2i\sqrt{3}-2i\sqrt{3})$

$\quad = 8$

Verify that $(-1-i\sqrt{3})^3 = 8$.

$(-1-i\sqrt{3})^3 = (-1-i\sqrt{3})(-1-i\sqrt{3})^2 = (-1-i\sqrt{3})[(-1)^2+2(-1)(-i\sqrt{3})+(-i\sqrt{3})^2]$

$\quad = (-1-i\sqrt{3})[1+2i\sqrt{3}+3i^2] = (-1-i\sqrt{3})[1+2i\sqrt{3}+3(-1)] = (-1-i\sqrt{3})[1+2i\sqrt{3}-3]$

$\quad = (-1-i\sqrt{3})(-2+2i\sqrt{3}) = -1(-2+2i\sqrt{3})-i\sqrt{3}(-2+2i\sqrt{3}) = 2-2i\sqrt{3}+2i\sqrt{3}-2i^2(\sqrt{3})^2$

$\quad = 2-2i\sqrt{3}+2i\sqrt{3}-2(-1)(3) = 2-2i\sqrt{3}+2i\sqrt{3}+6 = (2+6)+(-2i\sqrt{3}+2i\sqrt{3})$

$\quad = 8$

Copyright © Houghton Mifflin Company. All rights reserved.

79. $i + i^2 + i^3 + i^4 + \ldots + i^{28} = 7(i + i^2 + i^3 + i^4) = 7(i + (-1) + (-i) + 1) = 7(0) = 0$

●●●●●●●●●●●●●●●●●●●●●●●●●●●●●●●●●●●●

Prepare for Section 5.2

81. $(1+i)(2+i) = 2 + 3i + i^2 = 1 + 3i$

82. $\dfrac{2+i}{3-i} \cdot \dfrac{3+i}{3+i} = \dfrac{6 + 5i + i^2}{9 - i^2} = \dfrac{5 + 5i}{10} = \dfrac{1}{2} + \dfrac{1}{2}i$

83. $2 - 3i$

84. $3 + 5i$

85. $x = \dfrac{-1 \pm \sqrt{1^2 - 4(1)(1)}}{2(1)} = \dfrac{-1 \pm \sqrt{-3}}{2} = -\dfrac{1}{2} \pm \dfrac{\sqrt{3}}{2}i$

86. $x^2 + 9 = 0$

$x^2 = -9$

$x = \pm 3i$

Section 5.2

1.

Im, Re, $-2 - 2i \bullet$

$|z| = \sqrt{(-2)^2 + (-2)^2}$

$\quad = \sqrt{8} = 2\sqrt{2}$

3.

Im, Re, $\bullet \sqrt{3} - i$

$|z| = \sqrt{(\sqrt{3})^2 + (-1)^2}$

$\quad = \sqrt{3 + 1} = \sqrt{4}$

$\quad = 2$

5.

Im, Re, $\bullet -2i$

$|z| = \sqrt{0^2 + (-2)^2} = 2$

7.

Im, Re, $\bullet 3 - 5i$

$|z| = \sqrt{3^2 + (-5)^2}$

$\quad = \sqrt{34}$

9. $r = \sqrt{1^2 + (-1)^2}$

$r = \sqrt{2}$

$\alpha = \tan^{-1} \left| \dfrac{-1}{1} \right|$

$\quad = \tan^{-1} 1$

$\quad = 45°$

$\theta = 360° - 45° = 315°$

$z = \sqrt{2} \text{ cis } 315°$

11. $r = \sqrt{(\sqrt{3})^2 + (-1)^2}$

$r = 2$

$\alpha = \tan^{-1} \left| \dfrac{-1}{\sqrt{3}} \right|$

$\quad = \tan^{-1} \dfrac{1}{\sqrt{3}} = 30°$

$\alpha = 360° - 30° = 330°$

$z = 2 \text{ cis } 330°$

13. $r = \sqrt{0^2 + 3^2}$

$r = 3$

$\theta = 90°$

$z = 3 \text{ cis } 90°$

15. $r = \sqrt{(-5)^2 + 0^2}$

$r = 5$

$\theta = 180°$

$z = 5 \text{ cis } 180°$

17. $z = 2(\cos 45° + i \sin 45°)$

$z = 2\left(\dfrac{\sqrt{2}}{2} + \dfrac{\sqrt{2}}{2}i \right)$

$z = \sqrt{2} + i\sqrt{2}$

19. $z = \cos 315° + i \sin 315°$

$z = \dfrac{\sqrt{2}}{2} - \dfrac{\sqrt{2}}{2}i$

21. $z = 6 \text{ cis } 135°$

$z = 6(\cos 135° + i \sin 135°)$

$z = 6\left(-\dfrac{\sqrt{2}}{2} + \dfrac{\sqrt{2}}{2}i \right)$

$z = -3\sqrt{2} + 3i\sqrt{2}$

23. $z = 8 \text{ cis } 0°$

$z = 8(\cos 0° + i \sin 0°)$

$z = 8(1 + 0i)$

$z = 8$

Copyright © Houghton Mifflin Company. All rights reserved.

25.
$$z = 2\left(\cos\frac{5\pi}{6} + i\sin\frac{5\pi}{6}\right)$$
$$z = 2\left(-\frac{\sqrt{3}}{2} + \frac{1}{2}i\right)$$
$$z = -\sqrt{3} + i$$

27.
$$z = 3\left(\cos\frac{3\pi}{2} + i\sin\frac{3\pi}{2}\right)$$
$$z = 3(0 - i)$$
$$z = -3i$$

29.
$$z = 8\,\text{cis}\,\frac{3\pi}{4}$$
$$= 8\left(\cos\frac{3\pi}{4} + i\sin\frac{3\pi}{4}\right)$$
$$z = 8\left(-\frac{\sqrt{2}}{2} + \frac{i\sqrt{2}}{2}\right)$$
$$z = -4\sqrt{2} + 4i\sqrt{2}$$

31.
$$z = 9\,\text{cis}\,\frac{11\pi}{6}$$
$$z = 9\left(\cos\frac{11\pi}{6} + i\sin\frac{11\pi}{6}\right)$$
$$z = 9\left(\frac{\sqrt{3}}{2} - \frac{1}{2}i\right)$$
$$z = \frac{9\sqrt{3}}{2} - \frac{9}{2}i$$

33.
$$z = 2\,\text{cis}\,2$$
$$z = 2(\cos 2 + i\sin 2)$$
$$z \approx 2(-0.4161 + 0.9093i)$$
$$z \approx -0.832 + 1.819i$$

35.
$$z_1 z_2 = 2\,\text{cis}\,30° \cdot 3\,\text{cis}\,225°$$
$$z_1 z_2 = 6\,\text{cis}(30° + 225°)$$
$$z_1 z_2 = 6\,\text{cis}\,255°$$

37.
$$z_1 z_2 = 3(\cos 122° + i\sin 122°) \cdot 4(\cos 213° + i\sin 213°)$$
$$z_1 z_2 = 12[\cos(122° + 213°) + i\sin(122° + 213°)]$$
$$z_1 z_2 = 12(\cos 335° + i\sin 335°)$$
$$z_1 z_2 = 12\,\text{cis}\,335°$$

39.
$$z_1 z_2 = 5\left(\cos\frac{2\pi}{3} + i\sin\frac{2\pi}{3}\right) \cdot 2\left(\cos\frac{2\pi}{5} + i\sin\frac{2\pi}{5}\right)$$
$$z_1 z_2 = 10\left[\cos\left(\frac{2\pi}{3} + \frac{2\pi}{5}\right) + i\sin\left(\frac{2\pi}{3} + \frac{2\pi}{5}\right)\right]$$
$$z_1 z_2 = 10\left(\cos\frac{16\pi}{15} + i\sin\frac{16\pi}{15}\right)$$
$$z_1 z_2 = 10\,\text{cis}\,\frac{16\pi}{15}$$

41.
$$z_1 z_2 = 4\,\text{cis}\,2.4 \cdot 6\,\text{cis}\,4.1$$
$$z_1 z_2 = 24\,\text{cis}\,(2.4 + 4.1)$$
$$z_1 z_2 = 24\,\text{cis}\,6.5$$

43.
$$\frac{z_1}{z_2} = \frac{32\,\text{cis}\,30°}{4\,\text{cis}\,150°}$$
$$\frac{z_1}{z_2} = 8\,\text{cis}(30° - 150°)$$
$$\frac{z_1}{z_2} = 8\,\text{cis}(-120°)$$
$$\frac{z_1}{z_2} = 8(\cos 120° - i\sin 120°)$$
$$\frac{z_1}{z_2} = 8\left(-\frac{1}{2} - \frac{i\sqrt{3}}{2}\right) = -4 - 4i\sqrt{3}$$

45.
$$\frac{z_1}{z_2} = \frac{27(\cos 315° + i\sin 315°)}{9(\cos 225° + i\sin 225°)}$$
$$\frac{z_1}{z_2} = 3\,[\cos(315° - 225°) + i\sin(315° - 225°)]$$
$$\frac{z_1}{z_2} = 3(\cos 90° + i\sin 90°) = 3(0 + i) = 3i$$

47.
$$\frac{z_1}{z_2} = \frac{12\left(\cos\frac{2\pi}{3} + i\sin\frac{2\pi}{3}\right)}{4\left(\cos\frac{11\pi}{6} + i\sin\frac{11\pi}{6}\right)}$$
$$\frac{z_1}{z_2} = 3\left[\cos\left(\frac{2\pi}{3} - \frac{11\pi}{6}\right) + i\sin\left(\frac{2\pi}{3} - \frac{11\pi}{6}\right)\right]$$
$$\frac{z_1}{z_2} = 3\left(\cos\frac{7\pi}{6} - i\sin\frac{7\pi}{6}\right)$$
$$\frac{z_1}{z_2} = 3\left[-\frac{\sqrt{3}}{2} - \left(-\frac{1}{2}i\right)\right]$$
$$\frac{z_1}{z_2} = -\frac{3\sqrt{3}}{2} + \frac{3}{2}i$$

Copyright © Houghton Mifflin Company. All rights reserved.

49.

$$\frac{z_1}{z_2} = \frac{25\,\text{cis}\,3.5}{5\,\text{cis}\,1.5}$$

$$\frac{z_1}{z_2} = 5\,\text{cis}\,(3.5 - 1.5)$$

$$\frac{z_1}{z_2} = 5\,\text{cis}\,2$$

$$\frac{z_1}{z_2} = 5\,(\cos 2 + i\,\sin 2)$$

$$\frac{z_1}{z_2} \approx 5\,(-0.4161 + 0.9093i)$$

$$\frac{z_1}{z_2} \approx -2.081 + 4.546i$$

51.

$z_1 = 1 - i\sqrt{3}$

$r_1 = \sqrt{1^2 + \left(\sqrt{3}\right)^2}$ $\quad \alpha = \tan^{-1}\left|\dfrac{-\sqrt{3}}{1}\right| = 60°$

$r_1 = 2$

$\qquad\qquad \theta_1 = 300°$

$z_1 = 2(\cos 300° + i\,\sin 300°)$

$z_2 = 1 + i$

$r_2 = \sqrt{1^2 + 1^2}$ $\quad \alpha = \tan^{-1}\left|\dfrac{1}{1}\right| = 45°$

$r_2 = \sqrt{2}$

$\qquad\qquad \theta_2 = 45°$

$z_2 = \sqrt{2}(\cos 45° + i\,\sin 45°)$

$z_1 z_2 = 2(\cos 300° + i\,\sin 300°) \cdot \sqrt{2}(\cos 45° + i\,\sin 45°)$

$z_1 z_2 = 2\sqrt{2}[\cos(300° + 45°) + i\,\sin(300° + 45°)]$

$z_1 z_2 = 2\sqrt{2}(\cos 345° + i\,\sin 345°)$

$z_1 z_2 \approx 2.732 - 0.732i$

53.

$z_1 = 3 - 3i$

$r_1 = \sqrt{3^2 + (-3)^2}$ $\quad \alpha = \tan^{-1}\left|\dfrac{-3}{3}\right| = 45°$

$r_1 = 3\sqrt{2}$ $\qquad \theta_1 = 315°$

$z_1 = 3\sqrt{2}(\cos 315° + i\,\sin 315°)$

$z_2 = 1 + i$

$r_2 = \sqrt{1^2 + 1^2}$ $\quad \alpha = \tan^{-1}\left|\dfrac{1}{1}\right| = 45°$

$r_2 = \sqrt{2}$

$\qquad\qquad \theta_2 = 45°$

$z_2 = \sqrt{2}(\cos 45° + i\,\sin 45°)$

$z_1 z_2 = 3\sqrt{2}(\cos 315° + i\,\sin 315°) \cdot \sqrt{2}(\cos 45° + i\,\sin 45°)$

$z_1 z_2 = 6[\cos(315° + 45°) + i\,\sin(315° + 45°)]$

$z_1 z_2 = 6(\cos 360° + i\,\sin 360°)$

$z_1 z_2 = 6 + 0i$

$z_1 z_2 = 6$

55.

$z_1 = 1 + i\sqrt{3}$

$r_1 = \sqrt{1^2 + \left(\sqrt{3}\right)^2}$ $\quad \alpha_1 = \tan^{-1}\left|\dfrac{\sqrt{3}}{1}\right| = 60°$

$r_1 = 2$

$\qquad\qquad \theta_1 = 60°$

$z_1 = 2(\cos 60° + i\,\sin 60°)$

$z_2 = 1 - i\sqrt{3}$

$r_2 = \sqrt{1^2 + \left(\sqrt{3}\right)^2}$ $\quad \alpha = \tan^{-1}\left|\dfrac{-\sqrt{3}}{1}\right| = 60°$

$r_2 = 2$

$\qquad\qquad \theta_2 = 300°$

$z_2 = 2(\cos 300° + i\,\sin 300°)$

$$\frac{z_1}{z_2} = \frac{2(\cos 60° + i\,\sin 60°)}{2(\cos 300° + i\,\sin 300°)}$$

$$\frac{z_1}{z_2} = \cos(60° - 300°) + i\,\sin(60° - 300°)$$

$$\frac{z_1}{z_2} = \cos 240° - i\,\sin 240° = -\frac{1}{2} + \frac{\sqrt{3}}{2}i$$

Copyright © Houghton Mifflin Company. All rights reserved.

57.

$z_1 = \sqrt{2} - i\sqrt{2}$ $z_2 = 1 + i$

$r_1 = \sqrt{(\sqrt{2})^2 + (\sqrt{2})^2}$ $\alpha_1 = \tan^{-1}\left|\dfrac{-\sqrt{2}}{2}\right| = 45°$ $r_2 = \sqrt{1^2 + 1^2}$ $\alpha_2 = \tan^{-1}\left|\dfrac{1}{1}\right| = 45°$

$r_1 = 2$ $r_2 = \sqrt{2}$ $\theta_2 = 45°$

$\theta_1 = 315°$

$z_1 = 2(\cos 315° + i \sin 315°)$ $z_2 = \sqrt{2}(\cos 45° + i \sin 45°)$

$\dfrac{z_1}{z_2} = \dfrac{2(\cos 315° + i \sin 315°)}{\sqrt{2}(\cos 45° + i \sin 45°)}$

$\dfrac{z_1}{z_2} = \sqrt{2}[\cos(315° - 45°) + i \sin(315° - 45°)]$

$\dfrac{z_1}{z_2} = \sqrt{2}(\cos 270° + i \sin 270°)$

$\dfrac{z_1}{z_2} = \sqrt{2}[0 + i(-1)] = \sqrt{2}(0 - 1i) = 0 - \sqrt{2}i = -\sqrt{2}i \text{ or } -i\sqrt{2}$

••

59.

$z_1 = \sqrt{3} - 1$ $z_2 = 2 + 2i$ $z_3 = 2 - 2i\sqrt{3}$

$r_1 = \sqrt{(\sqrt{3})^2 + (-1)^2}$ $r_2 = \sqrt{2^2 + 2^2}$ $r_3 = \sqrt{2^2 + (-2\sqrt{3})^2}$

$r_1 = 2$ $r_2 = 2\sqrt{2}$ $r_3 = 4$

$\alpha_1 = \tan^{-1}\left|\dfrac{-1}{\sqrt{3}}\right| = 30°$ $\alpha_2 = \tan^{-1}\left|\dfrac{2}{2}\right| = 45°$ $\alpha_3 = \tan^{-1}\left|\dfrac{-2\sqrt{3}}{2}\right| = 60°$

$\theta_1 = 330°$ $\theta_2 = 45°$ $\theta_3 = 300°$

$z_1 = 2(\cos 330° + i \sin 330°)$ $z_2 = 2\sqrt{2}(\cos 45° + i \sin 45°)$ $z_3 = 4(\cos 300° + i \sin 300°)$

$z_1 z_2 z_3 = 2(\cos 330° + i \sin 330°) \cdot 2\sqrt{2}(\cos 45° + i \sin 45°) \cdot 4(\cos 300° + i \sin 300°)$

$z_1 z_2 z_3 = 16\sqrt{2}[\cos(330° + 45° + 300°) + i \sin(330° + 45° + 300°)]$

$z_1 z_2 z_3 = 16\sqrt{2}(\cos 675° + i \sin 675°)$

$z_1 z_2 z_3 = 16\sqrt{2}(\cos 315° + i \sin 315°)$

$z_1 z_2 z_3 = 16\sqrt{2}\left(\dfrac{1}{\sqrt{2}} - \dfrac{1}{\sqrt{2}}i\right) = 16 - 16i$

Copyright © Houghton Mifflin Company. All rights reserved.

61.

$z_1 = \sqrt{3} + i\sqrt{3}$

$r_1 = \sqrt{(\sqrt{3})^2 + (\sqrt{3})^2}$

$r_1 = \sqrt{6}$

$\alpha_1 = \tan^{-1}\left|\dfrac{\sqrt{3}}{\sqrt{3}}\right| = 45°$

$\theta_1 = 45°$

$z_1 = \sqrt{6}(\cos 45° + i \sin 45°)$

$z_2 = 1 - i\sqrt{3}$

$r_2 = \sqrt{1^2 + (-\sqrt{3})^2}$

$r_2 = 2$

$\alpha_2 = \tan^{-1}\left|\dfrac{-\sqrt{3}}{1}\right| = 60°$

$\theta_2 = 300°$

$z_2 = 2(\cos 300° + i \sin 300°)$

$z_3 = 2 - 2i$

$r_3 = \sqrt{2^2 + (-2)^2}$

$r_3 = 2\sqrt{2}$

$\alpha_3 = \tan^{-1}\left|\dfrac{-2}{2}\right| = 45°$

$\theta_3 = 315°$

$z_3 = 2\sqrt{2}(\cos 315° + i \sin 315°)$

$$\frac{z_1}{z_2 z_3} = \frac{\sqrt{6}(\cos 45° + i \sin 45°)}{2(\cos 300° + i \sin 300°) \cdot 2\sqrt{2}(\cos 315° + i \sin 315°)}$$

$$\frac{z_1}{z_2 z_3} = \frac{\sqrt{6}(\cos 45° + i \sin 45°)}{4\sqrt{2}[\cos(300° + 315°) + i \sin(300° + 315°)]}$$

$$\frac{z_1}{z_2 z_3} = \frac{\sqrt{6}(\cos 45° + i \sin 45°)}{4\sqrt{2}(\cos 255° + i \sin 255°)}$$

$$\frac{z_1}{z_2 z_3} = \frac{\sqrt{3}}{4}[\cos(45° - 255°) + i \sin(45° - 255°)]$$

$$\frac{z_1}{z_2 z_3} = \frac{\sqrt{3}}{4}(\cos 210° - i \sin 210°) = \frac{\sqrt{3}}{4}\left(-\frac{\sqrt{3}}{2} + \frac{i}{2}\right) = -\frac{3}{8} + \frac{\sqrt{3}}{8}i$$

63.

$z_1 = 1 - 3i$

$r_1 = \sqrt{1^2 + (-3)^2}$

$r_1 = \sqrt{10}$

$\alpha_1 = \tan^{-1}\left|\dfrac{-3}{1}\right| \approx 71.57°$

$\theta_1 = 288.43°$

$z_1 = \sqrt{10}(\cos 288.4° + i \sin 288.4°)$

$z_2 = 2 + 3i$

$r_2 = \sqrt{2^2 + 3^2}$

$r_2 = \sqrt{13}$

$\alpha_2 = \tan^{-1}\left|\dfrac{3}{2}\right| \approx 56.31°$

$\theta_2 = 56.31°$

$z_2 = \sqrt{13}(\cos 56.3° + i \sin 56.3°)$

$z_3 = 4 + 5i$

$r_3 = \sqrt{4^2 + 5^2}$

$r_3 = \sqrt{41}$

$\alpha_3 = \tan^{-1}\left|\dfrac{5}{4}\right| \approx 51.34°$

$\theta_3 = 51.34°$

$z_3 = \sqrt{41}(\cos 51.3° + i \sin 51.3°)$

$z_1 z_2 z_3 = \sqrt{10}(\cos 288.4° + i \sin 288.4°) \cdot \sqrt{13}(\cos 56.3° + i \sin 56.3°) \cdot \sqrt{41}(\cos 51.3° + i \sin 51.3°)$

$z_1 z_2 z_3 = \sqrt{10} \cdot \sqrt{13} \cdot \sqrt{41}[\cos(288.43° + 56.31° + 51.34°) + i \sin(288.43° + 56.31° + 51.34°)]$

$z_1 z_2 z_3 \approx 73.0(\cos 396.08° + i \sin 396.08°)$

$z_1 z_2 z_3 = 73.0(\cos 36.08° + i \sin 36.08°)$

$z_1 z_2 z_3 \approx 59.0 + 43.0i$

65.

$z = r(\cos\theta + i\sin\theta) \qquad \overline{z} = r(\cos\theta - i\sin\theta)$

$z \cdot \overline{z} = r(\cos\theta + i\sin\theta) \cdot r(\cos\theta - i\sin\theta)$

$z \cdot \overline{z} = r(\cos\theta + i\sin\theta) \cdot r[\cos(-\theta) + i\sin(-\theta)]$

$z \cdot \overline{z} = r^2[\cos(\theta - \theta) + i\sin(\theta - \theta)]$

$z \cdot \overline{z} = r^2(\cos 0 + i\sin 0)$

$z \cdot \overline{z} = r^2 \ \text{or} \ a^2 + b^2$

Copyright © Houghton Mifflin Company. All rights reserved.

67.

$$\left(\frac{\sqrt{2}}{2} + \frac{\sqrt{2}}{2}i\right)^2 = \frac{2}{4} + 2\frac{2}{4}i + \frac{2}{4}i^2 = i$$

68.

$x^3 - 8 = (x-2)(x^2 + 2x + 4)$

$(x^2 + 2x + 4)$ yields 2 complex solutions

$(x - 2)$ yields 1 real solution

The real root is 2.

69.

$x^5 - 243 = (x-3)(3x^4 + 3x^3 + 9x^2 + 27x + 81)$

$(3x^4 + 3x^3 + 9x^2 + 27x + 81)$ yields 4 complex solutions

$(x - 3)$ yields 1 real solution

The real root is 3.

70.

$$r = \sqrt{2^2 + 2^2} = 2\sqrt{2}$$

$$\alpha = \tan^{-1}\left|\frac{2}{2}\right|$$

$$= \tan^{-1} 1 = 45°$$

$$\theta = 45°$$

$$z = 2\sqrt{2} \text{ cis } 45° \text{ or } 2\sqrt{2} \text{ cis } \frac{\pi}{4}$$

71.

$$2(\cos 150° + i\sin 150°) = 2\left(-\frac{\sqrt{3}}{2} + \frac{1}{2}i\right) = -\sqrt{3} + i$$

72.

$$|z| = \sqrt{\left(\frac{\sqrt{2}}{2}\right)^2 + \left(-\frac{\sqrt{2}}{2}\right)^2}$$

$$= \sqrt{\frac{2}{4} + \frac{2}{4}}$$

$$= 1$$

Section 5.3

1.

$$[2(\cos 30° + i\sin 30°)]^8 = 2^8[\cos(8 \cdot 30°) + i\sin(8 \cdot 30°)]$$

$$= 256(\cos 240° + i\sin 240°)$$

$$= -128 - 128i\sqrt{3}$$

3.

$$[2(\cos 240° + i\sin 240°)]^5 = 2^5[\cos(5 \cdot 240°) + i\sin(5 \cdot 240°)]$$

$$= 32[\cos 1200° + i\sin 1200°]$$

$$= 32(\cos 120° + i\sin 120°)$$

$$= -16 + 16i\sqrt{3}$$

5.

$$[2\text{cis}(225°)]^5 = 2^5 \text{cis}(5 \cdot 225°)$$

$$= 32(\cos 1125° + i\sin 1125°)$$

$$= 32(\cos 45° + i\sin 45°)$$

$$= 16\sqrt{2} + 16i\sqrt{2}$$

7.

$$[2\text{cis }(120°)]^6 = 2^6 \text{cis}(6 \cdot 2\pi/3)$$

$$= 64(\cos 720° + i\sin 720°)$$

$$= 64(\cos 0° + i\sin 0°)$$

$$= 64$$

9.

$z = 1 - i$

$r = \sqrt{1^2 + (-1)^2}$ $\quad \alpha = \tan^{-1}\left|\frac{-1}{1}\right| = 45°$

$r = \sqrt{2}$

$\theta = 315°$

$z = \sqrt{2}(\cos 315° + i\sin 315°)$

$(1-i)^{10} = [\sqrt{2}(\cos 315° + i\sin 315°)]^{10}$

$\quad = (\sqrt{2})^{10}[\cos(10 \cdot 315°) + i\sin(10 \cdot 315°)]$

$\quad = 32(\cos 3150° + i\sin 3150°)$

$\quad = 32(\cos 270° + i\sin 270°)$

$\quad = 0 - 32i = -32i$

11.

$z = 2 + 2i$

$r = \sqrt{2^2 + 2^2}$ $\quad\quad \alpha = \tan^{-1}\left|\frac{2}{2}\right| = 45°$

$r = 2\sqrt{2}$

$\theta = 45°$

$z = 2\sqrt{2}(\cos 45° + i\sin 45°)$

$(2+2i)^7 = [2\sqrt{2}(\cos 45° + i\sin 45°)]^7$

$\quad = 1024\sqrt{2}[\cos(7 \cdot 45°) + i\sin(7 \cdot 45°)]$

$\quad = 1024\sqrt{2}(\cos 315° + i\sin 315°)$

$\quad = 1024 - 1024i$

Copyright © Houghton Mifflin Company. All rights reserved.

13.

$$z = \frac{\sqrt{2}}{2} + i\frac{\sqrt{2}}{2}$$

$$r = \sqrt{(\sqrt{2}/2)^2 + (\sqrt{2}/2)^2} \qquad \alpha = \tan^{-1}\left|\frac{\sqrt{2}/2}{\sqrt{2}/2}\right| = 45°$$

$$r = 1$$

$$\theta = 45°$$

$$z = \cos 45° + i\sin 45°$$

$$\left(\frac{\sqrt{2}}{2} - i\frac{\sqrt{2}}{2}\right)^6 = (\cos 45° + i\sin 45°)^6$$

$$= \cos(6 \cdot 45°) + i\sin(6 \cdot 45°)$$

$$= \cos 270° + i\sin 270°$$

$$= 0 - 1i = -i$$

15.

$$9 = 9(\cos 0° + i\sin 0°)$$

$$w_k = 9^{1/2}\left(\cos\frac{0° + 360°k}{2} + i\sin\frac{0° + 360°k}{2}\right) \quad k = 0,1$$

$$w_0 = 3(\cos 0° + i\sin 0°)$$

$$w_0 = 3 + 0i = 3$$

$$w_1 = 3\left(\frac{\cos 0° + 360°}{2} + i\sin\frac{0° + 360°}{2}\right)$$

$$w_1 = 3(\cos 180° + i\sin 180°)$$

$$w_1 = -3 + 0i = -3$$

17.

$$64 = 64(\cos 0° + i\sin 0°)$$

$$w_k = 64^{1/6}\left(\cos\frac{0° + 360°k}{6} + i\sin\frac{0° + 360°k}{6}\right) \quad k = 0, 1, 2, 3, 4, 5$$

$$w_0 = 2(\cos 0° + i\sin 0°)$$

$$w_0 = 2 + 0i = 2$$

$$w_1 = 2\left(\cos\frac{0 + 360°}{6} + i\sin\frac{0 + 360°}{6}\right)$$

$$w_1 = 2(\cos 60° + i\sin 60°)$$

$$w_1 = 2\left(\frac{1}{2} + i\frac{\sqrt{3}}{2}\right)$$

$$w_1 = 1 + i\sqrt{3}$$

$$w_2 = 2\left(\cos\frac{0° + 360° \cdot 2}{6} + i\sin\frac{0° + 360° \cdot 2}{6}\right)$$

$$w_2 = 2(\cos 120° + i\sin 120°)$$

$$w_2 = 2\left(-\frac{1}{2} + i\frac{\sqrt{3}}{2}\right)$$

$$w_2 = -1 + i\sqrt{3}$$

$$w_3 = 2\left(\cos\frac{0° + 360° \cdot 3}{6} + i\sin\frac{0° + 360° \cdot 3}{6}\right)$$

$$w_3 = 2(\cos 180° + i\sin 180°)$$

$$w_3 = 2(-1 + 0i)$$

$$w_3 = -2 + 0i = -2$$

$$w_4 = 2\left(\cos\frac{0° + 360° \cdot 4}{6} + i\sin\frac{0° + 360° \cdot 4}{6}\right)$$

$$w_4 = 2(\cos 240° + i\sin 240°)$$

$$w_4 = 2\left(-\frac{1}{2} - i\frac{\sqrt{3}}{2}\right)$$

$$w_4 = -1 - i\sqrt{3}$$

$$w_5 = 2\left(\cos\frac{0° + 360° \cdot 5}{6} + i\sin\frac{0° + 360° \cdot 5}{6}\right)$$

$$w_5 = 2(\cos 300° + i\sin 300°)$$

$$w_5 = 2\left(\frac{1}{2} - i\frac{\sqrt{3}}{2}\right)$$

$$w_5 = 1 - i\sqrt{3}$$

Copyright © Houghton Mifflin Company. All rights reserved.

19. $-1 = 1(\cos 180° + i\sin 180°)$

$$w_k = 1^{1/5}\left(\cos\frac{180° + 360°k}{5} + i\sin\frac{180° + 360°k}{5}\right) \quad k = 0, 1, 2, 3, 4$$

$w_0 = 1(\cos 36° + i\sin 36°)$
$w_0 \approx 0.809 + 0.588i$

$w_1 = \cos\dfrac{180° + 360°}{5} + i\sin\dfrac{180° + 360°}{5}$
$w_1 = \cos 108° + i\sin 108°$
$w_1 \approx -0.309 + 0.951i$

$w_2 = \cos\dfrac{180° + 360° \cdot 2}{5} + i\sin\dfrac{180° + 360° \cdot 2}{5}$
$w_2 = \cos 180° + i\sin 180°$
$w_2 = -1 + 0i = -1$

$w_3 = \cos\dfrac{180° + 360° \cdot 3}{5} + i\sin\dfrac{180° + 360° \cdot 3}{5}$
$w_3 = \cos 252° + i\sin 252°$
$w_3 \approx -0.309 - 0.951i$

$w_4 = \cos\dfrac{180° + 360° \cdot 4}{5} + i\sin\dfrac{180° + 360° \cdot 4}{5}$
$w_4 = \cos 324° + i\sin 324°$
$w_4 \approx 0.809 - 0.588i$

21. $1 = \cos 0° + i\sin 0°$

$$w_k = \cos\frac{0° + 360°k}{3} + i\sin\frac{0° + 360°k}{3} \quad k = 0, 1, 2$$

$w_0 = \cos\dfrac{0°}{3} + i\sin\dfrac{0°}{3}$
$w_0 = \cos 0° + i\sin 0°$
$w_0 = 1 + 0i = 1$

$w_1 = \cos\dfrac{0° + 360°}{3} + i\sin\dfrac{0° + 360°}{3}$
$w_1 = \cos 120° + i\sin 120°$
$w_1 = -\dfrac{1}{2} + \dfrac{\sqrt{3}}{2}i$

$w_2 = \cos\dfrac{0° + 360° \cdot 2}{3} + i\sin\dfrac{0° + 360° \cdot 2}{3}$
$w_2 = \cos 240° + i\sin 240°$
$w_2 = -\dfrac{1}{2} - \dfrac{\sqrt{3}}{2}i$

23. $1 + i = \sqrt{2}(\cos 45° + i\sin 45°)$

$$w_k = \left(\sqrt{2}\right)^{1/4}\left(\cos\frac{45° + 360°k}{4} + i\sin\frac{45° + 360°k}{4}\right) \quad k = 0, 1, 2, 3$$

$w_0 = 2^{1/8}\left(\cos\dfrac{45°}{4} + i\sin\dfrac{45°}{4}\right)$
$w_0 = 2^{1/8}(\cos 11.25° + i\sin 11.25°)$
$w_0 \approx 1.070 + 0.213i$

$w_1 = 2^{1/8}\left(\cos\dfrac{45° + 360°}{4} + i\sin\dfrac{45° + 360°}{4}\right)$
$w_1 = 2^{1/8}(\cos 101.25° = i\sin 101.25°)$
$w_1 \approx -0.213 - 1.070i$

$w_2 = 2^{1/8}\left(\cos\dfrac{45° + 360° \cdot 2}{4} + i\sin\dfrac{45° + 360° \cdot 2}{4}\right)$
$w_2 = 2^{1/8}(\cos 191.25° + i\sin 191.25°)$
$w_2 \approx -1.070 - 0.213i$

$w_3 = 2^{1/8}\left(\cos\dfrac{45° + 360° \cdot 3}{4} + i\sin\dfrac{45° + 360° \cdot 3}{4}\right)$
$w_3 = 2^{1/8}(\cos 281.25° + i\sin 281.25°)$
$w_3 \approx 0.213 - 1.070i$

Copyright © Houghton Mifflin Company. All rights reserved.

25. $\quad 2 - 2i\sqrt{3} = 4(\cos 300° + i\sin 300°) \quad k = 0, 1, 2$

$$w_k = 4^{1/3}\left(\cos\frac{300° + 360°k}{3} + i\sin\frac{300° + 360°k}{3}\right)$$

$$w_0 = 4^{1/3}\left(\cos\frac{300°}{3} + i\sin\frac{300°}{3}\right)$$

$w_0 = 4^{1/3}(\cos 100° + i\sin 100°)$

$w_0 \approx -0.276 + 1.563i$

$$w_1 = 4^{1/3}\left(\cos\frac{300° + 360°}{3} + i\sin\frac{300° + 360°}{3}\right)$$

$w_1 = 4^{1/3}(\cos 220° + i\sin 220°)$

$w_1 \approx -1.216 - 1.020i$

$$w_2 = 4^{1/3}\left(\cos\frac{300° + 360° \cdot 2}{3} + i\sin\frac{300° + 360° \cdot 2}{3}\right)$$

$w_2 = 4^{1/3}(\cos 340° + i\sin 340°)$

$w_2 \approx 1.492 - 0.543i$

27. $\quad -16 + 16i\sqrt{3} = 32(\cos 120° + i\sin 120°)$

$$w_k = 32^{1/2}\left(\cos\frac{120° + 360°k}{2} + i\sin\frac{120° + 360°k}{2}\right) \quad k = 0, 1$$

$$w_0 = 4\sqrt{2}\left(\cos\frac{120°}{2} + i\sin\frac{120°}{2}\right)$$

$w_0 = 4\sqrt{2}(\cos 60° + i\sin 60°)$

$w_0 \approx 2\sqrt{2} + 2i\sqrt{6}$

$$w_1 = 4\sqrt{2}\left(\cos\frac{120° + 360°}{2} + i\sin\frac{120° + 360°}{2}\right)$$

$w_1 = 4\sqrt{2}(\cos 240° + i\sin 240°)$

$w_1 \approx -2\sqrt{2} - 2i\sqrt{6}$

29. $\quad x^3 + 8 = 0$

$\qquad x^3 = -8$

Find the three cube roots of -8.

$-8 = 8(\cos 180° + i\sin 180°)$

$$x_k = 8^{1/3}\left(\cos\frac{180° + 360°k}{3} + i\sin\frac{180° + 360°k}{3}\right) \quad k = 0, 1, 2$$

$$w_0 = 2\left(\cos\frac{180°}{3} + i\sin\frac{180°}{3}\right)$$

$w_0 = 2(\cos 60° + i\sin 60°)$

$w_0 = 2 \text{ cis } 60°$

$$w_1 = 2\left(\cos\frac{180° + 360°}{3} + i\sin\frac{180° + 360°}{3}\right)$$

$w_1 = 2(\cos 180° + i\sin 180°)$

$w_1 = 2 \text{ cis } 180°$

$$w_2 = 2\left(\cos\frac{180° + 360° \cdot 2}{3} + i\sin\frac{180° + 360° \cdot 2}{3}\right)$$

$w_2 = 2(\cos 300° + i\sin 300°)$

$w_2 = 2 \text{ cis } 300°$

Copyright © Houghton Mifflin Company. All rights reserved.

31. $x^4 + i = 0$

$\qquad x^4 = -i$

Find the four fourth roots of $-i$.

$-i = (\cos 270° + i \sin 270°)$

$w_k = \cos \dfrac{270° + 360°k}{4} + i \sin \dfrac{270° + 360°k}{4} \quad k = 0, 1, 2, 3$

$w_0 = \text{cis} \dfrac{270°}{4} \qquad w_1 = \text{cis} \dfrac{270° + 360°}{4} \qquad\qquad w_2 = \text{cis} \dfrac{270° + 360° \cdot 2}{4} \qquad\qquad w_3 = \text{cis} \dfrac{270° + 360° \cdot 3}{4}$

$w_0 = \text{cis } 67.5° \qquad w_1 = \text{cis } 157.5° \qquad\qquad w_2 = \text{cis } 247.5° \qquad\qquad w_3 = \text{cis } 337.5°$

33. $x^3 - 27 = 0$

$\qquad x^3 = 27$

Find the three cube roots of 27.

$27 = 27(\cos 0° + i \sin 0°)$

$w_k = 3\left(\cos \dfrac{0° + 360°k}{3} + i \sin \dfrac{0° + 360°k}{3} \right) \quad k = 0, 1, 2$

$w_0 = 3 \text{ cis } \dfrac{0°}{3} \qquad\qquad w_1 = 3 \text{ cis } \dfrac{0° + 360°}{3} \qquad\qquad w_2 = 3 \text{ cis } \dfrac{0° + 360° \cdot 2}{3}$

$w_0 = 3 \text{ cis } 0° \qquad\qquad w_1 = 3 \text{ cis } 120° \qquad\qquad w_2 = 3 \text{ cis } 240°$

35. $x^4 + 81 = 0$

$\qquad x^4 = -81$

Find the four fourth roots of -81.

$-81 = 81(\cos 180° + i \sin 180°)$

$w_k = 81^{1/4}\left(\cos \dfrac{180° + 360°k}{4} + i \sin \dfrac{180° + 360°k}{4} \right) \quad k = 0, 1, 2, 3$

$w_0 = 3 \text{ cis } \dfrac{180°}{4} \qquad w_1 = 3 \text{ cis } \dfrac{180° + 360°}{4} \qquad\qquad w_2 = 2 \text{ cis } \dfrac{0° + 360° \cdot 2}{5} \qquad\qquad w_3 = 3 \text{ cis } \dfrac{180° + 360° \cdot 3}{4}$

$w_0 = 3 \text{ cis } 45° \qquad w_1 = 3 \text{ cis } 135° \qquad\qquad w_2 = 2 \text{ cis } 225° \qquad\qquad w_3 = 3 \text{ cis } 315°$

37. $x^4 - (1 - i\sqrt{3}) = 0$

$\qquad x^4 = 1 - i\sqrt{3}$

Find the four fourth roots of $1 - i\sqrt{3}$.

$1 - i\sqrt{3} = 2(\cos 300° + i \sin 300°)$

$w_k = 2^{1/4}\left(\cos \dfrac{300° + 360°k}{4} + i \sin \dfrac{300° + 360°k}{4} \right) \quad k = 0, 1, 2, 3$

$w_0 = \sqrt[4]{2} \text{ cis } \dfrac{300°}{4} \qquad w_1 = \sqrt[4]{2} \text{ cis } \dfrac{300° + 360°}{4} \qquad w_2 = \sqrt[4]{2} \text{ cis } \dfrac{300° + 360° \cdot 2}{4} \qquad w_3 = \sqrt[4]{2} \text{ cis } \dfrac{300° + 360° \cdot 3}{4}$

$w_0 = \sqrt[4]{2} \text{ cis } 75° \qquad w_1 = \sqrt[4]{2} \text{ cis } 165° \qquad w_2 = \sqrt[4]{2} \text{ cis } 255° \qquad w_3 = \sqrt[4]{2} \text{ cis } 345°$

Copyright © Houghton Mifflin Company. All rights reserved.

39. $x^3 + (1 + i\sqrt{3}) = 0$

$$x^3 = -1 - i\sqrt{3}$$

Find the three cube roots of $-1 - i\sqrt{3}$.

$$-1 - i\sqrt{3} = 2(\cos 240° + i \sin 240°)$$

$$w_k = 2^{1/3}\left(\cos\frac{240° + 360°k}{3} + i\sin\frac{240° + 360°k}{3}\right) \quad k = 0, 1, 2$$

$w_0 = \sqrt[3]{2}\ \text{cis}\ \dfrac{240°}{3}$ $\qquad$ $w_1 = \sqrt[3]{2}\ \text{cis}\ \dfrac{240° + 360°}{3}$ $\qquad$ $w_2 = \sqrt[3]{2}\ \text{cis}\ \dfrac{240° + 360° \cdot 2}{3}$

$w_0 = \sqrt[3]{2}\ \text{cis}\ 80°$ $\qquad\qquad$ $w_1 = \sqrt[3]{2}\ \text{cis}\ 200°$ $\qquad\qquad$ $w_2 = \sqrt[3]{2}\ \text{cis}\ 320°$

●●●●●●●●●●●●●●●●●●●●●●●●●●●●●●●●●●●●

Connecting Concepts

41.   Let $z = a + bi$. Then $\bar{z} = a - bi$ by definition.
Substitute $a = r\cos\theta$ and $b = r\sin\theta$.
Thus $\bar{z} = r\cos\theta - ri\sin\theta = r(\cos\theta - i\sin\theta)$

43. $z = r(\cos\theta + i\sin\theta)$

$z^2 = r^2(\cos 2\theta + i\sin 2\theta)$

$\dfrac{1}{z^2} = \dfrac{1}{r^2(\cos 2\theta + i\sin 2\theta)}$

$\qquad = \dfrac{\cos 2\theta - i\sin 2\theta}{r^2(\cos 2\theta + i\sin 2\theta)(\cos 2\theta - i\sin 2\theta)}$

$\qquad = \dfrac{\cos 2\theta - i\sin 2\theta}{r^2(\cos^2 2\theta - i^2 \sin^2 2\theta)} = \dfrac{\cos 2\theta - i\sin 2\theta}{r^2(\cos^2 2\theta + \sin^2 2\theta)}$

$z^{-2} = r^{-2}(\cos 2\theta - i\sin 2\theta)$

●●●●●●●●●●●●●●●●●●●●●●●●●●●●●●●●●●●●

Chapter 5 True/False Exercises

1.   False; $\sqrt{(-1)^2} = \sqrt{1} = 1$ $\qquad$ **2.** True $\qquad$ **3.** True $\qquad$ **4.** True

5.   False; $\dfrac{1+i}{1-i} = i$ $\qquad\qquad$ **6.** True $\qquad\qquad$ **7.** True

8.   False; $z^2 = r^2(\cos 2\theta + i\sin 2\theta)$ $\qquad$ **9.** True

10.   False; $\cos\pi + i\sin\pi = -1$ $\qquad$ **11.** False; $\cos\pi + i\sin\pi = -1$ $\qquad$ **12.** True

●●●●●●●●●●●●●●●●●●●●●●●●●●●●●●●●●●●●

Chapter Review

1. $3 - \sqrt{-64} = 3 - 8i$ [5.1]
$3 + 8i$

2. $\sqrt{-4} + 6 = 6 + 2i$ [5.1]
$6 - 2i$

3. $-2 + \sqrt{-5} = -2 + i\sqrt{5}$ [5.1]
$-2 - i\sqrt{5}$

4. $-5 + \sqrt{-27} = -5 + 3i\sqrt{3}$ [5.1]
$-5 - 3i\sqrt{3}$

5. $\left(\sqrt{-4}\right)\left(\sqrt{-4}\right) = (2i)(2i) = 4i^2 = -4$
[5.1]

6. $\left(-\sqrt{-27}\right)\left(\sqrt{-3}\right) = (-3i\sqrt{3})(i\sqrt{3})$ [5.1]
$\qquad\qquad = -9i^2 = 9$

7. $(3 + 7i) + (2 - 5i) = 5 + 2i$ [5.1]

8. $(3 - 4i) + (-6 + 8i) = -3 + 4i$ [5.1]

9. $(6 - 8i) - (9 - 11i) = -3 + 3i$ [5.1]

10. $(-3 - 5i) - (2 + 10i) = -5 - 15i$ [5.1]

11. $(5 + 3i)(2 - 5i) = 10 - 19i - 15i^2$ [5.1]
$\qquad\qquad = 25 - 19i$

12. $(-2 - 3i)(-4 + 7i) = 8 - 2i - 21i^2$ [5.1]
$\qquad\qquad = 29 - 2i$

Copyright © Houghton Mifflin Company. All rights reserved.

13. $\dfrac{-2i}{3-4i} \cdot \dfrac{3+4i}{3+4i} = \dfrac{-6i-8i^2}{9+16} = \dfrac{8}{25} - \dfrac{6}{25}i$ [5.1]

14. $\dfrac{4+i}{7-2i} \cdot \dfrac{7+2i}{7+2i} = \dfrac{28+15i+2i^2}{49+4} = \dfrac{26}{53} + \dfrac{15}{53}i$ [5.1]

15. $i(2i) - (1+i)^2 = 2i^2 - 1 - 2i - i^2$ [5.1]
$\qquad\qquad\qquad = -2 - 2i$

16. $(2-i)^3 = (4 - 4i + i^2)(2-i)$ [5.1]
$\qquad\quad = (3 - 4i)(2-i)$
$\qquad\quad = 6 - 11i + 4i^2$
$\qquad\quad = 2 - 11i$

17. $(3 + \sqrt{-4}) - (-3 - \sqrt{-16}) = 3 + 2i + 3 + 4i = 6 + 6i$ [5.1]

18. $(-2 + \sqrt{-9}) + (-3 - \sqrt{-81}) = -2 + 3i - 3 - 9i = -5 - 6i$ [5.1]

19. $(2 - \sqrt{-3})(2 + \sqrt{-3}) = 4 - (-3) = 7$ [5.1]

20. $(3 - \sqrt{-5})(2 + \sqrt{-5}) = 6 + i\sqrt{5} - (-5)$ [5.1]
$\qquad\qquad\qquad\qquad\qquad = 11 + i\sqrt{5}$

21. $i^{27} = i^3 = -i$ [5.1]   **22.** $i^{105} = i$ [5.1]   **23.** $\dfrac{i}{i^{17}} = \dfrac{1}{i^{16}} = \dfrac{1}{1} = 1$ [5.1]   **24.** $i^{62} = i^2 = -1$ [5.1]

25. $|-8i| = \sqrt{(0)^2 + (-8)^2}$ [5.2]
$\qquad\quad = 8$

26. $|2 - 3i| = \sqrt{(2)^2 + (-3)^2}$ [5.2]
$\qquad\qquad = \sqrt{4 + 9} = \sqrt{13}$

27. $|-4 + 5i| = \sqrt{(-4)^2 + (5)^2}$ [5.2]
$\qquad\qquad\quad = \sqrt{16 + 25} = \sqrt{41}$

28. $|-1 - i| = \sqrt{(-1)^2 + (-1)^2}$ [5.2]
$\qquad\qquad = \sqrt{1 + 1} = \sqrt{2}$

29. $r = \sqrt{(2)^2 + (-2)^2}$ [5.2]
$r = 2\sqrt{2}$

$\alpha = \tan^{-1}\left|\dfrac{-2}{2}\right|$

$\quad = \tan^{-1} 1 = 45°$

$\alpha = 360° - 45° = 315°$

$z = 2\sqrt{2} \text{ cis } 315°$

28. $|-1 - i| = \sqrt{(-1)^2 + (-1)^2}$ [5.2]
$\qquad\qquad = \sqrt{1 + 1} = \sqrt{2}$

29. $r = \sqrt{(2)^2 + (-2)^2}$ [5.2]
$r = 2\sqrt{2}$

$\alpha = \tan^{-1}\left|\dfrac{-2}{2}\right|$

$\quad = \tan^{-1} 1 = 45°$

$\alpha = 360° - 45° = 315°$

$z = 2\sqrt{2} \text{ cis } 315°$

30. $r = \sqrt{\left(-\sqrt{3}\right)^2 + (1)^2}$ [5.2]
$r = 2$

$\alpha = \tan^{-1}\left|\dfrac{1}{\sqrt{3}}\right|$

$\quad = \tan^{-1}\dfrac{1}{\sqrt{3}} = 30°$

$\alpha = 180° - 30° = 150°$

$z = 2 \text{ cis } 150°$

31. $r = \sqrt{(-3)^2 + (2)^2}$ [5.2]
$r = \sqrt{13}$

$\alpha = \tan^{-1}\left|\dfrac{2}{-3}\right|$

$\quad = \tan^{-1}\dfrac{2}{3} = 33.7°$

$\alpha = 180° - 33.7° = 146.3°$

$z = \sqrt{13} \text{ cis } 146.3°$

32. $r = \sqrt{(4)^2 + (-1)^2}$ [5.2]
$r = \sqrt{17}$

$\alpha = \tan^{-1}\left|\dfrac{-1}{4}\right|$

$\quad = \tan^{-1}\dfrac{1}{4} = 14.04°$

$\alpha = 360° - 14.04° = 345.96°$

$z = \sqrt{17} \text{ cis } 345.96°$

33. $z = 5(\cos 315° + i \sin 315°)$ [5.2]

$z = 5\left(\dfrac{\sqrt{2}}{2} - \dfrac{\sqrt{2}}{2}i\right)$

$z = \dfrac{5\sqrt{2}}{2} - \dfrac{5\sqrt{2}}{2}i$

$\approx 3.536 - 3.536i$

34. $z = 6\left(\cos\dfrac{4\pi}{3} + i \sin\dfrac{4\pi}{3}\right)$ [5.2]

$z = 6\left(-\dfrac{1}{2} - \dfrac{\sqrt{3}}{2}i\right)$

$z = -3 - 3\sqrt{3}i$

Copyright © Houghton Mifflin Company. All rights reserved.

35. $z = 2(\cos 2 + i \sin 2)$ [5.2]

$z \approx 2(-0.4161 + 0.9093i)$

$z \approx -0.832 + 1.819i$

36. $z = 3(\cos 115° + i \sin 115°)$ [5.2]

$z \approx 3(-0.4226 + 0.9063i)$

$z \approx -1.27 + 2.72i$

37. $z_1z_2 = 3(\cos 225° + i \sin 225°) \cdot 10(\cos 45° + i \sin 45°)$ [5.2]

$z_1z_2 = 30[\cos(225° + 45°) + i \sin(225° + 45°)]$

$z_1z_2 = 30(\cos 270° + i \sin 270°)$

$z_1z_2 = 30(0 - 1i)$

$z_1z_2 = -30i$

38. $z_1z_2 = 5(\cos 162° + i \sin 162°) \cdot 2(\cos 63° + i \sin 63°)$ [5.2]

$z_1z_2 = 10[\cos(162° + 63°) + i \sin(162° + 63°)]$

$z_1z_2 = 10(\cos 225° + i \sin 225°)$

$z_1z_2 = 10\left(-\dfrac{\sqrt{2}}{2} - \dfrac{\sqrt{2}}{2}i\right)$

$z_1z_2 = -5\sqrt{2} - 5i\sqrt{2}$

39. $z_1z_2 = 3(\cos 12° + i \sin 12°) \cdot 4(\cos 126° + i \sin 126°)$ [5.2]

$z_1z_2 = 12[\cos(12° + 126°) + i \sin(12° + 126°)]$

$z_1z_2 = 12(\cos 138° + i \sin 138°)$

$z_1z_2 \approx 12(-0.74314 + 0.66913i)$

$z_1z_2 \approx -8.918 + 8.030i$

40. $z_1z_2 = (\cos 23° + i \sin 23°) \cdot 4(\cos 233° + i \sin 233°)$ [5.2]

$z_1z_2 = 4[\cos(23° + 233°) + i \sin(23° + 233°)]$

$z_1z_2 = 4(\cos 256° + i \sin 256°)$

$z_1z_2 \approx 4(-0.24192 - 0.97030i)$

$z_1z_2 \approx -0.968 - 3.881i$

41. $z_1z_2 = 3(\cos 1.8 + i \sin 1.8) \cdot 5(\cos 2.5 + i \sin 2.5)$ [5.2]

$z_1z_2 = 15[\cos(1.8 + 2.5) + i \sin(1.8 + 2.5)]$

$z_1z_2 = 15(\cos 4.3 + i \sin 4.3)$

$z_1z_2 \approx 15(-0.4008 - 0.9162i)$

$z_1z_2 \approx -6.012 - 13.743i$

42. $z_1z_2 = 6(\cos 3.1 + i \sin 1.8) \cdot 5(\cos 4.3 + i \sin 4.3)$ [5.2]

$z_1z_2 = 30[\cos(3.1 + 4.3) + i \sin(3.1 + 4.3)]$

$z_1z_2 = 30(\cos 7.4 + i \sin 7.4)$

$z_1z_2 \approx 30(0.439 + 0.899i)$

$z_1z_2 \approx 13.2 + 27.0i$

43. $\dfrac{z_1}{z_2} = \dfrac{6(\cos 50° + i \sin 50°)}{2(\cos 150° + i \sin 150°)}$ [5.2]

$\dfrac{z_1}{z_2} = 3\,[\cos(50° - 150°) + i \sin(50° - 150°)]$

$\dfrac{z_1}{z_2} = 3(\cos -100° + i \sin -100°) = 3\text{cis}\,(-100°)$

44. $\dfrac{z_1}{z_2} = \dfrac{30(\cos 165° + i \sin 165°)}{10(\cos 55° + i \sin 55°)}$ [5.2]

$\dfrac{z_1}{z_2} = 3\,[\cos(165° - 55°) + i \sin(165° - 55°)]$

$\dfrac{z_1}{z_2} = 3(\cos 110° + i \sin 110°) = 3\text{cis}\,(110°)$

45. $\dfrac{z_1}{z_2} = \dfrac{40(\cos 66° + i \sin 66°)}{8(\cos 125° + i \sin 125°)}$ [5.2]

$\dfrac{z_1}{z_2} = 5\,[\cos(66° - 125°) + i \sin(66° - 125°)]$

$\dfrac{z_1}{z_2} = 5(\cos -59° + i \sin -59°) = 5\text{cis}\,(-59°)$

46. $\dfrac{z_1}{z_2} = \dfrac{2(\cos 150° + i \sin 150°)}{\sqrt{2}(\cos 200° + i \sin 200°)}$ [5.2]

$\dfrac{z_1}{z_2} = \sqrt{2}\,[\cos(150° - 200°) + i \sin(150° - 200°)]$

$\dfrac{z_1}{z_2} = \sqrt{2}(\cos -50° + i \sin -50°) = \sqrt{2}\text{cis}\,(-50°)$

47. $\dfrac{z_1}{z_2} = \dfrac{10(\cos 3.7 + i \sin 3.7)}{6(\cos 1.8 + i \sin 1.8)}$ [5.2]

$\dfrac{z_1}{z_2} = \dfrac{5}{3}\,[\cos(3.7 - 1.8) + i \sin(3.7 - 1.8)]$

$\dfrac{z_1}{z_2} = \dfrac{5}{3}(\cos 1.9 + i \sin 1.9) = \dfrac{5}{3}\text{cis}\,(1.9)$

48. $\dfrac{z_1}{z_2} = \dfrac{4(\cos 1.2 + i \sin 1.2)}{8(\cos 5.2 + i \sin 5.2)}$ [5.2]

$\dfrac{z_1}{z_2} = \dfrac{1}{2}\,[\cos(1.2 - 5.2) + i \sin(1.2 - 5.2)]$

$\dfrac{z_1}{z_2} = \dfrac{1}{2}(\cos -4 + i \sin -4) = \dfrac{1}{2}\text{cis}\,(-4)$

49. $[3(\cos 45° + i \sin 45°)]^5 = 3^5[\cos(5 \cdot 45°) + i \sin(5 \cdot 45°)]$ [5.3]

$= 243[\cos 225° + i \sin 225°]$

$= 243\left(-\dfrac{\sqrt{2}}{2} - \dfrac{\sqrt{2}}{2}i\right)$

$= -\dfrac{243\sqrt{2}}{2} - \dfrac{243\sqrt{2}}{2}i$

$\approx -171.827 - 171.827i$

50. $\left[\cos\dfrac{11\pi}{8} + i \sin\dfrac{11\pi}{8}\right]^8 = \cos\left(8 \cdot \dfrac{11\pi}{8}\right) + i \sin\left(8 \cdot \dfrac{11\pi}{8}\right)$ [5.3]

$= \cos(11\pi) + i \sin(11\pi)$

$= \cos \pi + i \sin \pi$

$= -1 + 0i$

$= -1$

Copyright © Houghton Mifflin Company. All rights reserved.

51. $z = 1 - i\sqrt{3}$ [5.3]

$r = \sqrt{1^2 + (-\sqrt{3})^2}$ $\qquad$ $\alpha = \tan^{-1}\left|\dfrac{-\sqrt{3}}{1}\right| = 60°$

$r = 2$

$\qquad\qquad\qquad$ $\theta = 300°$

$z = 2(\cos 300° + i\sin 300°)$

$(1 - i\sqrt{3})^7 = [2(\cos 300° + i\sin 300°)]^7$

$\qquad = 128[\cos(7 \cdot 300°) + i\sin(7 \cdot 300°)]$

$\qquad = 128(\cos 2100° + i\sin 2100°)$

$\qquad = 128(\cos 300° + i\sin 300°)$

$\qquad = 64 - 64i\sqrt{3}$

$\qquad \approx 64 - 110.851i$

52. $z = -2 - 2i$ [5.3]

$r = \sqrt{(-2)^2 + (-2)^2}$ $\qquad$ $\alpha = \tan^{-1}\left|\dfrac{-2}{-2}\right| = 45°$

$r = 2\sqrt{2}$

$\qquad\qquad\qquad$ $\theta = 225°$

$z = 2\sqrt{2}(\cos 225° + i\sin 225°)$

$(-2 - 2i)^{10} = [2\sqrt{2}(\cos 225° + i\sin 225°)]^{10}$

$\qquad = 32,768[\cos(10 \cdot 225°) + i\sin(10 \cdot 225°)]$

$\qquad = 32,768(\cos 2250° + i\sin 2250°)$

$\qquad = 32,768(\cos 90° + i\sin 90°)$

$\qquad = 0 + 32,768i$

53. $z = \sqrt{2} - i\sqrt{2}$ [5.3]

$r = \sqrt{(\sqrt{2})^2 + (-\sqrt{2})^2}$ $\qquad$ $\alpha = \tan^{-1}\left|\dfrac{-\sqrt{2}}{\sqrt{2}}\right| = 45°$

$r = 2$

$\qquad\qquad\qquad$ $\theta = 315°$

$z = 2(\cos 315° + i\sin 315°)$

$(\sqrt{2} - i\sqrt{2})^5 = [2(\cos 315° + i\sin 315°)]^5$

$\qquad = 32[\cos(5 \cdot 315°) + i\sin(5 \cdot 315°)]$

$\qquad = 32(\cos 1575° + i\sin 1575°)$

$\qquad = 32(\cos 135° + i\sin 135°)$

$\qquad = -16\sqrt{2} + 16i\sqrt{2}$

$\qquad \approx -22.627 + 22.627i$

54. $z = 3 - 4i$ [5.3]

$r = \sqrt{3^2 + (-4)^2}$ $\qquad$ $\alpha = \tan^{-1}\left|\dfrac{-4}{3}\right| = 53.13°$

$r = 5$

$\qquad\qquad\qquad$ $\theta = 306.87°$

$z = 5(\cos 306.87° + i\sin 306.87°)$

$(3 - 4i)^5 = [5(\cos 306.87° + i\sin 306.87°)]^5$

$\qquad = 3125[\cos(5 \cdot 306.87°) + i\sin(5 \cdot 306.87°)]$

$\qquad = 3125(\cos 1534.35° + i\sin 1534.35°)$

$\qquad = 3125(\cos 94.35° + i\sin 94.35°)$

$\qquad = -237 + 3116i$

55. $27i = 27\left(\cos 90° + i\sin 90°\right)$ [5.3]

$w_k = 27^{1/3}\left(\cos\dfrac{90° + 360°k}{3} + i\sin\dfrac{90° + 360°k}{3}\right)$ $\quad$ $k = 0, 1, 2$

$w_0 = 3\left(\cos\dfrac{90°}{3} + i\sin\dfrac{90°}{3}\right)$ $\qquad$ $w_1 = 3\left(\cos\dfrac{90° + 360°}{3} + i\sin\dfrac{90° + 360°}{3}\right)$ $\qquad$ $w_2 = 3\left(\cos\dfrac{90° + 360° \cdot 2}{3} + i\sin\dfrac{90° + 360° \cdot 2}{3}\right)$

$w_0 = 3(\cos 30° + i\sin 30°)$ $\qquad\qquad$ $w_1 = 3(\cos 150° + i\sin 150°)$ $\qquad\qquad$ $w_2 = 3(\cos 270° + i\sin 270°)$

$w_0 = 3 \text{ cis } 30°$ $\qquad\qquad\qquad$ $w_1 = 3 \text{ cis } 150°$ $\qquad\qquad\qquad$ $w_2 = 3 \text{ cis } 270°$

56. $8i = 8\left(\cos 90° + i\sin 90°\right)$ [5.3]

$w_k = 8^{1/4}\left(\cos\dfrac{90° + 360°k}{4} + i\sin\dfrac{9° + 360°k}{4}\right)$ $\quad$ $k = 0, 1, 2, 3$

$w_0 = \sqrt[4]{8}\left(\cos\dfrac{90°}{4} + i\sin\dfrac{90°}{4}\right)$ $\qquad\qquad\qquad$ $w_1 = \sqrt[4]{8}\left(\cos\dfrac{90° + 360°}{4} + i\sin\dfrac{90° + 360°}{4}\right)$

$w_0 = \sqrt[4]{8}(\cos 22.5° + i\sin 22.5°)$ $\qquad\qquad\qquad$ $w_1 = \sqrt[4]{8}(\cos 112.5° + i\sin 112.5°)$

$w_0 = \sqrt[4]{8} \text{ cis } 22.5°$ $\qquad\qquad\qquad\qquad\qquad$ $w_1 = \sqrt[4]{8} \text{ cis } 112.5°$

$w_2 = \sqrt[4]{8}\left(\cos\dfrac{90° + 360° \cdot 2}{4} + i\sin\dfrac{90° + 360° \cdot 2}{4}\right)$ $\qquad$ $w_3 = \sqrt[4]{8}\left(\cos\dfrac{90° + 360° \cdot 3}{4} + i\sin\dfrac{90° + 360° \cdot 3}{4}\right)$

$w_2 = \sqrt[4]{8}(\cos 202.5° + i\sin 202.5°)$ $\qquad\qquad\qquad$ $w_3 = \sqrt[4]{8}(\cos 292.5° + i\sin 292.5°)$

$w_2 = \sqrt[4]{8} \text{ cis } 202.5°$ $\qquad\qquad\qquad\qquad\qquad$ $w_3 = \sqrt[4]{8} \text{ cis } 292.5°$

Copyright © Houghton Mifflin Company. All rights reserved.

57. $256 = 256\left(\cos 0° + i\sin 0°\right)$ [5.3]

$$w_k = 81^{1/4}\left(\cos\frac{0° + 360°k}{4} + i\sin\frac{0° + 360°k}{4}\right) \quad k = 0, 1, 2, 3$$

$w_0 = 3(\cos 0° + i\sin 0°)$

$w_0 = 3 \text{ cis } 0°$

$w_1 = 3\left(\cos\frac{0° + 360°}{4} + i\sin\frac{0° + 360°}{4}\right)$

$w_1 = 3(\cos 90° + i\sin 90°)$

$w_1 = 3 \text{ cis } 90°$

$w_2 = 3\left(\cos\frac{0° + 360° \cdot 2}{4} + i\sin\frac{0° + 360° \cdot 2}{4}\right)$

$w_2 = 3(\cos 180° + i\sin 180°)$

$w_2 = 3 \text{ cis } 180°$

$w_3 = 3\left(\cos\frac{0° + 360° \cdot 3}{4} + i\sin\frac{0° + 360° \cdot 3}{4}\right)$

$w_3 = 3(\cos 270° + i\sin 270°)$

$w_3 = 3 \text{ cis } 270°$

58. $-16\sqrt{2} - 16i\sqrt{2} = 32(\cos 225° + i\sin 225°)$ [5.3]

$$w_k = 32^{1/5}\left(\cos\frac{225° + 360°k}{5} + i\sin\frac{225° + 360°k}{5}\right) \quad k = 0, 1, 2, 3, 4$$

$w_0 = 2\left(\cos\frac{225°}{5} + i\sin\frac{225°}{5}\right)$

$w_0 = 2(\cos 45° + i\sin 45°)$

$w_0 = 2 \text{ cis } 45°$

$w_1 = 2\left(\cos\frac{225° + 360°}{5} + i\sin\frac{225° + 360°}{5}\right)$

$w_1 = 2(\cos 117° + i\sin 117°)$

$w_1 = 2 \text{ cis } 117°$

$w_2 = 2\left(\cos\frac{225° + 360° \cdot 2}{5} + i\sin\frac{225° + 360° \cdot 2}{5}\right)$

$w_2 = 2(\cos 189° + i\sin 189°)$

$w_2 = 2 \text{ cis } 189°$

$w_3 = 2\left(\cos\frac{225° + 360° \cdot 3}{5} + i\sin\frac{225° + 360° \cdot 3}{5}\right)$

$w_3 = 2(\cos 261° + i\sin 261°)$

$w_3 = 2 \text{ cis } 261°$

$w_4 = 2\left(\cos\frac{225° + 360° \cdot 4}{5} + i\sin\frac{225° + 360° \cdot 4}{5}\right)$

$w_4 = 2(\cos 333° + i\sin 333°)$

$w_4 = 2 \text{ cis } 333°$

59. $81 = 81\left(\cos 0° + i\sin 0°\right)$ [5.3]

$$w_k = 256^{1/4}\left(\cos\frac{0° + 360°k}{4} + i\sin\frac{0° + 360°k}{4}\right) \quad k = 0, 1, 2, 3$$

$w_0 = 4(\cos 0° + i\sin 0°)$

$w_0 = 4 \text{ cis } 0°$

$w_1 = 4\left(\cos\frac{0° + 360°}{4} + i\sin\frac{0° + 360°}{4}\right)$

$w_1 = 4(\cos 90° + i\sin 90°)$

$w_1 = 4 \text{ cis } 90°$

$w_2 = 4\left(\cos\frac{0° + 360° \cdot 2}{4} + i\sin\frac{0° + 360° \cdot 2}{4}\right)$

$w_2 = 4(\cos 180° + i\sin 180°)$

$w_2 = 4 \text{ cis } 180°$

$w_3 = 4\left(\cos\frac{0° + 360° \cdot 3}{4} + i\sin\frac{0° + 360° \cdot 3}{4}\right)$

$w_3 = 4(\cos 270° + i\sin 270°)$

$w_3 = 4 \text{ cis } 270°$

60. $-125 = 125\left(\cos 180° + i\sin 180°\right)$ [5.3]

$$w_k = 125^{1/3}\left(\cos\frac{180° + 360°k}{3} + i\sin\frac{180° + 360°k}{3}\right) \quad k = 0, 1, 2$$

$w_0 = 5\left(\cos\frac{180°}{3} + i\sin\frac{180°}{3}\right)$

$w_0 = 5(\cos 60° + i\sin 60°)$

$w_0 = 5 \text{ cis } 60°$

$w_1 = 5\left(\cos\frac{180° + 360°}{3} + i\sin\frac{180° + 360°}{3}\right)$

$w_1 = 5(\cos 180° + i\sin 180°)$

$w_1 = 5 \text{ cis } 180° = -5$

$w_2 = 5\left(\cos\frac{180° + 360° \cdot 2}{3} + i\sin\frac{180° + 360° \cdot 2}{3}\right)$

$w_2 = 5(\cos 300° + i\sin 300°)$

$w_2 = 5 \text{ cis } 300°$

Copyright © Houghton Mifflin Company. All rights reserved.

•••••••••••••••••••••••••••••••••••••••

1. $6 + \sqrt{-9} = 6 + 3i$ [5.1]

2. $\sqrt{-18} = 3i\sqrt{2}$ [5.1]

3. $(3 + \sqrt{-4}) + (7 - \sqrt{-9}) = 3 + 2i + 7 - 3i$
$$= 10 - i$$
[5.1]

4. $(-1 + \sqrt{-25}) - (8 - \sqrt{-16}) = -1 + 5i - 8 + 4i$ [5.1]
$$= -9 + 9i$$

5. $(\sqrt{-12})(\sqrt{-3}) = (2i\sqrt{3})(i\sqrt{3})$ [5.1]
$$= 6i^2 = -6$$

6. $i^{263} = i^3 = -i$ [5.1]

7. $(3 + 7i) - (-2 - 9i) = 5 + 16i$ [5.1]

8. $(-6 - 9i)(4 + 3i) = -24 - 54i - 27i^2$
$$= 3 - 54i$$
[5.1]

9. $(3 - 5i)(-3 + 5i) = -9 + 30i - 25i^2$ [5.1]
$$= 16 + 30i$$

10. $\dfrac{4 - 5i}{i} \cdot \dfrac{-i}{-i} = \dfrac{-4i + 5i^2}{-i^2} = -5 - 4i$ [5.1]

11. $\dfrac{2 - 7i}{4 + 3i} \cdot \dfrac{4 - 3i}{4 - 3i} = \dfrac{8 - 34i + 21i^2}{16 + 9} = -\dfrac{13}{25} - \dfrac{34}{25}i$ [5.1]

12. $\dfrac{6 + 2i}{1 - i} \cdot \dfrac{1 + i}{1 + i} = \dfrac{6 + 8i + 2i^2}{1 + 1} = \dfrac{4}{2} + \dfrac{8}{2}i = 2 + 4i$ [5.1]

13. $|z| = \sqrt{(3)^2 + (-5)^2}$ [5.2]
$$= \sqrt{9 + 25} = \sqrt{34}$$

14. $r = \sqrt{(3)^2 + (-3)^2}$ [5.2]
$$r = 3\sqrt{2}$$
$$\alpha = \tan^{-1}\left|\dfrac{-3}{3}\right|$$
$$= \tan^{-1} 1 = 45°$$
$$\alpha = 360° - 45° = 315°$$
$$z = 3\sqrt{2} \text{ cis } 315°$$

15. $r = \sqrt{(0)^2 + (-6)^2}$ [5.2]
$$r = 6$$
$$\alpha = \tan^{-1}\left|\dfrac{-6}{0}\right| = 90°$$
$$\alpha = 360° - 90° = 270°$$
$$z = 6 \text{ cis } 270°$$

16. $z = 4(\cos 120° + i \sin 120°)$ [5.2]
$$z = 4\left(-\dfrac{1}{2} + \dfrac{\sqrt{3}}{2}i\right)$$
$$z = -2 + 2i\sqrt{3}$$

17. $z = 5(\cos 225° + i \sin 225°)$ [5.2]
$$z = 5\left(-\dfrac{\sqrt{2}}{2} - \dfrac{\sqrt{2}}{2}i\right)$$
$$z = -\dfrac{5\sqrt{2}}{2} - \dfrac{5\sqrt{2}}{2}i$$

18. $z_1 z_2 = 3(\cos 28° + i \sin 28°) \cdot 4(\cos 17° + i \sin 17°)$ [5.2]
$z_1 z_2 = 12[\cos(28° + 17°) + i \sin(28° + 17°)]$
$z_1 z_2 = 12(\cos 45° + i \sin 45°)$
$z_1 z_2 = 12\left(\dfrac{\sqrt{2}}{2} + \dfrac{\sqrt{2}}{2}i\right)$
$z_1 z_2 = 6\sqrt{2} + 6i\sqrt{2}$

19. $z_1 z_2 = 5(\cos 115° + i \sin 115°) \cdot 4(\cos 10° + i \sin 10°)$ [5.2]
$z_1 z_2 = 20[\cos(115° + 10°) + i \sin(115° + 10°)]$
$z_1 z_2 = 20(\cos 125° + i \sin 125°)$
$z_1 z_2 \approx 20(-0.5736 + 0.81915i)$
$z_1 z_2 \approx -11.472 + 16.383i$

20. $\dfrac{z_1}{z_2} = \dfrac{24(\cos 258° + i \sin 258°)}{6(\cos 78° + i \sin 78°)}$ [5.2]
$$\dfrac{z_1}{z_2} = 4\,[\cos(258° - 78°) + i \sin(258° - 78°)]$$
$$\dfrac{z_1}{z_2} = 4(\cos 180° + i \sin 180°) = -4 + 0i = -4$$

21. $\dfrac{z_1}{z_2} = \dfrac{18(\cos 50° + i \sin 50°)}{3(\cos 140° + i \sin 140°)}$ [5.2]
$$\dfrac{z_1}{z_2} = 6\,[\cos(50° - 140°) + i \sin(50° - 140°)]$$
$$\dfrac{z_1}{z_2} = 6(\cos -90° + i \sin -90°) = 0 - 6i = -6i$$

Copyright © Houghton Mifflin Company. All rights reserved.

22. $z = 2 - 2i\sqrt{3}$ [5.3]

$r = \sqrt{2^2 + \left(-2\sqrt{3}\right)^2}$ $\qquad$ $\alpha = \tan^{-1}\left|\dfrac{-2\sqrt{3}}{2}\right| = 60°$

$r = 4$

$\theta = 300°$

$z = 4(\cos 300° + i\sin 300°)$

$(2 - 3i\sqrt{3})^{12} = [4(\cos 300° + i\sin 300°)]^{12}$

$\qquad = 16,777,216[\cos(12 \cdot 300°) + i\sin(10 \cdot 300°)]$

$\qquad = 16,777,216(\cos 3600° + i\sin 3600°)$

$\qquad = 16,777,216(\cos 0° + i\sin 0°)$

$\qquad = 16,777,216 + 0i$

23. $64 = 64(\cos 0° + i\sin 0°)$ [5.3]

$w_k = 64^{1/6}\left(\cos\dfrac{0° + 360°k}{6} + i\sin\dfrac{0° + 360°k}{6}\right) \qquad k = 0, 1, 2, 3, 4, 5$

$w_0 = 2(\cos 0° + i\sin 0°)$ $\qquad\qquad\qquad$ $w_1 = 2\left(\cos\dfrac{0 + 360°}{6} + i\sin\dfrac{0 + 360°}{6}\right)$

$w_0 = 2$ cis $0°$ $\qquad\qquad\qquad\qquad\qquad$ $w_1 = 2(\cos 60° + i\sin 60°)$

$\qquad\qquad\qquad\qquad\qquad\qquad\qquad\qquad$ $w_1 = 2$ cis $60°$

$w_2 = 2\left(\cos\dfrac{0° + 360° \cdot 2}{6} + i\sin\dfrac{0° + 360° \cdot 2}{6}\right)$ $\qquad$ $w_3 = 2\left(\cos\dfrac{0° + 360° \cdot 3}{6} + i\sin\dfrac{0° + 360° \cdot 3}{6}\right)$

$w_2 = 2(\cos 120° + i\sin 120°)$ $\qquad\qquad\qquad\qquad$ $w_3 = 2(\cos 180° + i\sin 180°)$

$w_2 = 2$ cis $120°$ $\qquad\qquad\qquad\qquad\qquad\qquad$ $w_3 = 2$ cis $180°$

$w_4 = 2\left(\cos\dfrac{0° + 360° \cdot 4}{6} + i\sin\dfrac{0° + 360° \cdot 4}{6}\right)$ $\qquad$ $w_5 = 2\left(\cos\dfrac{0° + 360° \cdot 5}{6} + i\sin\dfrac{0° + 360° \cdot 5}{6}\right)$

$w_4 = 2(\cos 240° + i\sin 240°)$ $\qquad\qquad\qquad\qquad$ $w_5 = 2(\cos 300° + i\sin 300°)$

$w_4 = 2$ cis $240°$ $\qquad\qquad\qquad\qquad\qquad\qquad$ $w_5 = 2$ cis $300°$

24. $-1 + \sqrt{3}i = 2(\cos 120° + i\sin 120°)$ [5.3]

$w_k = 2^{1/3}\left(\cos\dfrac{120° + 360°k}{3} + i\sin\dfrac{120° + 360°k}{3}\right) \qquad k = 0, 1, 2$

$w_0 = \sqrt[3]{2}\left(\cos\dfrac{120°}{3} + i\sin\dfrac{120°}{3}\right)$ $\quad$ $w_1 = \sqrt[3]{2}\left(\cos\dfrac{120° + 360°}{3} + i\sin\dfrac{120° + 360°}{3}\right)$ $\quad$ $w_2 = \sqrt[3]{2}\left(\cos\dfrac{120° + 360° \cdot 2}{3} + i\sin\dfrac{120° + 360° \cdot 2}{3}\right)$

$w_0 = \sqrt[3]{2}(\cos 40° + i\sin 40°)$ $\qquad$ $w_1 = \sqrt[3]{2}(\cos 160° + i\sin 160°)$ $\qquad$ $w_2 = \sqrt[3]{2}(\cos 280° + i\sin 280°)$

$w_0 = \sqrt[3]{2}$ cis $40°$ $\qquad\qquad\quad$ $w_1 = \sqrt[3]{2}$ cis $160°$ $\qquad\qquad\quad$ $w_2 = \sqrt[3]{2}$ cis $280°$

25. $x^5 + 32 = 0$ [5.3]

$\qquad x^5 = -32$

Find the five fifth roots of -32.

$-32 = 32(\cos 180° + i\sin 180°)$

$w_k = 32^{1/5}\left(\cos\dfrac{180° + 360°k}{5} + i\sin\dfrac{180° + 360°k}{5}\right) \qquad k = 0, 1, 2, 3, 4$

$w_0 = 2$ cis $\dfrac{180°}{5}$ $\qquad\qquad$ $w_1 = 2$ cis $\dfrac{180° + 360°}{5}$ $\qquad\qquad$ $w_2 = 2$ cis $\dfrac{180° + 360° \cdot 2}{5}$

$w_0 = 2$ cis $36°$ $\qquad\qquad\quad$ $w_1 = 2$ cis $108°$ $\qquad\qquad\qquad$ $w_2 = 2$ cis $180°$

$w_3 = 2$ cis $\dfrac{180° + 360° \cdot 3}{5}$ $\qquad$ $w_4 = 2$ cis $\dfrac{180° + 360° \cdot 4}{5}$

$w_3 = 2$ cis $252°$ $\qquad\qquad\qquad$ $w_4 = 2$ cis $324°$

Copyright © Houghton Mifflin Company. All rights reserved.

1. $x^2 - x - 6 \le 0$ [1.1

$(x-3)(x+2) \le 0$

The product is negative or zero.
The critical values are –2 and 3.

$(x-3)(x+2)$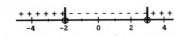

[–2, 3]

2. $x^2 - 4 = 0$

$(x-2)(x+2) = 0$

$x = -2,\ 2$

Domain: all real numbers except –2 and 2.

3. $f(c) = 2$ [1.3]

$2 = \dfrac{c}{c+1}$

$2(c+1) = c$

$2c + 2 = c$

$c = -2$

4. $(f \circ g)(x) = f[g(x)]$ [1.5]

$= f\left[\dfrac{x^2 - 1}{3}\right]$

$= \sin\left(3 \cdot \dfrac{x^2 - 1}{3}\right)$

$= \sin(x^2 - 1)$

5. $f(x) = \dfrac{x}{x-1}$ [1.6]

$y = \dfrac{x}{x-1}$

$x = \dfrac{y}{y-1}$

$x(y-1) = y$

$xy - x = y$

$xy - y = x$

$y(x-1) = x$

$y = \dfrac{x}{x-1}$

$f^{-1}(x) = \dfrac{x}{x-1}$

$f^{-1}(3) = \dfrac{3}{3-1} = \dfrac{3}{2}$

6. $\dfrac{3\pi}{2}\left(\dfrac{180°}{\pi}\right) = 270°$ [2.1]

7. $\cos 38° = \dfrac{20}{c}$ [2.2]

$c = \dfrac{20}{\cos 38°} \approx 25.4 \text{ cm}$

8. $-1 \le \sin t \le 1$ [2.4]

$a = -1,\ b = 1$

9. $y = 3\sin \pi x$ [2.5]

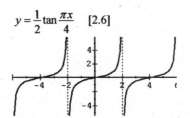

10. $y = \dfrac{1}{2}\tan \dfrac{\pi x}{4}$ [2.6]

11. $\dfrac{\cos x}{1 + \sin x} = \dfrac{\cos x}{1 + \sin x} \cdot \dfrac{1 - \sin x}{1 - \sin x}$ [3.1]

$= \dfrac{\cos x - \sin x \cos x}{1 - \sin^2 x}$

$= \dfrac{\cos x - \sin x \cos x}{\cos^2 x}$

$= \dfrac{1}{\cos x} - \dfrac{\sin x}{\cos x}$

$= \sec x - \tan x$

12. $\sin 2x \cos 3x - \cos 2x \sin 3x = \sin(2x - 3x)$ [3.2]

$= \sin(-x) \text{ or } -\sin x$

Copyright © Houghton Mifflin Company. All rights reserved.

13. $\sin\alpha = \dfrac{4}{5}$, $\cos\alpha = \dfrac{3}{5}$ [3.2]

$\cos\beta = \dfrac{12}{13}$, $\sin\beta = -\dfrac{5}{13}$

$\cos(\alpha+\beta) = \cos\alpha\cos\beta - \sin\alpha\sin\beta$

$\quad = \left(\dfrac{3}{5}\right)\left(\dfrac{12}{13}\right) - \left(\dfrac{4}{5}\right)\left(-\dfrac{5}{13}\right)$

$\quad = -\dfrac{48}{65} + \dfrac{15}{65} = \dfrac{56}{65}$

14. $y = \sin\left[\sin^{-1}\left(\dfrac{3}{5}\right) + \cos^{-1}\left(-\dfrac{5}{13}\right)\right]$ [3.5]

Let $\alpha = \sin^{-1}\dfrac{3}{5}$, $\sin\alpha = \dfrac{3}{5}$, $\cos\alpha = \sqrt{1 - \left(\dfrac{3}{5}\right)^2} = \dfrac{4}{5}$.

$\beta = \cos^{-1}\left(-\dfrac{5}{13}\right)$, $\cos\beta = -\dfrac{5}{13}$, $\sin\beta = \sqrt{1 - \left(-\dfrac{5}{13}\right)^2} = \dfrac{12}{13}$.

$y = \sin(\alpha+\beta)$

$\quad = \sin\alpha\cos\beta + \cos\alpha\sin\beta$

$\quad = \dfrac{3}{5}\left(-\dfrac{5}{13}\right) + \dfrac{4}{5}\left(\dfrac{12}{13}\right)$

$\quad = -\dfrac{15}{65} + \dfrac{48}{65} = \dfrac{33}{65}$

15. $\sin 2x = \sqrt{3}\sin x$ [3.6]

$2\sin x\cos x - \sqrt{3}\sin x = 0$

$\sin x(2\cos x - \sqrt{3}) = 0$

$\sin x = 0 \qquad 2\cos x - \sqrt{3} = 0$

$x = 0,\ \pi \qquad\qquad \cos x = \dfrac{\sqrt{3}}{2}$

$\qquad\qquad\qquad x = \dfrac{\pi}{6},\ \dfrac{11\pi}{6}$

The solutions are $0,\ \dfrac{\pi}{6},\ \pi,\ \dfrac{11\pi}{6}$.

16. $c^2 = a^2 + b^2 - 2ab\cos C$ [4.2]

$\quad = (140)^2 + (130)^2 - 2(140)(130)\cos 78°$

$\quad = 36{,}500 - 36{,}400\cos 78°$

$\quad \approx 28{,}932$

$c \approx 170$ cm

17. $\cos\theta = \dfrac{\mathbf{v}\cdot\mathbf{w}}{\|\mathbf{v}\|\,\|\mathbf{w}\|}$ [4.3]

$\cos\theta = \dfrac{(3\mathbf{i}+2\mathbf{j})\cdot(5\mathbf{i}-3\mathbf{j})}{\sqrt{3^2+2^2}\,\sqrt{5^2+(-3)^2}}$

$\cos\theta = \dfrac{3(5)+(2)(-3)}{\sqrt{13}\sqrt{34}}$

$\cos\theta = \dfrac{9}{\sqrt{442}} = 0$

$\theta = 64.7°$

18. $W = \mathbf{F}\cdot\mathbf{s}$ [4.3]

$W = \|\mathbf{F}\|\,\|\mathbf{s}\|\cos\alpha$

$W = (100)(15)(\cos 15°)$

$W \approx 1449$ foot-pounds

19. $r = \sqrt{(2)^2 + (2)^2}$ [5.2]

$r = 2\sqrt{2}$

$\alpha = \tan^{-1}\left|\dfrac{2}{2}\right|$

$\quad = \tan^{-1} 1 = 45°$

$z = 2\sqrt{2}\ \text{cis}\ 45°$

20. $-27 = 27\left(\cos 180° + i\sin 180°\right)$ [5.3]

$w_k = 27^{1/3}\left(\cos\dfrac{180°+360°k}{3} + i\sin\dfrac{180°+360°k}{3}\right) \qquad k = 0, 1, 2$

$w_0 = 3\left(\cos\dfrac{180°}{3} + i\sin\dfrac{180°}{3}\right)$

$w_0 = 3(\cos 60° + i\sin 60°)$

$w_0 = 3\left(\dfrac{1}{2} + \dfrac{\sqrt{3}}{2}i\right) = \dfrac{3}{2} + \dfrac{3\sqrt{3}}{2}$

$w_1 = 3\left(\cos\dfrac{180°+360°}{3} + i\sin\dfrac{180°+360°}{3}\right)$

$w_1 = 3(\cos 180° + i\sin 180°)$

$w_1 = 3(-1+0i) = -3$

$w_2 = 3\left(\cos\dfrac{180°+360°\cdot 2}{3} + i\sin\dfrac{180°+360°\cdot 2}{3}\right)$

$w_2 = 3(\cos 300° + i\sin 300°)$

$w_2 = 3\left(\dfrac{1}{2} - \dfrac{\sqrt{3}}{2}i\right) = \dfrac{3}{2} - \dfrac{3\sqrt{3}}{2}i$

Copyright © Houghton Mifflin Company. All rights reserved.

Chapter 6
Topics in Analytic Geometry

1.
$$x^2 = -4y$$
$$4p = -4$$
$$p = -1$$
vertex $= (0, 0)$
focus $= (0, -1)$
directrix: $y = 1$

3.
$$y^2 = \frac{1}{3}x$$
$$4p = \frac{1}{3}$$
$$p = \frac{1}{12}$$
vertex $= (0, 0)$
focus $= \left(\frac{1}{12}, 0\right)$
directrix: $x = -\frac{1}{12}$

5.
$$(x-2)^2 = 8(y+3)$$
vertex $= (2, -3)$
$$4p = 8 \quad p = 2$$
$$(h, k+p) = (2, -3+2) = (2, -1)$$
focus $= (2, -1)$
$$k - p = -3 - 2 = -5$$
directrix : $y = -5$

7.
$$(y+4)^2 = -4(x-2)$$
vertex $= (2, -4)$
$$4p = -4 \quad p = -1$$
$$(h+p, k) = (2-1, -4) = (1, -4)$$
focus $= (1, -4)$
$$h - p = 2 + 1 = 3$$
directrix : $x = 3$

9.
$$(y-1)^2 = 2(x+4)$$
vertex $= (-4, 1)$
$$4p = 2 \quad p = \frac{1}{2}$$
$$(h+p, k) = \left(-4+\frac{1}{2}, 1\right) = \left(-\frac{7}{2}, 1\right)$$
focus $= \left(-\frac{7}{2}, 1\right)$
$$h - p = -4 - \frac{1}{2} = -\frac{9}{2}$$
directrix: $x = -\frac{9}{2}$

11.
$$(x-2)^2 = 2(y-2)$$
vertex $= (2, 2)$
$$4p = 2 \quad p = \frac{1}{2}$$
$$(h, k+p) = \left(2, 2+\frac{1}{2}\right) = \left(2, \frac{5}{2}\right)$$
focus $= \left(2, \frac{5}{2}\right)$
$$k - p = 2 - \frac{1}{2} = \frac{3}{2}$$
directrix: $y = \frac{3}{2}$

13.
$$x^2 + 8x - y + 6 = 0$$
$$x^2 + 8x = y - 6$$
$$x^2 + 8x + 16 = y - 6 + 16$$
$$(x+4)^2 = y + 10$$
vertex $= (-4, -10)$
$$4p = 1, \quad p = \frac{1}{4}$$
focus $= \left(-4, -\frac{39}{4}\right)$
directrix: $y = -\frac{41}{4}$

15.
$$x + y^2 - 3y + 4 = 0$$
$$y^2 - 3y = -x - 4$$
$$y^2 - 3y + 9/4 = -x - 4 + 9/4$$
$$\left(y - \frac{3}{2}\right)^2 = -\left(x + \frac{7}{4}\right)$$
vertex $= \left(-\frac{7}{4}, \frac{3}{2}\right)$
$$4p = -1, \quad p = -\frac{1}{4}$$
focus $= \left(-2, \frac{3}{2}\right)$
directrix: $x = -\frac{3}{2}$

17.
$$2x - y^2 - 6y + 1 = 0$$
$$-y^2 - 6y = -2x - 1$$
$$y^2 + 6y = 2x + 1$$
$$y^2 + 6y + 9 = 2x + 1 + 9$$
$$(y+3)^2 = 2(x+5)$$
vertex $= (-5, -3)$
$$4p = 2, \quad p = \frac{1}{2}$$
focus $= \left(-\frac{9}{2}, -3\right)$
directrix: $x = -\frac{11}{2}$

Copyright © Houghton Mifflin Company. All rights reserved.

19.
$$x^2 + 3x + 3y - 1 = 0$$
$$x^2 + 3x = -3y + 1$$
$$x^2 + 3x + 9/4 = -3y + 1 + 9/4$$
$$\left(x + \tfrac{3}{2}\right)^2 = -3\left(y - \tfrac{13}{12}\right)$$
vertex $\left(-\tfrac{3}{2}, \tfrac{13}{12}\right)$
$$4p = -3, \quad p = -\tfrac{3}{4}$$
focus $= \left(-\tfrac{3}{2}, \tfrac{1}{3}\right)$
directrix: $y = \tfrac{11}{6}$

21.
$$2x^2 - 8x - 4y + 3 = 0$$
$$2(x^2 - 4x) = 4y - 3$$
$$2(x^2 - 4x + 4) = 4y - 3 + 8$$
$$2(x - 2)^2 = 4y + 5$$
$$(x - 2)^2 = 2y + \tfrac{5}{2}$$
$$(x - 2)^2 = 2\left(y + \tfrac{5}{4}\right)$$
vertex $= \left(2, -\tfrac{5}{4}\right), \ 4p = 2, \ p = \tfrac{1}{2}$
focus $= \left(2, -\tfrac{3}{4}\right)$
directrix $y = -\tfrac{7}{4}$

23.
$$2x + 4y^2 + 8y - 5 = 0$$
$$4y^2 + 8y = -2x + 5$$
$$4(y^2 + 2y) = -2x + 5$$
$$4(y^2 + 2y + 1) = -2x + 5 + 4$$
$$4(y + 1)^2 = -2x + 9$$
$$(y + 1)^2 = -\tfrac{1}{2}x + \tfrac{9}{4}$$
$$(y + 1)^2 = -\tfrac{1}{2}\left(x - \tfrac{9}{2}\right)$$
vertex $= \left(\tfrac{9}{2}, -1\right)$
$$4p = -\tfrac{1}{2}, \quad p = -\tfrac{1}{8}$$
focus $= \left(\tfrac{35}{8}, -1\right)$
directrix $x = \tfrac{37}{8}$

25.
$$(x - 1)^2 = 3\left(y - \tfrac{1}{9}\right)$$
vertex $= \left(1, \tfrac{1}{9}\right), \ 4p = 3, \ p = \tfrac{3}{4}$
focus $= \left(1, \tfrac{31}{36}\right)$
directrix $y = -\tfrac{23}{36}$

27. vertex $(0, 0)$, focus $(0, -4)$
$$x^2 = 4py$$
$$p = -4 \text{ since focus is } (0, p)$$
$$x^2 = 4(-4)y$$
$$x^2 = -16y$$

29. vertex $(-1, 2)$, focus $(-1, 3)$
$$(x - h)^2 = 4p(y - k)$$
$$h = -1, \ k = 2.$$
The distance p from the vertex to the focus is 1.
$$(x + 1)^2 = 4(1)(y - 2)$$
$$(x + 1)^2 = 4(y - 2)$$

31. focus $(3, -3)$, directrix $y = -5$
The vertex is the midpoint of the line segment joining $(3, -3)$ and the point $(3, -5)$ on the directrix.
$$(h, k) = \left(\frac{3 + 2}{2}, \frac{-3 + (-5)}{2}\right) = (3, -4)$$
The distance p from the vertex to the focus is 1.
$$4p = 4(1) = 4$$
$$(x - h)^2 = 4p(y - k)$$
$$(x - 3)^2 = 4(y + 4)$$

33. vertex $= (-4, 1)$, point: $(-2, 2)$ on the parabola.

Axis of symmetry $x = -4$.

If $P_1 = (-2, 2)$ and axis of symmetry is $x = -4$, then we must have $(x + 4)^2 = 4p(y - 1)$. Since $(-2, 2)$ is on the curve, we get
$$(-2 + 4)^2 = 4p(2 - 1)$$
$$4 = 4p \Rightarrow p = 1$$

Thus, the equation in standard form is $(x + 4)^2 = 4(y - 1)$

Copyright © Houghton Mifflin Company. All rights reserved.

35. Find the vertex.

$$x = -0.325y^2 + 13y + 120$$
$$x - 120 = -0.325y^2 + 13y$$
$$x - 120 = -0.325(y^2 - 40y)$$
$$x - 120 - 130 = -0.325(y^2 - 40y + 400)$$
$$x - 250 = -0.325(y - 20)^2$$
$$-\frac{40}{13}(x - 250) = (y - 20)^2$$

The vertex is (250, 20).

Find the focus.

$$4p = -\frac{40}{13}, \quad p = -\frac{10}{13}$$

The focus is $\left(\dfrac{3240}{13}, 20\right)$.

37. Place the satellite dish on an xy-coordinate system with its vertex at $(0, -1)$ as shown.

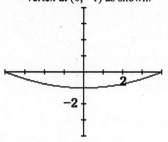

The equation of the parabola is $x^2 = 4p(y+1)$ $-1 \le y \le 0$

Because $(4, 0)$ is a point on this graph, $(4, 0)$ must be a solution of the equation of the parabola. Thus,

$$16 = 4p(0+1)$$
$$16 = 4p$$
$$4 = p$$

Because p is the distance from the vertex to the focus, the focus is on the-axis 4 feet above the vertex.

39. The focus of the parabola is $(p, 0)$ where $y^2 = 4px$.
Half of 18.75 inches is 9.375 inches.
Therefore, the point (3.66, 9.375) is on the parabola.

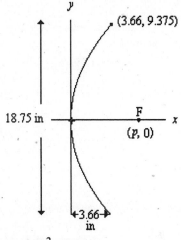

$$(9.375)^2 = 4p(3.66)$$
$$87.890625 = 14.64p$$
$$\frac{87.890625}{14.64} = p$$
$$p \approx 6.0 \text{ inches}$$

41. $$S = \frac{\pi r}{6d^2}\left[\left(r^2 + 4d^2\right)^{3/2} - r^3\right]$$

a. $r = 40.5$ feet
$d = 16$ feet

$$S = \frac{\pi(40.5)}{6(16)^2}\left[\left([40.5]^2 + 4[16]^2\right)^{3/2} - (40.5)^3\right]$$
$$= \frac{40.5\pi}{1536}\left[(266.25)^{3/2} - 66430.125\right]$$
$$= \frac{40.5\pi}{1536}[137518.9228 - 66430.125]$$
$$= \frac{40.5\pi}{1536}[71088.79775]$$
$$\approx 5900 \text{ square feet}$$

b. $r = 125$ feet
$d = 52$ feet

$$S = \frac{\pi(125)}{6(52)^2}\left[\left([125]^2 + 4[52]^2\right)^{3/2} - (125)^3\right]$$
$$= \frac{125\pi}{16224}\left[(26441)^{3/2} - 1953125\right]$$
$$= \frac{125\pi}{16224}[4299488.724 - 1953125]$$
$$= \frac{125\pi}{16224}[2346363.724]$$
$$\approx 56,800 \text{ square feet}$$

Copyright © Houghton Mifflin Company. All rights reserved.

43. The equation of the mirror is given by

$$x^2 = 4py \quad -60 \le x \le 60$$

Because p is the distance from the vertex to the focus and the coordinates of the focus are $(0, 600)$, $p = 600$.

Therefore,

$$x^2 = 4(600)y$$

$$x^2 = 2400y$$

To determine a, substitute $(60, a)$ into the equation

$x^2 = 2400y$ and solve for a.

$$x^2 = 2400y$$

$$60^2 = 2400a$$

$$3600 = 2400a$$

$$1.5 = a.$$

The concave depth of the mirror is 1.5 inches.

47. $(-1.5616, 3.8769), \ (2.5616, 12.1231)$

45. $(-0.3660, -0.3660), \ (1.3660, 1.3660)$

● ●

Connecting Concepts

49.
$$x^2 = 4y$$

$$4p = 4$$

$$p = 1$$

focus $= (0, 1)$

Substituting the vertical coordinate of the focus for y to obtain x-coordinates of endpoints (x_1, y_1), (x_2, y_2), we have

$$x^2 = 4(1), \ \text{or} \ x^2 = 4$$

$$x = \pm\sqrt{4}$$

$$x_1 = -2 \quad x_2 = 2$$

Length of latus rectum $= |x_2 - x_1|$
$$= 2 - (-2) = 4.$$

51. $(x - h)^2 = 4p(y - k)$

focus $= (h, k + p)$

Substituting the vertical coordinate of the focus for y to obtain x-coordinates of endpoints (x_1, y_1), (x_2, y_2), we have

$$(x - h)^2 = 4p(k + p - k)$$
$$(x - h)^2 = 4p^2$$
$$x - h = \pm 2p$$

$$x_1 = h - 2p \quad x_2 = h + 2p$$

Solving for $|x_2 - x_1|$, we obtain

$$\Delta x = |x_2 - x_1| = |h + 2p - h + 2p| = 4|p|$$

or $(y - k)^2 = 4p(x - h)$

focus $= (h + p, k)$

Substituting the horizontal coordinate of the focus for x to obtain the y-coordinates of the endpoints (x_1, y_1), (x_2, y_2), we have

$$(y - k)^2 = 4p(h + p - h)$$
$$(y - k)^2 = 4p^2$$
$$y - k = \pm 2p$$

$$y_1 = k - 2p \quad y_2 = k + 2p$$

Solving for $|y_2 - y_1|$, we obtain $\Delta y = |y_2 - y_1|$
$$= |k + 2p - k + 2p|$$
$$= 4|p|$$

Thus, the length of the latus rectum is $4|p|$.

Copyright © Houghton Mifflin Company. All rights reserved.

53.

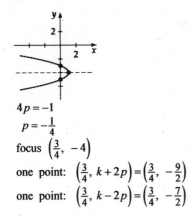

$4p = -1$

$p = -\dfrac{1}{4}$

focus $\left(\dfrac{3}{4}, \ -4\right)$

one point: $\left(\dfrac{3}{4}, \ k+2p\right) = \left(\dfrac{3}{4}, \ -\dfrac{9}{2}\right)$

one point: $\left(\dfrac{3}{4}, \ k-2p\right) = \left(\dfrac{3}{4}, \ -\dfrac{7}{2}\right)$

55.   Graph $y = \dfrac{7}{4} + \dfrac{1}{4}x|x|$.

57.   By definition, any point on the curve (x, y) will be equidistant from both the focus $(1, 1)$ and the directrix, $\left(y_2 = -x_2 - 2\right)$.

If we let d_1 equal the distance from the focus to the point (x, y), we get $d_1 = \sqrt{(x-1)^2 + (y-1)^2}$

To determine the distance d_2 from the point (x, y) to the line $y = -x - 2$, draw a line segment from (x, y) to the directrix so as to meet the directrix at a $90°$ angle.

Now drop a line segment parallel to the y-axis from (x, y) to the directrix. This segment will meet the directrix at a $45°$ angle, thus forming a right isosceles triangle with the directrix and the line segment perpendicular to the directrix from (x, y). The length of this segment, which is the hypotenuse of the triangle, is the difference between y and the y-value of the directrix at x, or $-x - 2$. Thus, the hypotenuse has a length of $y + x + 2$, and since the right triangle is also isosceles, each leg has a length of $\dfrac{y+x+2}{\sqrt{2}}$.

But since d_2 is the length of the leg drawn from (x, y) to the directrix, $d_2 = \dfrac{y+x+2}{\sqrt{2}}$.

Thus, $d_1 = \sqrt{(x-1)^2 + (y-1)^2}$ and $d_2 = \dfrac{x+y+2}{\sqrt{2}}$.

By definition, $d_1 = d_2$. So, by substitution,

$$\sqrt{(x-1)^2 + (y-1)^2} = \frac{x+y+2}{\sqrt{2}}$$

$$\sqrt{2}\sqrt{(x-1)^2 + (y-1)^2} = x+y+2$$

$$2\left[(x-1)^2 + (y-1)^2\right] = x^2 + y^2 + 4x + 4y + 2xy + 4$$

$$2\left(x^2 - 2x + 1 + y^2 - 2y + 1\right) = x^2 + y^2 + 4x + 4y + 2xy + 4$$

$$2x^2 - 4x + 2y^2 - 4y + 4 = x^2 + y^2 + 4x + 4y + 2xy + 4$$

$$x^2 + y^2 - 8x - 8y - 2xy = 0$$

<hr>

Prepare for Section 6.2

58.   midpoint: $\dfrac{x_1 + x_2}{2} \qquad \dfrac{y_1 + y_2}{2}$

$\qquad\qquad\quad \dfrac{5 + -1}{2} = 2 \qquad \dfrac{1+5}{2} = 3$

The midpoint is $(2, 3)$.

length: $\sqrt{(x_2 - x_1)^2 + (y_2 - y_1)^2}$

$\qquad\quad \sqrt{(-1-5)^2 + (5-1)^2} = \sqrt{36 + 16} = \sqrt{52} = 2\sqrt{13}$

The length is $2\sqrt{13}$.

59. $x^2 + 6x - 16 = 0$

$(x+8)(x-2) = 0$

$x + 8 = 0 \qquad x - 2 = 0$

$\qquad x = -8 \qquad\quad x = 2$

The solutions are $-8, 2$.

Copyright © Houghton Mifflin Company. All rights reserved.

60.
$$x^2 - 2x = 2$$
$$x^2 - 2x + 1 = 2 + 1$$
$$(x-1)^2 = 3$$
$$x - 1 = \pm\sqrt{3}$$
$$x = 1 \pm \sqrt{3}$$

61. $x^2 - 8x + 16 = (x-4)^2$

62.
$$(x-2)^2 + y^2 = 4$$
$$y^2 = 4 - (x-2)^2$$
$$y = \pm\sqrt{4 - (x-2)^2}$$

63.

$$(x-2)^2 + (y+3)^2 = 16$$
Center: (2, –3), Radius 4

Section 6.2

1. $\dfrac{x^2}{16} + \dfrac{y^2}{25} = 1$

$a^2 = 25 \rightarrow a = 5$

$b^2 = 16 \rightarrow b = 4$

$c = \sqrt{a^2 - b^2}$

$\quad = \sqrt{25 - 16}$

$\quad = \sqrt{9}$

$\quad = 3$

Center $(0, 0)$

Vertices $(0, \pm 5)$

Foci $(0, \pm 3)$

3. $\dfrac{x^2}{9} + \dfrac{y^2}{4} = 1$

$a^2 = 9 \rightarrow a = 3$

$b^2 = 4 \rightarrow b = 2$

$c = \sqrt{a^2 - b^2}$

$\quad = \sqrt{9 - 4}$

$\quad = \sqrt{5}$

Center $(0, 0)$

Vertices $(\pm 3, 0)$

Foci $\left(\pm\sqrt{5}, 0\right)$

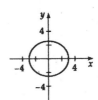

5. $\dfrac{x^2}{7} + \dfrac{y^2}{9} = 1$

$a^2 = 9 \rightarrow a = 3$

$b^2 = 7 \rightarrow b = \sqrt{7}$

$c = \sqrt{a^2 - b^2}$

$\quad = \sqrt{9 - 7}$

$\quad = \sqrt{2}$

Center $(0, 0)$

Vertices $(0, \pm 3)$

Foci $\left(0, \pm\sqrt{2}\right)$

7. $\dfrac{4x^2}{9} + \dfrac{y^2}{16} = 1$

Rewrite as

$\dfrac{x^2}{9/4} + \dfrac{y^2}{16} = 1$

$a^2 = 16 \rightarrow a = 4$

$b^2 = 9/4 \rightarrow b = 3/2$

$c = \sqrt{a^2 - b^2}$

$\quad = \sqrt{16 - 9/4}$

$\quad = \sqrt{55}/2$

Center $(0, 0)$

Vertices $(0, \pm 4)$

Foci $\left(0, \pm\dfrac{\sqrt{55}}{2}\right)$

9. $\dfrac{(x-3)^2}{25} + \dfrac{(y+2)^2}{16} = 1$

Center $(3, -2)$

Vertices $(3 \pm 5, -2) = (8, -2), (-2, -2)$

Foci $(3 \pm 3, -2) = (6, -2), (0, -2)$

11. $\dfrac{(x+2)^2}{9} + \dfrac{y^2}{16} = 1$

Center $(-2, 0)$

Vertices $(-2, 5), (-2, -5)$

Foci $(-2, 4), (-2, -4)$

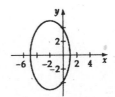

Copyright © Houghton Mifflin Company. All rights reserved.

13.

$$\frac{(x-1)^2}{21} + \frac{(y-3)^2}{4} = 1$$

Center $(1, 3)$

Vertices $\left(1 \pm \sqrt{21}, 3\right)$

Foci $\left(1 \pm \sqrt{17}, 3\right)$

15.

$$\frac{9(x-1)^2}{16} + \frac{(y+1)^2}{9} = 1$$

Center $(1, -1)$

Vertices $(1, -1 \pm 3) = (1, 2), (1, -4)$

Foci $\left(1, -1 \pm \frac{\sqrt{65}}{3}\right)$

17.

$$3x^2 + 4y^2 = 12$$

$$\frac{x^2}{4} + \frac{y^2}{3} = 1$$

Center $(0, 0)$

Vertices $(\pm 2, 0)$

Foci $(\pm 1, 0)$

19.

$$25x^2 + 16y^2 = 400$$

$$\frac{x^2}{16} + \frac{y^2}{25} = 1$$

Center $(0, 0)$

Vertices $(0, \pm 5)$

Foci $(0, \pm 3)$

21.

$$64x^2 + 25y^2 = 400$$

$$\frac{x^2}{\frac{25}{4}} + \frac{y^2}{16} = 1$$

Center $(0, 0)$

Vertices $(0, \pm 4)$

Foci $\left(0, \pm \frac{\sqrt{39}}{2}\right)$

23.

$$4x^2 + y^2 - 24x - 8y + 48 = 0$$

$$4\left(x^2 - 6x\right) + \left(y^2 - 8y\right) = -48$$

$$4\left(x^2 - 6x + 9\right) + \left(y^2 - 8y + 16\right) = -48 + 36 + 16$$

$$4(x-3)^3 + (y-4)^2 = 4$$

$$\frac{(x-3)^2}{1} + \frac{(y-2)^2}{4} = 1$$

Center $(3, 4)$

Vertices $(3, 4 \pm 2) = (3, 6), (3, 2)$

Foci $\left(3, 4 \pm \sqrt{3}\right)$

25.

$$5x^2 + 9y^2 - 20x + 54y + 56 = 0$$

$$5\left(x^2 - 4x\right) + 9\left(y^2 + 6y\right) = -56$$

$$5\left(x^2 - 4x + 4\right) + 9\left(y^2 + 6y + 9\right) = -56 + 20 + 81$$

$$5(x-2)^2 + 9(y+3)^2 = 45$$

$$\frac{(x-2)^2}{9} + \frac{(y+3)^2}{5} = 1$$

Center $(2, -3)$

Vertices $(2 \pm 3, -3) = (-1, -3), (5, -3)$

Foci $(2 \pm 2, 3) = (0, -3), (4, -3)$

Copyright © Houghton Mifflin Company. All rights reserved.

27.
$$16x^2 + 9y^2 - 64x - 80 = 0$$
$$16(x^2 - 4x) + 9y^2 = 80$$
$$16(x^2 - 4x + 4) + 9y^2 = 80 + 64$$
$$16(x-2)^2 + 9y^2 = 144$$
$$\frac{(x-2)^2}{9} + \frac{y^2}{16} = 1$$
Center $(2, 0)$
Vertices $(2, \pm 4) = (2, 4), (2, -4)$
Foci $(2 \pm \sqrt{7})$

29.
$$25x^2 + 16y^2 + 50x - 32y - 359 = 0$$
$$25(x^2 + 2x) + 16(y^2 - 2y) = 359$$
$$25(x^2 + 2x + 1) + 16(y^2 - 2y + 1) = 359 + 25 + 16$$
$$25(x+1)^2 + 16(y-1)^2 = 400$$
$$\frac{(x+1)^2}{16} + \frac{(y-1)^2}{25} = 1$$
Center $(-1, 1)$
Vertices $(-1, 1 \pm 5) = (-1, 6), (-1, -4)$
Foci $(-1, 1 \pm 3) = (-1, 4), (-1, -2)$

31.
$$8x^2 + 25y^2 - 48x + 50y + 47 = 0$$
$$8(x^2 - 6x) + 25(y^2 + 2y) = -47$$
$$8(x^2 - 6x + 9) + 25(y^2 + 2y + 1) = -47 + 72 + 25$$
$$8(x-3)^2 + 25(y+1)^2 = 50$$
$$\frac{(x-3)^2}{25/4} + \frac{(y+1)^2}{2} = 1$$
Center: $(3, -1)$
Vertices: $\left(3 \pm \frac{5}{2}, -1\right) = \left(\frac{11}{2}, -1\right), \left(\frac{1}{2}, -1\right)$
Foci: $\left(3 \pm \frac{\sqrt{17}}{2}, -1\right)$

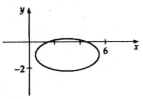

33.
$$2a = 10$$
$$a = 5$$
$$a^2 = 25$$
$$c = 4$$
$$c^2 = a^2 - b^2$$
$$16 = 25 - b^2$$
$$b^2 = 9$$
$$\frac{x^2}{25} + \frac{y^2}{9} = 1$$

35.
$$a = 6$$
$$a^2 = 36$$
$$b = 4$$
$$b^2 = 16$$
$$\frac{x^2}{36} + \frac{y^2}{16} = 1$$

37.
$$2a = 12$$
$$a = 6$$
$$a^2 = 36$$
$$\frac{x^2}{36} + \frac{y^2}{b^2} = 1$$
$$\frac{(2)^2}{36} + \frac{(-3)^2}{b^2} = 1$$
$$\frac{4}{36} + \frac{9}{b^2} = 1$$
$$\frac{9}{b^2} = \frac{8}{9}$$
$$8b^2 = 81$$
$$b^2 = \frac{81}{8}$$
$$\frac{x^2}{36} + \frac{y^2}{81/8} = 1$$

39.
$$c = 3$$
$$2a = 8$$
$$a = 4$$
$$a^2 = 16$$
$$c^2 = a^2 - b^2$$
$$9 = 16 - b^2$$
$$b^2 = 7$$
$$\frac{(x+2)^2}{16} + \frac{(y-4)^2}{7} = 1$$

Copyright © Houghton Mifflin Company. All rights reserved.

41. $2a = 10$

$a = 5$

$a^2 = 25$

Since the center of the ellipse is $(2, 4)$ and the point $(3, 3)$ is on the ellipse, we have

$\dfrac{(x-2)^2}{b^2} + \dfrac{(y-4)^2}{a^2} = 1$

$\dfrac{(3-2)^2}{b^2} + \dfrac{(3-4)^2}{25} = 1$

$\dfrac{1}{b^2} = 1 - \dfrac{1}{25}$

$b^2 = \dfrac{25}{24}$

$\dfrac{(x-2)^2}{\frac{25}{24}} + \dfrac{(y-4)^2}{25} = 1$

43. center $(5, 1)$

$c = 3$

$2a = 10$

$a = 5$

$a^2 = 25$

$c^2 = a^2 - b^2$

$9 = 25 - b^2$

$b^2 = 16$

$\dfrac{(x-5)^2}{16} + \dfrac{(y-1)^2}{25} = 1$

45. $2a = 10$

$a = 5$

$a^2 = 25$

$\dfrac{c}{a} = \dfrac{2}{5}$

$\dfrac{c}{5} = \dfrac{2}{5}$

$c = 2$

$c^2 = a^2 - b^2$

$4 = 25 - b^2$

$b^2 = 21$

$\dfrac{x^2}{25} + \dfrac{y^2}{21} = 1$

47. center $(0, 0)$

$c = 4$

$\dfrac{c}{a} = \dfrac{2}{3}$

$\dfrac{4}{a} = \dfrac{2}{3}$

$a = 6$

$c^2 = a^2 - b^2$

$16 = 36 - b^2$

$b^2 = 20$

$\dfrac{x^2}{20} + \dfrac{y^2}{36} = 1$

49. center $(1, 3)$

$c = 2$

$\dfrac{c}{a} = \dfrac{2}{5}$

$\dfrac{2}{a} = \dfrac{2}{5}$

$a = 5$

$c^2 = a^2 - b$

$4 = 25 - b$

$b^2 = 21$

$\dfrac{(x-1)^2}{25} + \dfrac{(y-3)^2}{21} = 1$

51. $2a = 24$

$a = 12$

$\dfrac{c}{a} = \dfrac{2}{3}$

$\dfrac{c}{12} = \dfrac{2}{3}$

$c = 8$

$c^2 = a^2 - b^2$

$64 = 144 - b^2$

$b^2 = 80$

$\dfrac{x^2}{80} + \dfrac{y^2}{144} = 1$

53. $484 = 64 + c^2$

$c^2 = 420$

$c = 20.494$

$2c = 40.9878 \approx 41$

The emitter should be placed 41 cm away.

55. Aphelion $= 2a -$ perihelion

$934.34 = 2a - 835.14$

$a = 884.74$ million miles

Aphelion $= a + c = 934.34$

$884.74 + c = 934.34$

$c = 49.6$ million miles

$b = \sqrt{a^2 - c^2}$

$= \sqrt{884.74^2 - 49.6^2}$

≈ 883.35 million miles

An equation of the orbit of Saturn is

$\dfrac{x^2}{884.74^2} + \dfrac{y^2}{883.35^2} = 1$

57. $a =$ semimajor axis $= 50$ feet

$b =$ height $= 30$ feet

$c^2 = a^2 - b^2$

$c^2 = 50^2 - 30^2$

$c = \sqrt{1600} = 40$

The foci are located 40 feet to the right and to the left of center.

59. $2a = 36 \qquad 2b = 9$

$a = 18 \qquad b = \dfrac{9}{2}$

$c^2 = a^2 - b^2$

$c^2 = 18^2 - \left(\dfrac{9}{2}\right)^2$

$c^2 = 324 - \dfrac{81}{4}$

$c^2 = \dfrac{1215}{4}$

$c = \dfrac{9\sqrt{15}}{2}$

Since one focus is at $(0, 0)$, the center of the ellipse is at $(9\sqrt{15}/2, 0)$ $(17.43, 0)$. The equation of the path of Halley's Comet in astronomical units is

$\dfrac{\left(x - 9\sqrt{15}/2\right)^2}{324} + \dfrac{y^2}{81/4} = 1$

Copyright © Houghton Mifflin Company. All rights reserved.

61. $2a = 16 \Rightarrow a = 8$

$2b = 10 \Rightarrow b = 5$

$$p = \pi\sqrt{2\left(a^2 + b^2\right)}$$

$$= \pi\sqrt{2\left(8^2 + 5^2\right)}$$

$$= \pi\sqrt{2(64 + 25)}$$

$$= \pi\sqrt{2(89)}$$

$$= \pi\sqrt{178} \text{ inches}$$

1 mile = 5280 feet = 5280(12) inches = 63360 inches

$$\frac{63360 \text{ inches}}{\pi\sqrt{178} \text{ inches per revolution}} \approx 1512 \text{ revolutions}$$

63. **a.** $2a = 615 \Rightarrow a = 307.5$

$2b = 510 \Rightarrow b = 255$

$$\frac{x^2}{307.5^2} + \frac{y^2}{255^2} = 1$$

b. $A = \pi ab$

$$= \pi(307.5)(255)$$

$$= 78412.5\pi$$

$$\approx 246,300 \text{ square feet}$$

65. $9y^2 + 36y + 16x^2 - 108 = 0$

$$y = \frac{-36 \pm \sqrt{36^2 - 4(9)(16x^2 - 108)}}{2(9)}$$

$$= \frac{-36 \pm \sqrt{1296 - 36(16x^2 - 108)}}{18}$$

$$= \frac{-36 \pm \sqrt{1296 - 576x^2 + 3888}}{18}$$

$$= \frac{-36 \pm \sqrt{-576x^2 + 5184}}{18}$$

$$= \frac{-36 \pm \sqrt{576(-x^2 + 9)}}{18}$$

$$= \frac{-36 \pm 24\sqrt{(-x^2 + 9)}}{18}$$

$$= \frac{-6 \pm 4\sqrt{(-x^2 + 9)}}{3}$$

67. $9y^2 - 54y + 16x^2 - 64x + 1 = 0$

$$y = \frac{-(-54) \pm \sqrt{(-54)^2 - 4(9)(16x^2 - 64x + 1)}}{2(9)}$$

$$= \frac{54 \pm \sqrt{2916 - 36(16x^2 - 64x + 1)}}{18}$$

$$= \frac{54 \pm \sqrt{2916 - 576x^2 + 2304x - 36}}{18}$$

$$= \frac{54 \pm \sqrt{-576x^2 + 2304x + 2880}}{18}$$

$$= \frac{54 \pm \sqrt{576(-x^2 + 4x + 5)}}{18}$$

$$= \frac{54 \pm 24\sqrt{-x^2 + 4x + 5}}{18}$$

$$= \frac{9 \pm 4\sqrt{-x^2 + 4x + 5}}{3}$$

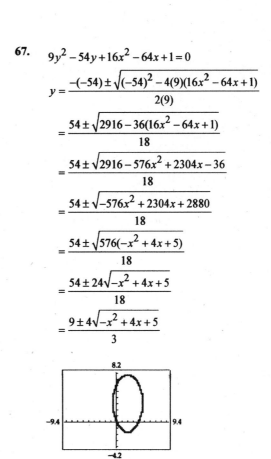

Copyright © Houghton Mifflin Company. All rights reserved.

69.

$$9y^2 + 18y + 4x^2 + 24x + 44 = 0$$

$$y = \frac{-18 \pm \sqrt{18^2 - 4(9)(4x^2 + 24x + 44)}}{2(9)}$$

$$= \frac{-18 \pm \sqrt{324 - 36(4x^2 + 24x + 44)}}{18}$$

$$= \frac{-18 \pm \sqrt{324 - 144x^2 - 864x - 1584}}{18}$$

$$= \frac{-18 \pm \sqrt{-144x^2 - 864x - 1260}}{18}$$

$$= \frac{-18 \pm \sqrt{36(-4x^2 - 24x - 35)}}{18}$$

$$= \frac{-18 \pm 6\sqrt{-4x^2 - 24x - 35}}{18}$$

$$= \frac{-3 \pm \sqrt{-4x^2 - 24x - 35}}{3}$$

71. The sum of the distances between the two foci and a point on the ellipse is $2a$.

$$2a = \sqrt{\left(\frac{9}{2} - 0\right)^2 + (3-3)^2} + \sqrt{\left(\frac{9}{2} - 0\right)^2 + (3+3)^2}$$

$$= \sqrt{\left(\frac{9}{2}\right)^2} + \sqrt{\frac{225}{4}}$$

$$= \frac{9}{2} + \frac{15}{2}$$

$$= 12$$

$$a = 6$$
$$c = 3$$
$$c^2 = a^2 - b^2$$
$$9 = 36 - b^2$$
$$b^2 = 27$$

$$\frac{x^2}{36} + \frac{y^2}{27} = 1$$

Connecting Concepts

73. The sum of the distances between the two foci and a point on the ellipse is $2a$.

$$2a = \sqrt{(5-2)^2 + (3+1)^2} + \sqrt{(5-2)^2 + (3-3)^2}$$

$$= \sqrt{25} + \sqrt{3^2}$$

$$= 5 + 3$$

$$= 8$$

$$a = 4$$
$$c = 2$$

$$c^2 = a^2 - b^2$$
$$4 = 16 - b^2$$
$$b^2 = 12$$

$$\frac{(x-1)^2}{16} + \frac{(y-2)^2}{12} = 1$$

Copyright © Houghton Mifflin Company. All rights reserved.

75. Center $(1, -1)$

$c^2 = a^2 - b^2$

$c^2 = 16 - 9$

$c^2 = 7$

$c = \sqrt{7}$

The latus rectum is on the graph of $y = -1 + \sqrt{7}$, or $y = -1 - \sqrt{7}$

$$\frac{(x-1)^2}{9} + \frac{(y+1)^2}{16} = 1$$

$$\frac{(x-1)^2}{9} + \frac{\left(-1+\sqrt{7}+1\right)^2}{16} = 1 \quad \text{or} \quad \frac{(x-1)^2}{9} + \frac{\left(-1-\sqrt{7}+1\right)^2}{16} = 1$$

$$\frac{(x-1)^2}{9} + \frac{7}{16} = 1$$

$$\frac{(x-1)^2}{9} = \frac{9}{16}$$

$$16(x-1)^2 = 81$$

$$(x-1)^2 = \frac{81}{16}$$

$$x - 1 = \pm\sqrt{\frac{81}{16}}$$

$$x - 1 = \pm\frac{9}{4}$$

$$x = \frac{13}{4} \text{ and } -\frac{5}{4}$$

The x-coordinates of the endpoints of the latus rectum are $\frac{13}{4}$ and $-\frac{5}{4}$.

$$\left| \frac{13}{4} - \left(-\frac{5}{4}\right) \right| = \frac{9}{2}$$

The length of the latus rectum is $\frac{9}{2}$.

77. Let us transform the general equation of an ellipse into an $x'y'$ - coordinate system where the center is at the origin by replacing $(x - h)$ by x' and $(y - k)$ by y'.

We have $\dfrac{x'^2}{a^2} + \dfrac{y'^2}{b^2} = 1$.

Letting $x' = c$ and solving for y' yields

$$\frac{(c)^2}{a^2} + \frac{y'^2}{b^2} = 1$$

$$b^2 c^2 + a^2 y'^2 = a^2 b^2$$

$$a^2 y'^2 = a^2 b^2 - b^2 c^2$$

$$a^2 y'^2 = b^2 (a^2 - c^2)$$

But since $c^2 = a^2 - b^2$, $b^2 = a^2 - c^2$, we can substitute to obtain

$$a^2 y'^2 = b^2 (b^2)$$

$$y'^2 = \frac{b^4}{a^2}$$

$$y' = \pm\sqrt{\frac{b^4}{a^2}} = \pm\frac{b^2}{a}$$

The endpoints of the latus rectum, then, are $\left(c, \dfrac{b^2}{a} \right)$ and $\left(c, -\dfrac{b^2}{a} \right)$.

The distance between these points is $\dfrac{2b^2}{a}$.

Copyright © Houghton Mifflin Company. All rights reserved.

79.

$$a^2 = 9$$

$$b^2 = 4$$

$$c^2 = a^2 - b^2 = 9 - 4 = 5$$

$$c = \sqrt{5}$$

$$x = \pm \frac{a^2}{c} = \pm \frac{9}{\sqrt{5}} = \pm \frac{9\sqrt{5}}{5}$$

The directrices are $x = \dfrac{9\sqrt{5}}{5}$ and $x = -\dfrac{9\sqrt{5}}{5}$.

81.

The eccentricity is $e = \dfrac{c}{a} = \dfrac{2}{\sqrt{12}} = \dfrac{1}{\sqrt{3}}$. Let $P(x, y)$ be a point on the ellipse $\dfrac{x^2}{12} + \dfrac{y^2}{8} = 1$ and $F(2, 0)$ be a focus.

Then $d(P, F) = \sqrt{(x-2)^2 + y^2}$. The distance from the line $x = 6$ to $P(x, y)$ is $|x - 6|$. Thus, solving the equation of the ellipse for y^2, we have

$$\frac{\sqrt{(x-2)^2 + y^2}}{|x-6|} = \frac{\sqrt{(x-2)^2 + 8\left(1 - \dfrac{x^2}{12}\right)}}{|x-6|}$$

$$= \frac{\sqrt{x^2 - 4x + 4 + 8 - \dfrac{2}{3}x^2}}{|x-6|}$$

$$= \frac{\sqrt{\dfrac{1}{3}x^2 - 4x + 12}}{|x-6|} = \frac{\sqrt{\dfrac{x^2 - 12x + 36}{3}}}{|x-6|}$$

$$= \frac{\sqrt{\dfrac{(x-6)^2}{3}}}{|x-6|} = \frac{\dfrac{|x-6|}{\sqrt{3}}}{|x-6|} = \frac{1}{\sqrt{3}} = e \text{ (eccentricity)}$$

• •

83.

$$\frac{4 + -2}{2} = 1$$

$$\frac{-3 + 1}{2} = -1$$

Midpoint: $(1, -1)$

$$\sqrt{(-2-4)^2 + (1--3)^2} = \sqrt{52} = 2\sqrt{13}$$

Distance: $2\sqrt{13}$

84.

$$(x-1)(x+3) = 5$$

$$x^2 + 2x - 3 = 5$$

$$x^2 + 2x - 8 = 0$$

$$(x+4)(x-2) = 0$$

$$x + 4 = 0 \quad x - 2 = 0$$

$$x = -4 \quad\quad x = 2$$

85.

$$\frac{4}{\sqrt{8}} = \frac{4\sqrt{8}}{8} = \frac{8\sqrt{2}}{8} = \sqrt{2}$$

86.

$$4x^2 + 24x = 4(x^2 + 6x)$$

$$= 4(x^2 + 6x + 9)$$

$$= 4(x+3)^2$$

87.

$$\frac{x^2}{4} - \frac{y^2}{9} = 1$$

$$-\frac{y^2}{9} = 1 - \frac{x^2}{4}$$

$$y^2 = \frac{9x^2}{4} - 9$$

$$y = \pm\sqrt{\frac{9x^2}{4} - 9}$$

$$y = \pm\frac{3}{2}\sqrt{x^2 - 4}$$

88.

Copyright © Houghton Mifflin Company. All rights reserved.

Section 6.3

1.
$$\frac{x^2}{16} - \frac{y^2}{25} = 1$$

Center $(0, 0)$

Vertices $(\pm 4, 0)$

Foci $\left(\pm\sqrt{41}, 0\right)$

Asymptotes $y = \pm\frac{5}{4}x$

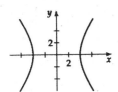

3.
$$\frac{y^2}{4} - \frac{x^2}{25} = 1$$

Center $(0, 0)$

Vertices $(0, \pm 2)$

Foci $\left(0, \pm\sqrt{29}\right)$

Asymptotes $y = \pm\frac{2}{5}x$

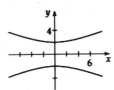

5.
$$\frac{x^2}{7} - \frac{y^2}{9} = 1$$

Center $(0, 0)$

Vertices $\left(\pm\sqrt{7}, 0\right)$

Foci $(\pm 4, 0)$

Asymptotes $y = \pm\frac{3\sqrt{7}}{7}x$

7.
$$\frac{4x^2}{9} - \frac{y^2}{16} = 1$$

Center $(0, 0)$

Vertices $\left(\pm\frac{3}{2}, 0\right)$

Foci $\left(\pm\frac{\sqrt{73}}{2}, 0\right)$

Asymptotes $y = \pm\frac{8}{3}x$

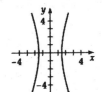

9.
$$\frac{(x-3)^2}{16} - \frac{(y+4)^2}{9} = 1$$

Center $(3, -4)$

Vertices $(3 \pm 4, -4) = (7, -4), (-1, -4)$

Foci $(3 \pm 5, -4) = (8, -4), (-2, -4)$

Asymptotes $y + 4 = \pm\frac{3}{4}(x - 3)$

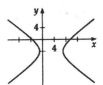

11.
$$\frac{(y+2)^2}{4} - \frac{(x-1)^2}{16} = 1$$

Center $(1, -2)$

Vertices $(1, -2 \pm 2) = (1, 0), (1, -4)$

Foci $\left(1, -2 \pm 2\sqrt{5}\right) = \left(1, -2 + 2\sqrt{5}\right), \left(1, -2 - 2\sqrt{5}\right)$

Asymptotes $y + 2 = \pm\frac{1}{2}(x - 1)$

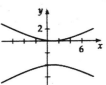

13.
$$\frac{(x+2)^2}{9} - \frac{y^2}{25} = 1$$

Center $(-2, 0)$

Vertices $(-2 \pm 3, 0) = (1, 0), (-5, 0)$

Foci $\left(-2 \pm \sqrt{34}, 0\right)$

Asymptotes $y = \pm\frac{5}{3}(x + 2)$

15.
$$\frac{9(x-1)^2}{16} - \frac{(y+1)^2}{9} = 1$$

$$\frac{(x-1)^2}{16/9} - \frac{(y+1)^2}{9}$$

Center $(1, -1)$

Vertices $\left(1 \pm \frac{4}{3}, -1\right) = \left(\frac{7}{3}, -1\right), \left(-\frac{1}{3}, -1\right)$

Foci $\left(1 \pm \frac{\sqrt{97}}{3}, -1\right)$

Asymptotes $(y + 1) = \pm\frac{9}{4}(x - 1)$

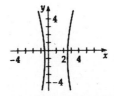

Copyright © Houghton Mifflin Company. All rights reserved.

17. $x^2 - y^2 = 9$

$\dfrac{x^2}{9} - \dfrac{y^2}{9} = 1$

Center $(0, 0)$

Vertices $(\pm 3, 0)$

Foci $(\pm 3\sqrt{2}, 0)$

Asymptotes $y = \pm x$

19. $16y^2 - 9x^2 = 144$

$\dfrac{y^2}{9} - \dfrac{x^2}{16} = 1$

Center $(0, 0)$

Vertices $(0, \pm 3)$

Foci (0 ± 5)

Asymptotes $y = \pm \dfrac{3}{4}x$

21. $9y^2 - 36x^2 = 4$

$\dfrac{y^2}{4/9} - \dfrac{x^2}{1/9} = 1$

Center $(0, 0)$

Vertices $\left(0, \pm \dfrac{2}{3}\right)$

Foci $\left(0, \pm \dfrac{\sqrt{5}}{3}\right)$

Asymptotes $y = \pm 2x$

23.
$$x^2 - y^2 - 6x + 8y = 3$$
$$\left(x^2 - 6x\right) - \left(y^2 - 8y\right) = 3$$
$$\left(x^2 - 6x + 9\right) - \left(y^2 - 8y + 16\right) = 3 + 9 - 16$$
$$(x - 3)^2 - (y - 4)^2 = -4$$
$$\dfrac{(y - 4)^2}{4} - \dfrac{(x - 3)^2}{4} = 1$$

Center $(3, 4)$

Vertices $(3, 4 \pm 2) = (3, 6), (3, 2)$

Foci $\left(3, 4 \pm 2\sqrt{2}\right) = \left(3, 4 + 2\sqrt{2}\right), \left(3, 4 - 2\sqrt{2}\right)$

Asymptotes $y - 4 = \pm (x - 3)$

25.
$$9x^2 - 4y^2 + 36x - 8y + 68 = 0$$
$$9x^2 + 36x - 4y^2 - 8y = -68$$
$$9\left(x^2 + 4x\right) - 4\left(y^2 + 2y\right) = -68$$
$$9\left(x^2 + 4x + 4\right) - 4\left(y^2 + 2y + 1\right) = -68 + 36 - 4$$
$$9(x + 2)^2 - 4(y + 1)^2 = -36$$
$$\dfrac{(y + 1)^2}{9} - \dfrac{(x + 2)^2}{4} = 1$$

Center $(-2, -1)$

Vertices $(-2, -1 \pm 3) = (-2, 2), (-2, -4)$

Foci $\left(-2, -1 \pm \sqrt{13}\right) = \left(-2, -1 + \sqrt{13}\right), \left(-2, -1 - \sqrt{13}\right)$

Asymptotes $y + 1 = \pm \dfrac{3}{2}(x + 2)$

Copyright © Houghton Mifflin Company. All rights reserved.

27.

$$y = \frac{-6 \pm \sqrt{6^2 - 4(-1)\left(4x^2 + 32x + 39\right)}}{2(-1)}$$

$$= \frac{-6 \pm \sqrt{36 + 4\left(4x^2 + 32x + 39\right)}}{-2}$$

$$= \frac{-6 \pm \sqrt{16x^2 + 128x + 192}}{-2}$$

$$= \frac{-6 \pm \sqrt{16(x^2 + 8x + 12)}}{-2}$$

$$= \frac{-6 \pm 4\sqrt{x^2 + 8x + 12}}{-2}$$

$$= 3 \pm 2\sqrt{x^2 + 8x + 12}$$

29.

$$y = \frac{64 \pm \sqrt{(-64)^2 - 4(-16)\left(9x^2 - 36x + 116\right)}}{2(-16)}$$

$$= \frac{64 \pm \sqrt{4096 + 64(9x^2 - 36x + 116)}}{-32}$$

$$= \frac{64 \pm \sqrt{64(9x^2 - 36x + 116 + 64)}}{-32}$$

$$= \frac{64 \pm 8\sqrt{(9x^2 - 36x + 180)}}{-32}$$

$$= \frac{64 \pm 8\sqrt{9(x^2 - 4x + 20)}}{-32}$$

$$= \frac{64 \pm 24\sqrt{x^2 - 4x + 20}}{-32}$$

$$= \frac{-8 \pm 3\sqrt{x^2 - 4x + 20}}{4}$$

31.

$$y = \frac{18 \pm \sqrt{(-18)^2 - 4(-9)\left(4x^2 + 8x - 6\right)}}{2(-9)}$$

$$= \frac{18 \pm \sqrt{324 + 36\left(4x^2 + 8x - 6\right)}}{-18}$$

$$= \frac{18 \pm \sqrt{36\left(4x^2 + 8x - 6 + 9\right)}}{-18}$$

$$= \frac{18 \pm 6\sqrt{\left(4x^2 + 8x + 3\right)}}{-18}$$

$$= \frac{-3 \pm \sqrt{\left(4x^2 + 8x + 3\right)}}{3}$$

33. vertices $(3, 0)$ and $(-3, 0)$, foci $(4, 0)$ and $(-4, 0)$

Traverse axis is on x-axis. For a standard hyperbola, the vertices are at $(h + a, k)$ and $(h - a, k)$, $h + a = 3$, $h - a = -3$, and $k = 0$..

If $h + a = 3$ and $h - a = -3$, then $h = 0$ and $a = 3$.

The foci are located at $(4, 0)$ and $(-4, 0)$. Thus, $h = 0$ and $c = 4$.

Since $c^2 = a^2 + b^2$, $b^2 = c^2 - a^2$

$$b^2 = (4)^2 - (3)^2 = 16 - 9 = 7$$

$$\frac{(x-h)^2}{a^2} - \frac{(y-k)^2}{b^2} = 1$$

$$\frac{(x-0)^2}{(3)^2} - \frac{(y-0)^2}{7} = 1$$

$$\frac{x^2}{9} - \frac{y^2}{7} = 1$$

Copyright © Houghton Mifflin Company. All rights reserved.

35. foci $(0, 5)$ and $(0, -5)$, asymptotes $y = 2x$ and $y = -2x$

Transverse axis is on y-axis. Since foci are at $(h, k + c)$ and $(h, k - c)$, $k + c = 5$, $k - c = -5$, and $h = 0$.

Therefore, $k = 0$ and $c = 5$.

Since one of the asymptotes is $y = \frac{a}{b}x$, $\frac{a}{b} = 2$ and $a = 2b$.

$a^2 + b^2 = c^2$; then substituting $a = 2b$ and $c = 5$ yields $(2b)^2 + b^2 = (5)^2$, or $5b^2 = 25$.

Therefore, $b^2 = 5$ and $b = \sqrt{5}$.
Since $a = 2b$, $a = 2(\sqrt{5}) = 2\sqrt{5}$.

$$\frac{(y-k)^2}{a^2} - \frac{(x-h)^2}{b^2} = 1$$
$$\frac{y^2}{(2\sqrt{5})^2} - \frac{x^2}{5} = 1$$
$$\frac{y^2}{20} - \frac{x^2}{5} = 1$$

37. vertices $(0, 3)$ and $(0, -3)$, point $(2, 4)$

The distance between the two vertices is the length of the transverse axis, which is $2a$.

$$2a = |3 - (-3)| = 6 \text{ or } a = 3.$$

Since the midpoint of the transverse axis is the center of the hyperbola, the center is given by
$$\left(\frac{0+0}{2}, \frac{3+(-3)}{2}\right), \text{ or } (0, 0)$$

Since both vertices lie on the y-axis, the transverse axis must be on the y-axis.

Taking the standard form of the hyperbola, we have
$$\frac{y^2}{a^2} - \frac{x^2}{b^2} = 1$$

Substituting the point $(2, 4)$ for x and y, and 3 for a, we have
$$\frac{16}{9} - \frac{4}{b^2} = 1$$

Solving for b^2 yields $b^2 = \frac{36}{7}$.

Therefore, the equation is
$$\frac{y^2}{9} - \frac{x^2}{36/7} = 1$$

39. vertices $(0, 4)$ and $(0, -4)$, asymptotes $y = \frac{1}{2}x$ and $y = -\frac{1}{2}x$.

The length of the transverse axis, or the distance between the vertices, is equal to $2a$.

$2a = 4 - (-4) = 8$, or $a = 4$

The center of the hyperbola, or the midpoint of the line segment joining the vertices, is
$$\left(\frac{0+0}{2}, \frac{4+(-4)}{2}\right), \text{ or } (0, 0)$$

Since both vertices lie on the y-axis, the transverse axis must lie on the y-axis. Therefore, the asymptotes are given by $y = \frac{a}{b}x$ and $y = -\frac{a}{b}x$. One asymptote is $y = \frac{1}{2}x$. Thus $\frac{a}{b} = \frac{1}{2}$ or $b = 2a$.
Since $b = 2a$ and $a = 4$, $b = 2(4) = 8$.

Thus, the equation is
$$\frac{y^2}{4^2} - \frac{x^2}{8^2} = 1 \text{ or } \frac{y^2}{16} - \frac{x^2}{64} = 1$$

41. vertices $(6, 3)$ and $(2, 3)$, foci $(7, 3)$ and $(1, 3)$
Length of transverse axis = distance between vertices
$$2a = |6 - 2|$$
$$a = 2$$

The center of the hyperbola (h, k) is the midpoint of the line segment joining the vertices, or the point $\left(\frac{6+2}{2}, \frac{3+3}{2}\right)$.

Thus, $h = \frac{6+2}{2}$, or 4, and $k = \frac{3+3}{3}$, or 3.

Since both vertices lie on the horizontal line $y = 3$, the transverse axis is parallel to the x-axis. The location of the foci is given by $(h + c, k)$ and $(h - c, k)$, or specifically $(7, 3)$ and $(1, 3)$. Thus $h + c = 7$, $h - c = 1$, and $k = 3$. Solving for h and c simultaneously yields $h = 4$ and $c = 3$.

Since $c^2 = a^2 + b^2$, $b^2 = c^2 - a^2$.

Substituting, we have $b^2 = 3^2 - 2^2 = 9 - 4 = 5$.

Substituting $a = 2$, $b^2 = 5$, $h = 4$, and $k = 3$ in the standard equation $\frac{(x-h)^2}{a^2} - \frac{(y-k)^2}{b^2} = 1$ yields $\frac{(x-4)^2}{4} - \frac{(y-3)^2}{5} = 1$.

Copyright © Houghton Mifflin Company. All rights reserved.

43. foci $(1, -2)$ and $(7, -2)$, slope of an asymptote $= \frac{5}{4}$

Both foci lie on the horizontal line $y = -2$; therefore, the transverse axis is parallel to the x-axis.

The foci are given by $(h + c, k)$ and $(h - c, k)$.

Thus, $h - c = 1, h + c = 7$, and $k = -2$. Solving simultaneously for h and c yields $h = 4$ and $c = 3$.

Since $y - k = \frac{b}{a}(x - h)$ is the equation for an asymptote, and the slope of an asymptote is given as $\frac{5}{4}, \frac{b}{a} = \frac{5}{4}, b = \frac{5a}{4}$,

and $b^2 = \frac{25a^2}{16}$.

Because $a^2 + b^2 = c^2$, substituting $c = 3$ and $b^2 = \frac{25a^2}{16}$ yields $a^2 = \frac{144}{41}$.

Therefore, $b^2 = \frac{3600}{656} = \frac{225}{41}$.

Substituting in the standard equation for a hyperbola yields $\dfrac{(x - 4)^2}{144/41} - \dfrac{(y + 2)^2}{225/41} = 1$

45. Because the transverse axis is parallel to the y-axis and the center is $(7, 2)$, the equation of the hyperbola is

$$\frac{(y - 2)^2}{a^2} - \frac{(x - 7)^2}{b^2} = 1$$

Because $(9, 4)$ is a point on the hyperbola,

$$\frac{(4 - 2)^2}{a^2} - \frac{(9 - 7)^2}{b^2} = 1$$

The slope of the asymptote is $\frac{1}{2}$. Therefore $\frac{1}{2} = \frac{a}{b}$ or $b = 2a$.

Substituting, we have

$$\frac{4}{a^2} - \frac{4}{4a^2} = 1$$

$$\frac{4}{a^2} - \frac{1}{a^2} = 1, \text{ or } a^2 = 3$$

Since $b = 2a$, $b^2 = 4a^2$, or $b^2 = 12$. The equation is $\dfrac{(y - 2)^2}{3} - \dfrac{(x - 7)^2}{12} = 1$.

47. vertices $(1, 6)$ and $(1, 8)$, eccentricity $= 2$

Length of transverse axis = distance between vertices

$$2a = |6 - 8| = 2$$

$$a = 1 \text{ and } a^2 = 1$$

Center (midpoint of transverse axis) is $\left(\dfrac{1 + 1}{2}, \dfrac{6 + 8}{2}\right)$, or $(1, 7)$.

Therefore, $h = 1$ and $k = 7$.

Since both vertices lie on the vertical line $x = 1$, the transverse axis is parallel to the y-axis.

Since $e = \frac{c}{a}$, $c = ae = (1)(2) = 2$.

Because $b^2 = c^2 - a^2$, $b^2 = (2)^2 - (1)^2 = 4 - 1 = 3$.

Substituting h, k, a^2, and b^2 into the standard equation yields $\dfrac{(y - 7)^2}{1} - \dfrac{(x - 1)^2}{3} = 1$

Copyright © Houghton Mifflin Company. All rights reserved.

49.   foci $(4, 0)$ and $(-4, 0)$, eccentricity $= 2$

Center (midpoint of line segment joining foci) is $\left(\dfrac{4+(-4)}{2}, \dfrac{0+0}{2}\right)$, or $(0, 0)$

Thus, $h = 0$ and $k = 0$.

Since both foci lie on the horizontal line $y = 0$, the transverse axis is parallel to the x-axis. The locations of the foci are given by $(h + c, k)$ and $(h - c, k)$, or specifically $(4, 0)$ and $(-4, 0)$

Since $h = 0$, $c = 4$.

Because $e = \dfrac{c}{a}$, $a = \dfrac{c}{e} = \dfrac{4}{2} = 2$ and $a^2 = 4$.

Because $b^2 = c^2 - a^2$, $b^2 = 4^2 - 2^2 = 16 - 4 = 12$.

Substituting h, k, a^2 and b^2 into the standard formula for a hyperbola yields $\dfrac{x^2}{4} - \dfrac{y^2}{12} = 1$

51.   conjugate axis length $= 4$, center $(4, 1)$, eccentricity $= \frac{4}{3}$

$2b =$ conjugate axis length $= 4$

$b = 2$ and $b^2 = 4$

Since

$e = \dfrac{c}{a} = \dfrac{4}{3}$, $c = \dfrac{4a}{3}$ and $c^2 = \dfrac{16a^2}{9}$. Since $a^2 + b^2 = c^2$, substituting $b^2 = 4$ and $c^2 = \dfrac{16a^2}{9}$ and solving for a^2 yields $a^2 = \dfrac{36}{7}$.

Substituting into the two standard equations of a hyperbola yields $\dfrac{(x-4)^2}{36/7} - \dfrac{(y-1)^2}{4} = 1$ and $\dfrac{(y-1)^2}{36/7} - \dfrac{(x-4)^2}{4} = 1$

53.   **a.**   Because the transmitters are 250 miles apart,

$2c = 250$ and $c = 125$.

$2a =$ rate $\times$ time

$2a = 0.186 \times 500 = 93$

Thus, $a = 46.5$ miles.

$b = \sqrt{c^2 - a^2} = \sqrt{125^2 - 46.5^2} = \sqrt{13,462.75}$ miles

The ship is located on the hyperbola given by

$\dfrac{x^2}{2,162.25} - \dfrac{y^2}{13,462.75} = 1$

  b. $x = 100$

$\dfrac{10,000}{2,162.25} - \dfrac{y^2}{13,462.75} = 1$

$\dfrac{-y^2}{13,462.75} \approx -3.6248121$

$y^2 \approx 48,799.939$

$y \approx 221$

The ship is 221 miles from the coastline.

55.   When the wave hits Earth, $z = 0$.

$y^2 = x^2 + (z - 10,000)^2$

$y = x^2 + (0 - 10,000)^2$

$y - x^2 = 10,000^2$

It is a hyperbola.

Copyright © Houghton Mifflin Company. All rights reserved.

57.

$$4x^2 + 9y^2 - 16x - 36y + 16 = 0$$

$$4\left(x^2 - 4x\right) + 9\left(y^2 - 4y\right) = -16$$

$$4\left(x^2 - 4x + 4\right) + 9\left(y^2 - 4y + 4\right) = -16 + 16 + 36$$

$$4(x-2)^2 + 9(y-2)^2 = 36$$

$$\frac{(x-2)^2}{9} + \frac{(y-2)^2}{4} = 1$$

ellipse

center $(2, 2)$

vertices $(2 \pm 3, 2) = (5, 2), (-1, 2)$

foci $\left(2 \pm \sqrt{5}, 2\right) = \left(2 + \sqrt{5}, 2\right), \left(2 - \sqrt{5}, 2\right)$

59.

$$5x - 4y^2 + 24y - 11 = 0$$

$$-4\left(y^2 - 6y\right) = -5x + 11$$

$$-4\left(y^2 - 6y + 9\right) = -5x + 11 - 36$$

$$-4(y-3)^2 = -5(x - 25)$$

$$-4(y-3)^2 = -5(x + 5)$$

$$(y-3)^2 = \frac{5}{4}(x + 5)$$

parabola

vertex $(-5, 3)$

focus $\left(-5 + \frac{5}{16}, 3\right) = \left(-\frac{75}{16}, 3\right)$

directrix $x = -5 - \frac{5}{16}$, or $x = \frac{-85}{16}$

61.

$$x^2 + 2y - 8x = 0$$

$$x^2 - 8x = -2y$$

$$x^2 - 8x + 16 = -2y + 16$$

$$(x-4)^2 = -2(y - 8)$$

parabola

vertex $(4, 8)$

foci $\left(4, 8 - \frac{1}{2}\right) = \left(4, \frac{15}{2}\right)$

directrix $y = 8 + \frac{1}{2}$, or $y = \frac{17}{2}$

63.

$$25x^2 + 9y^2 - 50x - 72y - 56 = 0$$

$$25\left(x^2 - 2x\right) + 9\left(y^2 - 8y\right) = 56$$

$$25\left(x^2 - 2x + 1\right) + 9\left(y^2 - 8y + 16\right) = 56 + 25 + 144$$

$$25(x-1)^2 + 9(y-4)^2 = 225$$

$$\frac{(x-1)^2}{9} + \frac{(y-4)^2}{25} = 1$$

ellipse

center $(1, 4)$

vertices $(1, 4 \pm 5) = (1, 9), (1, -1)$

foci $(1, 4 \pm 4) = (1, 8), (1, 0)$

Copyright © Houghton Mifflin Company. All rights reserved.

65. foci $F_1(2,0)$, $F_2(-2,0)$ passing through $P_1(2,3)$

$$d(P_1, F_2) - d(P_1, F_1) = \sqrt{(2+2)^2 + 3^2} - \sqrt{(2-2)^2 + 3^2} = 5 - 3 = 2$$

Let $P(x, y)$ be any point on the hyperbola. Since the difference between F_1P and F_2P is the same as the difference between F_1P_1 and F_2P_1, we have

$$\sqrt{(x-2)^2 + y^2} - \sqrt{(x+2)^2 + y^2} = 2$$
$$\sqrt{(x-2)^2 + y^2} = 2 + \sqrt{(x+2)^2 + y^2}$$
$$x^2 - 4x + 4 + y^2 = 4 + 4\sqrt{(x+2)^2 + y^2} + x^2 + 4x + 4 + y^2$$
$$-8x - 4 = 4\sqrt{(x+2)^2 + y^2}$$
$$-2x - 1 = \sqrt{(x+2)^2 + y^2}$$
$$4x^2 + 4x + 1 = x^2 + 4x + 4 + y^2$$
$$3x^2 - y^2 = 3$$
$$\frac{x^2}{1} - \frac{y^2}{3} = 1$$

67. foci $(0,4)$ and $(0,-4)$, point $\left(\frac{7}{3}, 4\right)$

Difference in distances from (x, y) to foci = difference of distances from $\left(\frac{7}{3}, 4\right)$ to foci

$$\sqrt{(x-0)^2 + (y-4)^2} - \sqrt{(x-0)^2 + (y+4)^2} = \sqrt{\left(\frac{7}{3} - 0\right)^2 + (4-4)^2} - \sqrt{\left(\frac{7}{3} - 0\right)^2 + (4+4)^2}$$
$$\sqrt{x^2 + y^2 - 8y + 16} - \sqrt{x^2 + y^2 + 8y + 16} = \frac{7}{3} - \frac{25}{3} = -6$$
$$\sqrt{x^2 + y^2 - 8y + 16} = \sqrt{x^2 + y^2 + 8y + 16} - 6$$
$$x^2 + y^2 - 8y + 16 = x^2 + y^2 + 8y + 16 - 12\sqrt{x^2 + y^2 + 8y + 16} + 36$$
$$-16y - 36 = -12\sqrt{x^2 + y^2 + 8y + 16}$$
$$4y + 9 = 3\sqrt{x^2 + y^2 + 8y + 16}$$
$$16y^2 + 72y + 81 = 9x^2 + 9y^2 + 72y + 144$$
$$7y^2 - 9x^2 = 63$$
$$\frac{y^2}{9} - \frac{x^2}{7} = 1$$

69.
$$\frac{x^2}{16} - \frac{y^2}{25} = 1$$
$$a^2 = 16$$
$$b^2 = 25$$
$$c^2 = a^2 + b^2 = 16 + 25 = 41$$
$$c = \sqrt{41}$$
$$\frac{a^2}{c} = \frac{16}{\sqrt{41}} = \frac{16\sqrt{41}}{41}.$$

Thus, the directrices are $x = \pm\dfrac{16\sqrt{41}}{41}$.

Copyright © Houghton Mifflin Company. All rights reserved.

71. $\dfrac{x^2}{9} - \dfrac{y^2}{16} = 1$, focus $(5, 0)$, directrix $x = \dfrac{9}{5}$

Solving for y^2 gives us

$$16x^2 - 9y^2 = 144$$

$$9y^2 = 16x^2 - 144$$

$$y^2 = \dfrac{16}{9}x^2 - 16$$

Let $k = \dfrac{\text{distance from } P(x, y) \text{ to focus } (5, 0)}{\text{distance from } P(x, y) \text{ to directrix } (x = 9/5)}$

$$k = \dfrac{\sqrt{(x-5)^2 + (y-0)^2}}{|x - 9/5|} = \dfrac{\sqrt{x^2 - 10x + 25 + y^2}}{|x - 9/5|}$$

But since $P(x, y)$ lies on the curve, $y^2 = \dfrac{16}{9}x^2 - 16$.

Substituting gives us $k = \dfrac{\sqrt{x^2 - 10x + 25 + 16x^2/9 - 16}}{|x - 9/5|} = \dfrac{\sqrt{\dfrac{25x^2}{9} - \dfrac{90x}{9} + \dfrac{81}{9}}}{|x - 9/5|} = \dfrac{\sqrt{25x^2 - 90x + 81}}{3|x - 9/5|}$.

Solving for k^2 yields

$$k^2 = \dfrac{25\left(x^2 - \dfrac{18}{5} + \dfrac{81}{25}\right)}{9\left(x^2 - \dfrac{18}{5} + \dfrac{81}{25}\right)} = \dfrac{25}{9}$$

$$k = \pm\sqrt{\dfrac{25}{9}} = \pm\dfrac{5}{3}$$

But since $k = \dfrac{\text{distance from } P \text{ to focus}}{\text{distance from } P \text{ to directrix}}$, and the ratio of two distances must be positive, $k = \dfrac{5}{3}$.

$$a^2 = 9$$
$$b^2 = 16$$
$$c^2 = a^2 + b^2 = 9 + 16 = 25$$
$$c = 5 \text{ and } a = 3$$

Since $e = \dfrac{c}{a} = \dfrac{5}{3}$ and $k = \dfrac{5}{3}$, $e = k$.

73.

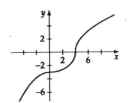

· ·

75. $\cos(\alpha + \beta) = \cos\alpha\cos\beta - \sin\alpha\sin\beta$

76. $\sin(\alpha + \beta) = \sin\alpha\cos\beta + \cos\alpha\sin\beta$

77.
$$\cot 2\alpha = \dfrac{\sqrt{3}}{3}$$
$$\tan 2\alpha = \dfrac{3}{\sqrt{3}}$$
$$2\alpha = \tan^{-1}\left(\dfrac{3}{\sqrt{3}}\right) = \dfrac{\pi}{3}$$
$$\alpha = \dfrac{\pi}{6}$$

78.
$$\sin\alpha = \dfrac{1}{2}, \ \alpha = 30° \text{ or } 150°$$
$$\cos\alpha = -\dfrac{\sqrt{3}}{2} \ \alpha = 150° \text{ or } 210°$$
$$\alpha = 150°$$

Copyright © Houghton Mifflin Company. All rights reserved.

79. $4x^2 - 6y^2 + 9x + 16y - 8 = 0$

$A = 4, B = 0, C = -6$

Since $B^2 - 4AC = 0^2 - 4(4)(-6) = 96 > 0$,

the graph is a hyperbola.

80.

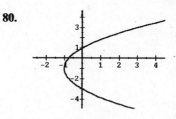

Section 6.4

1. $xy = 3$

$A = 0, B = 1, C = 0$

$\cot 2\alpha = \dfrac{A - C}{B}$

$\cot 2\alpha = \dfrac{0 - 0}{1}$

$\cot 2\alpha = 0$

$2\alpha = 90°$

$\alpha = 45°$

3. $9x^2 - 24xy + 16y^2 - 320x - 240y = 0$

$A = 9, B = -24, C = 16$

$\cot 2\alpha = \dfrac{A - C}{B}$

$\cot 2\alpha = \dfrac{9 - 16}{-24}$

$\cot 2\alpha = \dfrac{-7}{-24}$

$\cot 2\alpha = \dfrac{7}{24}$

$2\alpha \approx 73.74°$

$\alpha \approx 36.9°$

5. $5x^2 - 6\sqrt{3}xy - 11y^2 + 4x - 3y + 2 = 0$

$A = 5, B = -6\sqrt{3}, C = -11$

$\cot 2a = \dfrac{A - C}{B}$

$\cot 2a = \dfrac{5 - (-11)}{-6\sqrt{3}}$

$\cot 2a = \dfrac{5 + 11}{-6\sqrt{3}}$

$\cot 2a = \dfrac{16}{-6\sqrt{3}}$

$\cot 2a = -\dfrac{8}{3\sqrt{3}}$

$\cot 2a = -\dfrac{8\sqrt{3}}{9}$

$2\alpha \approx 147°$

$\alpha \approx 73.5°$

Copyright © Houghton Mifflin Company. All rights reserved.

7.
$$xy = 4$$
$$xy - 4 = 0$$

$$A = 0, B = 1, C = 0, F = -4$$

$$\cot 2\alpha = \frac{A-C}{B} = \frac{0-0}{1} = 0$$

$$\csc^2 2\alpha = \cot^2 2\alpha + 1$$
$$\csc^2 2\alpha = 0^2 + 1 = 1$$
$$\csc 2\alpha = +1 \quad (2\alpha \text{ is in the first quadrant.})$$

$$\sin 2\alpha = \frac{1}{\csc 2\alpha} = \frac{1}{1} = 1$$

$$\sin^2 2\alpha + \cos^2 2\alpha = 1$$
$$\cos^2 2\alpha = 1 - \sin^2 2\alpha$$
$$\cos^2 2\alpha = 1 - (1)^2$$
$$\cos^2 2\alpha = 0$$
$$\cos 2\alpha = 0$$

$$\sin\alpha = \sqrt{\frac{1-(0)}{2}} = \frac{\sqrt{2}}{2} \qquad \cos\alpha = \sqrt{\frac{1+(0)}{2}} = \frac{\sqrt{2}}{2}$$

$$\alpha = 45°$$

$$A' = A\cos^2\alpha + B\cos\alpha\sin\alpha + C\sin^2\alpha = 0\left(\frac{\sqrt{2}}{2}\right)^2 + 1\left(\frac{\sqrt{2}}{2}\right)\left(\frac{\sqrt{2}}{2}\right) + 0\left(\frac{\sqrt{2}}{2}\right)^2 = \frac{1}{2}$$

$$C' = A\sin^2\alpha - B\cos\alpha\sin\alpha + C\cos^2\alpha = 0\left(\frac{\sqrt{2}}{2}\right)^2 - 1\left(\frac{\sqrt{2}}{2}\right)\left(\frac{\sqrt{2}}{2}\right) + 0\left(\frac{\sqrt{2}}{2}\right)^2 = -\frac{1}{2}$$

$$F' = F = -4$$

$$\frac{1}{2}x'^2 - \frac{1}{2}y'^2 - 4 = 0 \text{ or } \frac{x'^2}{8} - \frac{y'^2}{8} = 1$$

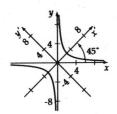

Copyright © Houghton Mifflin Company. All rights reserved.

9. $6x^2 - 6xy + 14y^2 - 45 = 0$

$A = 6, B = -6, C = 14, F = -45$

$\cot 2\alpha = \dfrac{A-C}{B} = \dfrac{6-14}{-6} = \dfrac{4}{3}$

$\csc^2 2\alpha = \cot^2 2\alpha + 1$

$\csc^2 2\alpha = \left(\dfrac{4}{3}\right)^2 + 1 = \dfrac{25}{9}$

$\csc 2\alpha = +\sqrt{\dfrac{25}{9}} = \dfrac{5}{3}$ $(2\alpha$ is in the first quadrant.$)$

$\sin 2\alpha = \dfrac{1}{\csc 2\alpha} = \dfrac{3}{5}$

$\sin^2 \alpha + \cos^2 2\alpha = 1$

$\cos^2 2\alpha = 1 - \sin^2 2\alpha$

$\cos^2 2\alpha = 1 - \left(\dfrac{3}{5}\right)^2 = \dfrac{16}{25}$

$\cos 2\alpha = +\sqrt{\dfrac{16}{25}} = \dfrac{4}{5}$ $(2\alpha$ is in the first quadrant.$)$

$\sin\alpha = \sqrt{\dfrac{1-\left(\frac{4}{5}\right)}{2}} = \dfrac{\sqrt{10}}{10}$ $\cos\alpha = \sqrt{\dfrac{1+\left(\frac{4}{5}\right)}{2}} = \dfrac{3\sqrt{10}}{10}$

$\alpha = 18.4°$

$A' = A\cos^2\alpha + B\cos\alpha\sin\alpha + C\sin^2\alpha = 6\left(\dfrac{3\sqrt{10}}{10}\right)^2 - 6\left(\dfrac{3\sqrt{10}}{10}\right)\left(\dfrac{\sqrt{10}}{10}\right) + 14\left(\dfrac{\sqrt{10}}{10}\right)^2 = 5$

$C' = A\sin^2\alpha - B\cos\alpha\sin\alpha + C\cos^2\alpha = 6\left(\dfrac{\sqrt{10}}{10}\right)^2 + 6\left(\dfrac{3\sqrt{10}}{10}\right)\left(\dfrac{\sqrt{10}}{10}\right) + 14\left(\dfrac{3\sqrt{10}}{10}\right)^2 = 15$

$F' = F = -45$

$5x'^2 + 15y'^2 - 45 = 0$ or $\dfrac{(x')^2}{9} + \dfrac{(y')^2}{3} = 1$

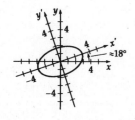

Copyright © Houghton Mifflin Company. All rights reserved.

11.

$$x^2 - 4xy + 2y^2 - 1 = 0$$

$$A = 1, B = 4, C = -2, F = -1$$

$$\cot 2\alpha = \frac{A-C}{B} = \frac{1-(-2)}{4} = \frac{3}{4}$$

$$\csc^2 2\alpha = \cot^2 2\alpha + 1$$

$$\csc^2 2\alpha = \left(\frac{3}{4}\right)^2 + 1 = \frac{25}{16}$$

$$\csc 2\alpha = +\sqrt{\frac{25}{16}} = \frac{5}{4} \qquad (2\alpha \text{ is in the first quadrant.})$$

$$\sin 2\alpha = \frac{1}{\csc 2\alpha} = \frac{4}{5}$$

$$\sin^2 \alpha + \cos^2 2\alpha = 1$$

$$\cos^2 2\alpha = 1 - \sin^2 \alpha$$

$$\cos^2 2\alpha = 1 - \left(\frac{4}{5}\right)^2 = \frac{9}{25}$$

$$\cos 2\alpha = +\sqrt{\frac{9}{25}} = \frac{3}{5} \qquad (2\alpha \text{ is in first quadrant})$$

$$\sin \alpha = \sqrt{\frac{1-\left(\frac{3}{5}\right)}{2}} = \frac{\sqrt{5}}{5} \qquad\qquad \cos \alpha = \sqrt{\frac{1+\left(\frac{3}{5}\right)}{2}} = \frac{2\sqrt{5}}{5}$$

$$\alpha \approx 26.6°$$

$$A' = A\cos^2 \alpha + B\cos \alpha \sin \alpha + C\sin^2 \alpha = 1\left(\frac{2\sqrt{5}}{5}\right)^2 + 4\left(\frac{2\sqrt{5}}{5}\right)\left(\frac{\sqrt{5}}{5}\right) - 2\left(\frac{\sqrt{5}}{5}\right)^2 = 2$$

$$C' = A\sin^2 \alpha - B\cos \alpha \sin \alpha + C\cos^2 \alpha = 1\left(\frac{\sqrt{5}}{5}\right)^2 - 4\left(\frac{2\sqrt{5}}{5}\right)\left(\frac{\sqrt{5}}{5}\right) - 2\left(\frac{2\sqrt{5}}{5}\right)^2 = -3$$

$$F' = F = -1$$

$$2(x')^2 + 3(y')^2 = 1$$

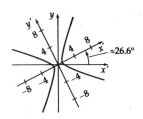

Copyright © Houghton Mifflin Company. All rights reserved.

13. $3x^2 + 2\sqrt{3}xy + y^2 + 2x - 2\sqrt{3}y + 16 = 0$

$A = 3, B = 2\sqrt{3}, C = 1, D = 2, E = -2\sqrt{3}, F = 16$

$\cot 2\alpha = \dfrac{A-C}{B} = \dfrac{3-1}{2\sqrt{3}} = \dfrac{\sqrt{3}}{3}$

$\csc^2 2\alpha = \cot^2 2\alpha + 1$

$\csc^2 2\alpha = \left(\dfrac{\sqrt{3}}{3}\right)^2 + 1 = \dfrac{4}{3}$

$\csc 2\alpha = +\sqrt{\dfrac{4}{3}} = \dfrac{2\sqrt{3}}{3}$ (2α is in the first quadrant.)

$\sin 2\alpha = \dfrac{1}{\csc 2\alpha} = \dfrac{\sqrt{3}}{2}$

$\sin^2 \alpha + \cos^2 2\alpha = 1$

$\quad \cos^2 2\alpha = 1 - \sin^2 2\alpha$

$\quad \cos^2 2\alpha = 1 - \left(\dfrac{\sqrt{3}}{2}\right)^2 = \dfrac{1}{4}$

$\quad \cos 2\alpha = +\sqrt{\dfrac{1}{4}} = \dfrac{1}{2}$ (2α is in the first quadrant.)

$\sin\alpha = \sqrt{\dfrac{1-\left(\frac{1}{2}\right)}{2}} = \dfrac{1}{2}$ $\cos\alpha = \sqrt{\dfrac{1+\left(\frac{1}{2}\right)}{2}} = \dfrac{\sqrt{3}}{2}$

$\alpha = 30°$

$A' = A\cos^2\alpha + B\cos\alpha\sin\alpha + C\sin^2\alpha = 3\left(\dfrac{\sqrt{3}}{2}\right)^2 + 2\sqrt{3}\left(\dfrac{\sqrt{3}}{2}\right)\left(\dfrac{1}{2}\right) + 1\left(\dfrac{1}{2}\right)^2 = 4$

$C' = A\sin^2\alpha - B\cos\alpha\sin\alpha + C\cos^2\alpha = 3\left(\dfrac{1}{2}\right)^2 - 2\sqrt{3}\left(\dfrac{\sqrt{3}}{2}\right)\left(\dfrac{1}{2}\right) + 1\left(\dfrac{\sqrt{3}}{2}\right)^2 = 0$

$D' = D\cos\alpha + E\sin\alpha = 2\left(\dfrac{\sqrt{3}}{2}\right) - 2\sqrt{3}\left(\dfrac{1}{2}\right) = 0$

$E' = -D\sin\alpha + E\cos\alpha = -2\left(\dfrac{1}{2}\right) - 2\sqrt{3}\left(\dfrac{\sqrt{3}}{2}\right) = -4$

$F' = F = 16$

$4(x')^2 - 4y' + 16 = 0$ or $y' = (x')^2 + 4$

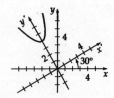

Copyright © Houghton Mifflin Company. All rights reserved.

15. $9x^2 + 24xy + 16y^2 - 40x - 30y + 100 = 0$

$A = 9, B = -24, C = 16, D = -40, E = -30 \ F = 100$

$$\cot 2\alpha = \frac{A-C}{B} = \frac{9-16}{-24} = \frac{7}{24}$$

$$\csc^2 2\alpha = \cot^2 2\alpha + 1$$

$$\csc^2 2\alpha = \left(\frac{7}{24}\right)^2 + 1 = \frac{625}{576}$$

$$\csc 2\alpha = +\sqrt{\frac{625}{576}} = \frac{25}{24}\qquad (2\alpha \text{ is in the first quadrant.})$$

$$\sin 2\alpha = \frac{1}{\csc 2\alpha} = \frac{24}{25}$$

$$\sin^2 \alpha + \cos^2 2\alpha = 1$$

$$\cos^2 2\alpha = 1 - \sin^2 2\alpha$$

$$\cos^2 2\alpha = 1 - \left(\frac{24}{25}\right)^2 = \frac{49}{625}$$

$$\cos 2\alpha = +\sqrt{\frac{49}{625}} = \frac{7}{25}\qquad (2\alpha \text{ is in first quadrant.})$$

$$\sin \alpha = \sqrt{\frac{1-\left(\frac{7}{25}\right)}{2}} = \frac{3}{5}\qquad \cos \alpha = \sqrt{\frac{1+\left(\frac{7}{25}\right)}{2}} = \frac{4}{5}$$

$\alpha \approx 36.9°$

$$A' = A\cos^2 \alpha + B\cos \alpha \sin \alpha + C\sin^2 \alpha = 9\left(\frac{4}{5}\right)^2 - 24\left(\frac{4}{5}\right)\left(\frac{3}{5}\right) + 16\left(\frac{3}{5}\right)^2 = 0$$

$$C' = A\sin^2 \alpha - B\cos \alpha \sin \alpha + C\cos^2 \alpha = 9\left(\frac{3}{5}\right)^2 + 24\left(\frac{4}{5}\right)\left(\frac{3}{5}\right) + 16\left(\frac{4}{5}\right)^2 = 25$$

$$D' = D\cos \alpha + E\sin \alpha = -40\left(\frac{4}{5}\right) - 30\left(\frac{3}{5}\right) = -50$$

$$E' = -D\sin \alpha + E\cos \alpha = 40\left(\frac{3}{5}\right) - 30\left(\frac{4}{5}\right) = 0$$

$$F' = F = 100$$

$$25(x')^2 - 50x' + 100 = 0 \text{ or } (y')^2 = 2(x-2)$$

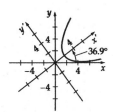

Copyright © Houghton Mifflin Company. All rights reserved.

17.
$$6x^2 + 24xy - y^2 - 12x + 26y + 11 = 0$$
$$A = 6, B = 24, C = -1, D = -12, E = 26 \ F = 11$$

$$\cot 2\alpha = \frac{A-C}{B} = \frac{6-(-1)}{24} = \frac{7}{24}$$

$$\csc^2 2\alpha = \cot^2 2\alpha + 1$$

$$\csc^2 2\alpha = \left(\frac{7}{24}\right)^2 + 1 = \frac{625}{576}$$

$$\csc 2\alpha = +\sqrt{\frac{625}{576}} = \frac{25}{24} \qquad (2\alpha \text{ is in the first quadrant.})$$

$$\sin 2\alpha = \frac{1}{\csc 2\alpha} = \frac{24}{25}$$

$$\sin^2 2\alpha + \cos^2 2\alpha = 1$$

$$\cos^2 2\alpha = 1 - \sin^2 2\alpha$$

$$\cos^2 2\alpha = 1 - \left(\frac{24}{25}\right)^2 = \frac{49}{625}$$

$$\cos 2\alpha = +\sqrt{\frac{49}{625}} = \frac{7}{25} \qquad (2\alpha \text{ is in the first quadrant.})$$

$$\sin\alpha = \sqrt{\frac{1-\left(\frac{7}{25}\right)}{2}} = \frac{3}{5} \qquad\qquad \cos\alpha = \sqrt{\frac{1+\left(\frac{7}{25}\right)}{2}} = \frac{4}{5}$$

$$\alpha \approx 36.9°$$

$$A' = A\cos^2\alpha + B\cos\alpha\sin\alpha + C\sin^2\alpha = 6\left(\frac{4}{5}\right)^2 + 24\left(\frac{4}{5}\right)\left(\frac{3}{5}\right) - 1\left(\frac{3}{5}\right)^2 = 15$$

$$C' = A\sin^2\alpha - B\cos\alpha\sin\alpha + C\cos^2\alpha = 6\left(\frac{3}{5}\right)^2 - 24\left(\frac{4}{5}\right)\left(\frac{3}{5}\right) - 1\left(\frac{4}{5}\right)^2 = -10$$

$$D' = D\cos\alpha + E\sin\alpha = -12\left(\frac{4}{5}\right) + 26\left(\frac{3}{5}\right) = 6$$

$$E' = -D\sin\alpha + E\cos\alpha = 12\left(\frac{3}{5}\right) + 26\left(\frac{4}{5}\right) = 28$$

$$F' = F = 11$$

$$15(x')^2 - 10(y')^2 + 6x' + 28y' + 11 = 0$$

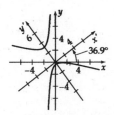

19. $A = 6, B = -1, C = 2, D = 4, E = -12, F = 7$

Graph $y = \dfrac{-(-x-12) \pm \sqrt{(-x-12)^2 - 8(6x^2 + 4x + 7)}}{4}$

The graph will appear disconnected at the endpoints of the minor axes on a graphing utility.

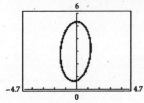

21. $A = 1, B = -6, C = 1, D = -2, E = -5, F = 4$

Graph $y = \dfrac{-(-6x-5) \pm \sqrt{(-6x-5)^2 - 4(x^2 - 2x + 4)}}{2}$

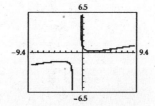

Copyright © Houghton Mifflin Company. All rights reserved.

23. $A = 3, B = -6, C = 3, D = 10, E = -8, F = -2$

Graph $y = \dfrac{-(-6x - 8) \pm \sqrt{(-6x - 8)^2 - 12(3x^2 + 10x - 2)}}{6}$

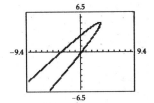

25. $\dfrac{2x'^2}{1} - \dfrac{3y'^2}{1} = 1 \qquad \sin\alpha = \dfrac{\sqrt{5}}{5} \qquad \cos\alpha = \dfrac{2\sqrt{5}}{5}$

$a^2 = \dfrac{1}{2}; \quad a = \dfrac{\sqrt{2}}{2}$

$b^2 = \dfrac{1}{3}; \quad b = \dfrac{\sqrt{3}}{3}$

Asymptotes: $y' = \pm\dfrac{b}{a}x'$ or $y' = \pm\dfrac{\sqrt{6}}{3}x'$

Using the transformation formulas for x' and y' yields

$y\cos\alpha - x\sin\alpha = \pm\dfrac{\sqrt{6}}{3}\left(x\cos\alpha + y\sin\alpha\right)$

$\dfrac{2\sqrt{5}}{5}y - \dfrac{\sqrt{5}}{5}x = \pm\dfrac{\sqrt{6}}{3}\left(\dfrac{2\sqrt{5}}{5}x + \dfrac{\sqrt{5}}{5}y\right)$

$\dfrac{2\sqrt{5}}{5}y - \dfrac{\sqrt{5}}{5}x = \pm\left(\dfrac{2\sqrt{30}}{15} + \dfrac{\sqrt{30}}{15}y\right)$

Multiplying both sides of the equation by $15/\sqrt{5}$ yields

$6y - 3x = \pm\left(2\sqrt{6}x + \sqrt{6}y\right)$

$6y - 3x = 2\sqrt{6}x + \sqrt{6}y \qquad \text{and} \qquad 6y - 3x = -(2\sqrt{6}x + \sqrt{6}y)$

$6y - \sqrt{6}y = 3x + 2\sqrt{6}x \qquad\qquad 6y + \sqrt{6}y = 3x - 2\sqrt{6}x$

$\qquad y = \dfrac{3 + 2\sqrt{6}}{6 - \sqrt{6}}x \qquad\qquad\qquad y = \dfrac{3 - 2\sqrt{6}}{6 + \sqrt{6}}x$

Rationalizing the denominators, we obtain

$\qquad y = \dfrac{2 + \sqrt{6}}{2}x \qquad \text{and} \qquad y = \dfrac{2 - \sqrt{6}}{2}x$

27.

From Exercise 9, $\dfrac{x'^2}{9} + \dfrac{y'^2}{3} = 1, \sin\alpha = \dfrac{\sqrt{10}}{10}, \cos\alpha = \dfrac{3\sqrt{10}}{10}, a^2 = 9, b^2 = 3, c^2 = 9 - 3 = 6.$

Thus $c = \sqrt{6}$.

Foci in $x'y'$-coordinates are $(\pm\sqrt{6}, 0)$. Thus $x' = \pm\sqrt{6}, y' = 0$.

$x = x'\cos\alpha - y'\sin\alpha \qquad\qquad y = y'\cos\alpha + x'\sin\alpha$

$\quad = \pm\sqrt{6}\left(\dfrac{3\sqrt{10}}{10}\right) - 0\cdot\dfrac{\sqrt{10}}{10} \qquad\quad = 0\cdot\cos\alpha \pm\sqrt{6}\left(\dfrac{\sqrt{10}}{10}\right)$

$\quad = \pm\dfrac{3\sqrt{15}}{5} \qquad\qquad\qquad\qquad\quad = \pm\dfrac{\sqrt{15}}{5}$

Foci in the xy-coordinate sysetm are $\left(\dfrac{3\sqrt{15}}{5}, \dfrac{\sqrt{15}}{5}\right)$ and $\left(-\dfrac{3\sqrt{15}}{5}, -\dfrac{\sqrt{15}}{5}\right)$.

Copyright © Houghton Mifflin Company. All rights reserved.

29. $x^2 + xy - y^2 - 40 = 0$

$A = 1, B = 1, C = -1$

Since $B^2 - 4AC = 1^2 - 4(1)(-1) = 5 > 0$,
the graph is a hyperbola.

31. $3x^2 + 2\sqrt{3}xy + y^2 - 3x + 2y + 20 = 0$

$A = 3, B = 2\sqrt{3}, C = 1$

Since $B^2 - 4AC = (2\sqrt{3})^2 - 4(3)(1) = 0$,
the graph is a parabola.

33. $4x^2 - 4xy + y^2 - 12y - 20 = 0$

$A = 4, B = -4, C = 1$

Since $B^2 - 4AC = (-4)^2 - 4(4)(1) = 0$, the graph is a parabola.

35. $5x^2 - 6\sqrt{3}xy - 11y^2 + 4x - 3y + 2 = 0$

$A = 5, B = -6\sqrt{3}, C = -11$

Since $B^2 - 4AC = (-6\sqrt{3})^2 - 4(5)(-11) = 328 > 0$,
the graph is a hyperbola.

Connecting Concepts

37. $x^2 + y^2 = r^2$

Substitute $x = x'\cos\alpha - y'\sin\alpha$ and $y = y'\cos\alpha + x'\sin\alpha$.

$$(x'\cos\alpha - y'\sin\alpha)^2 + (y'\cos\alpha + x'\sin\alpha)^2 = r^2$$
$$x'^2\cos^2\alpha - 2x'y'\cos\alpha\sin\alpha + y'^2\sin^2\alpha + y'^2\cos^2\alpha + 2x'y'\cos\alpha\sin\alpha + x'^2\sin^2\alpha = r^2$$
$$x'^2(\cos^2\alpha + \sin^2\alpha) + x'y'(2\cos\alpha\sin\alpha - 2\cos\alpha\sin\alpha) + y'^2(\sin^2\alpha + \cos^2\alpha) = r^2$$
$$x'^2(1) + x'y'(0) + y'^2(1) = r^2$$
$$x'^2 + y'^2 = r^2$$

39. Vertices $(2, 4)$ and $(-2, -4)$ imply that the major axis is on the line $y = 2x$. Consider an $x'y'$-coordinate system rotated an angle α, where $\tan\alpha = 2$. From this equation, using identities, $\cos\alpha = \dfrac{\sqrt{5}}{5}$ and $\sin\alpha = \dfrac{2\sqrt{5}}{5}$.

The equation of the ellipse in the $x'y'$-coordinate system is

$$\frac{x'^2}{20} + \frac{y'^2}{10} = 1 \qquad (1)$$

This follows from the fact that $(2, 4)$ in xy-coordinates is $\left(\sqrt{20}, 0\right)$ in $x'y'$-coordinates.

Therefore, $\alpha = \sqrt{20}$. Also, $\left(\sqrt{2}, 2\sqrt{2}\right)$ in xy-coordinates is $\left(\sqrt{10}, 0\right)$ in $x'y'$-coordinates.

Therefore, $c = \sqrt{10}$. Thus $b = \sqrt{10}$.

Now let $x' = \dfrac{\sqrt{5}}{5}x + \dfrac{2\sqrt{5}}{5}y$ and $y' = \dfrac{\sqrt{5}}{5}y - \dfrac{2\sqrt{5}}{5}x$ and substitute into Equation (1):

$$\frac{\left(\frac{\sqrt{5}}{5}x + \frac{2\sqrt{5}}{5}y\right)^2}{20} + \frac{\left(\frac{\sqrt{5}}{5}y - \frac{2\sqrt{5}}{5}x\right)^2}{10} = 1$$

Simplifying, we have

$$\frac{\left(\frac{1}{5}x^2 + \frac{4}{5}xy + \frac{4}{5}y^2\right)}{20} + \frac{\left(\frac{4}{5}x^2 - \frac{4}{5}xy + \frac{1}{5}y^2\right)}{10} = 1$$
$$\frac{\frac{9}{5}x^2 - \frac{4}{5}xy + \frac{6}{5}y^2}{20} = 1$$
$$\frac{9x^2 - 4xy + 6y^2}{100} = 1$$
$$\text{or } 9x^2 - 4xy + 6y^2 = 100$$

41. $A' + C' = A\cos^2\alpha + B\cos\alpha\sin\alpha + C\sin^2\alpha + A\sin^2\alpha - B\cos\alpha\sin\alpha + C\cos^2\alpha$

$\qquad = A(\cos^2\alpha + \sin^2\alpha) + B(\cos\alpha\sin\alpha - \cos\alpha\sin\alpha) + C(\sin^2\alpha + \cos^2\alpha)$

$\qquad = A + C$

Copyright © Houghton Mifflin Company. All rights reserved.

43.   Ellipse with major axis parallel to x-axis:

$$\frac{(x-h)^2}{a^2} + \frac{(y-k)^2}{b^2} = 1$$

$$b^2(x-h)^2 + a^2(y-k)^2 = a^2b^2$$
$$b^2(x^2 - 2hx + h^2) + a^2(y^2 - 2ky + k^2) = a^2b^2$$
$$b^2x^2 - 2b^2hx + b^2h^2 + a^2y^2 - 2a^2ky + a^2k^2 = a^2b^2$$
$$b^2x^2 + a^2y^2 - 2b^2hx - 2a^2ky + b^2h^2 + a^2k^2 - a^2b^2 = 0$$

$A = b^2, B = 0, C = a^2$

$B^2 - 4AC = 0^2 - 4b^2a^2 = 4a^2b^2 < 0$

$B^2 - 4AC < 0$ for an ellipse whose major axis is parallel to the x-axis.

Ellipse with major axis parallel to y-axis:

$$\frac{(y-k)^2}{a^2} + \frac{(x-h)^2}{b^2} = 1$$

$$b^2(y-k)^2 + a^2(x-h)^2 = a^2b^2$$
$$b^2(y^2 - 2ky + k^2) + a^2(x^2 - 2hx + h^2) = a^2b^2$$
$$b^2y^2 - 2b^2ky + b^2k^2 + a^2x^2 - 2a^2hx + a^2h^2 = a^2b^2$$
$$b^2y^2 + a^2x^2 - 2b^2ky - 2a^2hx + b^2k^2 + a^2h^2 - a^2b^2 = 0$$

$A = b^2, B = 0, C = a^2$

$B^2 - 4AC = 0^2 - 4b^2a^2 = -4a^2b^2 < 0$

$B^2 - 4AC < 0$ for an ellipse whose major axis is parallel to the y-axis.

Parabola with axis of symmetry parallel to y-axis:

$$(x-h)^2 = 4p(y-k)$$
$$x^2 - 2hx + h^2 = 4py - 4pk$$
$$x^2 - 2hx - 4py + h^2 + 4pk = 0$$

$A = 1, B = 0, C = 0$

$B^2 - 4AC = 0^2 - 4(1)(0) = 0$

$B^2 - 4AC = 0$ for a parabola with axis of symmetry parallel to the y-axis.

Parabola with axis of symmetry parallel to x-axis:

$$(y-k)^2 = 4p(x-h)$$
$$y^2 - 2ky + k^2 = 4px - 4ph$$
$$y^2 - 2ky - 4px + k^2 + 4ph = 0$$

$A = 1, B = 0, C = 0$

$B^2 - 4AC = 0^2 - 4(1)(0) = 0$

$B^2 - 4AC = 0$ for a parabola with axis of symmetry parallel to the x-axis.

Hyperbola with the transverse axis parallel to x-axis:

$$\frac{(x-h)^2}{a^2} - \frac{(y-k)^2}{b^2} = 1$$

$$b^2(x-h)^2 - a^2(y-k)^2 = a^2b^2$$
$$b^2(x^2 - 2hx + h^2) - a^2(y^2 - 2ky + k^2) = a^2b^2$$
$$b^2x^2 - 2b^2hx + b^2h^2 - a^2y^2 + 2a^2ky - a^2k^2 = a^2b^2$$
$$b^2x^2 - a^2y^2 - 2b^2hx + 2a^2ky + b^2h^2 - a^2k^2 - a^2b^2 = 0$$

$A = b^2, B = 0, C = -a^2$

$B^2 - 4AC = 0^2 - 4b^2(-a^2) = 4a^2b^2 > 0$

$B^2 - 4AC > 0$ for a hyperbola whose transverse axis is parallel to the x-axis.

Copyright © Houghton Mifflin Company. All rights reserved.

Hyperbola with transverse axis parallel to y-axis:

$$\frac{(y-k)^2}{a^2} - \frac{(x-h)^2}{b^2} = 1$$

$$b^2(y-k)^2 - a^2(x-h)^2 = a^2b^2$$
$$b^2(y^2 - 2ky + k^2) - a^2(x^2 - 2hx + h^2) = a^2b^2$$
$$b^2y^2 - 2b^2ky + b^2k^2 - a^2x^2 + 2a^2hx - a^2h^2 = a^2b^2$$
$$b^2y^2 - a^2x^2 - 2b^2ky + 2a^2hx + b^2k^2 - a^2h^2 - a^2b^2 = 0$$

$A = -a^2$, $B = 0$, $C = b^2$

$B^2 - 4AC = 0^2 - 4b^2(-a^2) = 4a^2b^2 > 0$

$B^2 - 4AC > 0$ for a hyperbola whose transverse axis is parallel to the y-axis.

• •

Prepare for Section 6.5

45. $\sin(-x) = -\sin x$ odd function

46. $\cos(-x) = \cos x$ even function

47. $\tan\alpha = -\sqrt{3}$

$\alpha = \dfrac{2\pi}{3}, \dfrac{5\pi}{3}$

48.

$\sin\alpha = -\dfrac{\sqrt{3}}{2}$, $\alpha = 240°$ or $300°$

$\cos\alpha = -\dfrac{1}{2}$ $\alpha = 120°$ or $240°$

$\alpha = 240°$

49. $(r\cos\theta)^2 + (r\sin\theta)^2 = r^2\cos^2\theta + r^2\sin^2\theta$
$$= r^2(\cos^2\theta + \sin^2\theta)$$
$$= r^2$$

50.

$\sin 32° = \dfrac{y}{5}$ $\quad y = 5\sin 32° \approx 2.6$

$\cos 32° = \dfrac{x}{5}$ $\quad x = 5\cos 32° \approx 4.2$

$(4.2, 2.6)$

Section 6.5

1.

3.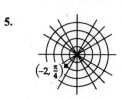

5.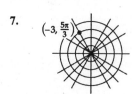

7. $\left(-3, \dfrac{5\pi}{3}\right)$

9. $0 \le \theta \le 2\pi$
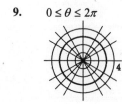

11.

13. $0 \le \theta \le \pi$
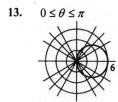

15. $0 \le \theta \le 2\pi$
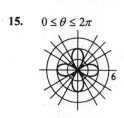

17. $0 \le \theta \le \pi$
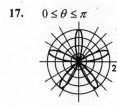

19. $0 \le \theta \le 2\pi$
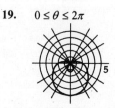

21. $0 \le \theta \le 2\pi$
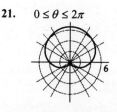

23. $0 \le \theta \le 2\pi$
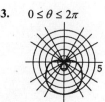

Copyright © Houghton Mifflin Company. All rights reserved.

25. Graph for $0 \le \theta \le 2\pi$.

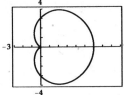

27. Graph for $0 \le \theta \le \pi$.

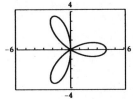

29. Graph for $0 \le \theta \le \pi$.

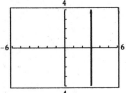

31. Graph for $0 \le \theta \le \pi$.

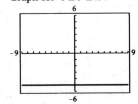

33. Graph for $0 \le \theta \le 4\pi$.

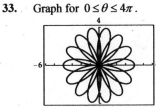

35. Graph for $0 \le \theta \le 6\pi$.

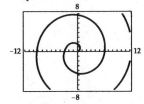

37. Graph for $0 \le \theta \le 2\pi$.

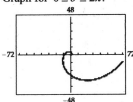

39. Graph for $0 \le \theta \le 2\pi$ with $\theta_{step} = \pi/200$.

(Some graphing utilities may draw a false asymptote in "connected" mode.)

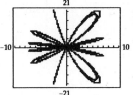

41.
$$r = \sqrt{x^2 + y^2}$$
$$= \sqrt{1^2 + (-\sqrt{3})^2}$$
$$= \sqrt{1+3}$$
$$= \sqrt{4}$$
$$= 2$$

$$\theta = \tan^{-1}\frac{y}{x}$$
$$= \tan^{-1}\left(\frac{-\sqrt{3}}{1}\right)$$
$$= \tan^{-1}\left(-\frac{\sqrt{3}}{1}\right)$$
$$= -60°$$

The polar coordinates of the point are $(2, -60°)$.

43.
$$x = r\cos\theta$$
$$= (-3)\left(\cos\frac{2\pi}{3}\right)$$
$$= (-3)\left(-\frac{1}{2}\right)$$
$$= \frac{3}{2}$$

$$y = r\sin\theta$$
$$= (-3)\left(\sin\frac{2\pi}{3}\right)$$
$$= (-3)\left(\frac{\sqrt{3}}{2}\right)$$
$$= -\frac{3\sqrt{3}}{2}$$

The rectangular coordinates of the point are $\left(\frac{3}{2}, -\frac{3\sqrt{3}}{2}\right)$.

45.
$$x = r\cos\theta$$
$$= 0\cos\left(-\frac{\pi}{2}\right)$$
$$= 0$$

$$y = r\sin\theta$$
$$= 0\sin\left(-\frac{\pi}{2}\right)$$
$$= 0$$

The rectangular coordinates of the point are $(0, 0)$

47.
$$r = \sqrt{x^2 + y^2}$$
$$= \sqrt{(3)^2 + (4)^2}$$
$$= \sqrt{9+16}$$
$$= \sqrt{25}$$
$$= 5$$

$$\theta = \tan^{-1}\frac{y}{x}$$
$$= \tan^{-1}\frac{4}{3}$$
$$\approx 53.1°$$

The approximate polar coordinates of the point are $(5, 53.1°)$.

49.
$$r = 3\cos\theta$$
$$r - 3\cos\theta = 0$$
$$r^2 - 3r\cos\theta = 0$$
$$x^2 + y^2 - 3x = 0$$

51.
$$r = 3\sec\theta$$
$$r = \frac{3}{\cos\theta}$$
$$r\cos\theta = 3$$
$$x = 3$$

53.
$$r = 4$$
$$\sqrt{x^2 + y^2} = 4$$
$$x^2 + y^2 = 16$$

Copyright © Houghton Mifflin Company. All rights reserved.

55.
$$r = \tan\theta$$
$$r = \frac{\sin\theta}{\cos\theta}$$
$$r\cos\theta = \sin\theta$$
$$r\cos\theta - \sin\theta = 0$$
$$r^2\cos\theta - r\sin\theta = 0$$
$$\sqrt{x^2+y^2}(x) - y = 0$$
$$\sqrt{x^2+y^2} = \frac{y}{x}$$
$$x^2+y^2 = \frac{y^2}{x^2}$$
$$x^4 - y^2 + x^2y^2 = 0$$

57.
$$r = \frac{2}{1+\cos\theta}$$
$$r + r\cos\theta = 2$$
$$\sqrt{x^2+y^2} + x = 2$$
$$\sqrt{x^2+y^2} = 2 - x$$
$$x^2 + y^2 = 4 - 4x + x^2$$
$$y^2 + 4x - 4 = 0$$

59.
$$r(\sin\theta - 2\cos\theta) = 6$$
$$r\sin\theta - 2r\cos\theta = 6$$
$$y - 2x = 6$$
$$y = 2x + 6$$

61.
$$y = 2$$
$$r\sin\theta = 2$$
$$r = 2\csc\theta$$

63.
$$x^2 + y^2 = 4$$
$$r^2 = 4$$
$$r = 2$$

65.
$$x^2 = 8y$$
$$r^2\cos^2\theta = 8r\sin\theta$$
$$r\cos^2\theta = 8\sin\theta$$

67.
$$x^2 - y^2 = 25$$
$$r^2\cos^2\theta - r^2\sin^2\theta = 25$$
$$r^2(\cos^2\theta - \sin^2\theta) = 25$$
$$r^2(\cos 2\theta) = 25$$

69.
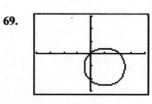

71.
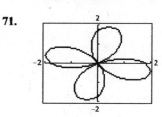

73.
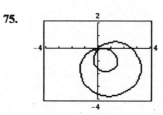

75.
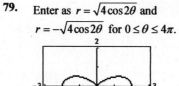

• •

Connecting Concepts

77. $\cos\theta = \pm\sqrt{\cos^2\theta}$ is *not* an identity.

79. Enter as $r = \sqrt{4\cos 2\theta}$ and $r = -\sqrt{4\cos 2\theta}$ for $0 \le \theta \le 4\pi$.

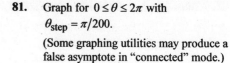

81. Graph for $0 \le \theta \le 2\pi$ with $\theta_{\text{step}} = \pi/200$.

(Some graphing utilities may produce a false asymptote in "connected" mode.)

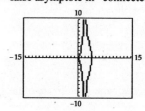

83. Graph $r = 2/\theta$ for $-4\pi < \theta < 4\pi$.

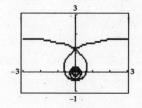

85. Graph for $-30 \le \theta \le 30$.

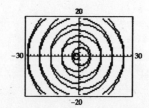

Copyright © Houghton Mifflin Company. All rights reserved.

87. a. $0 \le \theta \le 5\pi$ **b.** $0 \le \theta \le 20\pi$

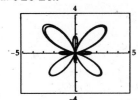   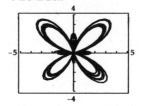

P<small>REPARE FOR</small> S<small>ECTION</small> **6.6**

88.
$$\frac{x^2}{25} + \frac{y^2}{16} = 1$$
$$a^2 = 25, \ a = 5$$
$$b^2 = 16$$
$$c^2 = a^2 - b^2 = 25 - 16 = 9$$
$$c = 3$$
$$e = \frac{c}{a} = \frac{3}{5}$$

The eccentricity is $\frac{3}{5}$.

89.
$$y^2 = 4x$$
$$4p = 4$$
$$p = 1$$
directrix: $x = -1$

90.
$$y = 2(1 + yx)$$
$$y = 2 + 2yx$$
$$y - 2yx = 2$$
$$y(1 - 2x) = 2$$
$$y = \frac{2}{1 - 2x}$$

91.
$$1 + \sin x = 0$$
$$\sin x = -1$$
$$x = \frac{3\pi}{2}$$

92. For a hyperbola, $e > 1$.

93.
$$\frac{4\sec x}{2\sec x - 1} = \frac{4\dfrac{1}{\cos x}}{2\dfrac{1}{\cos x} - 1}$$
$$= \frac{\cos x}{\cos x} \cdot \frac{4\dfrac{1}{\cos x}}{2\dfrac{1}{\cos x} - 1}$$
$$= \frac{4}{2 - \cos x}$$

Copyright © Houghton Mifflin Company. All rights reserved.

Section 6.6

1. $r = \dfrac{12}{3-6\cos\theta} = \dfrac{4}{1-2\cos\theta}$

$e = 2$ The graph is a hyperbola.

The transverse axis is on the polar axis.

Let $\theta = 0$.

$r = \dfrac{12}{3-6\cos 0} = \dfrac{12}{3-6} = -4$

Let $\theta = \pi$.

$r = \dfrac{12}{3-6\cos\pi} = \dfrac{12}{3+6} = \dfrac{4}{3}$

The vertices are at $(-4,\ 0)$ and $\left(\dfrac{4}{3},\ \pi\right)$.

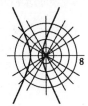

3. $r = \dfrac{8}{4+3\sin\theta} = \dfrac{2}{1+\frac{3}{4}\sin\theta}$

$e = \dfrac{3}{4}$ The graph is an ellipse.

The major axis is on the line $\theta = \dfrac{\pi}{2}$.

Let $\theta = \dfrac{\pi}{2}$.

$r = \dfrac{8}{4+3\sin\frac{\pi}{2}} = \dfrac{8}{4+3} = \dfrac{8}{7}$

Let $\theta = \dfrac{3\pi}{2}$.

$r = \dfrac{8}{4+3\sin\frac{3\pi}{2}} = \dfrac{8}{4-3} = 8$

Vertices on major axis are at $\left(\dfrac{8}{7},\ \dfrac{\pi}{2}\right)$ and $\left(8,\ \dfrac{3\pi}{2}\right)$.

Let $\theta = 0$.

$r = \dfrac{8}{4+3\sin 0} = \dfrac{8}{4+0} = 2$

Let $\theta = \pi$.

$r = \dfrac{8}{4+3\sin\pi} = \dfrac{8}{4+0} = 2$

The curve also goes through $(2, 0)$ and $(2, \pi)$.

5. $r = \dfrac{9}{3-3\sin\theta} = \dfrac{3}{1-\sin\theta}$

$e = 1$ The graph is a parabola.

The axis of symmetry is $\theta = \dfrac{\pi}{2}$.

When $\theta = \dfrac{\pi}{2}$, r is undefined.

Let $\theta = \dfrac{3\pi}{2}$.

$r = \dfrac{9}{3-3\sin\frac{3\pi}{2}} = \dfrac{9}{3+3} = \dfrac{3}{2}$

Vertex is at $\left(\dfrac{3}{2}, \dfrac{3\pi}{2}\right)$.

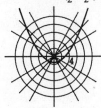

7. $r = \dfrac{10}{5+6\cos\theta} = \dfrac{2}{1+\frac{6}{5}\cos\theta}$

$e = \dfrac{6}{5}$ The graph is a hyperbola.

The transverse axis is on the polar axis.

Let $\theta = 0$.

$r = \dfrac{10}{5+6\cos 0} = \dfrac{10}{5+6} = \dfrac{10}{11}$

Let $\theta = \pi$.

$r = \dfrac{10}{5+6\cos\pi} = \dfrac{10}{5-6} = -10$

The vertices are at $\left(\dfrac{10}{11}, 0\right)$ and $(-10, \pi)$.

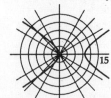

Copyright © Houghton Mifflin Company. All rights reserved.

9.

$$r = \frac{4\sec\theta}{2\sec\theta - 1}$$

$$= \frac{\dfrac{4}{\cos\theta}}{\dfrac{2}{\cos\theta} - 1} = \frac{4}{2 - \cos\theta}$$

$$= \frac{2}{1 - \frac{1}{2}\cos\theta}$$

$e = \frac{1}{2}$ The graph is an ellipse.

The major axis is on the polar axis.

However, the original equation is undefined at $\frac{\pi}{2}$ and at $\frac{3\pi}{2}$.

Thus, the ellipse will have holes at those angles.

Let $\theta = 0$.

$$r = \frac{4}{2 - \cos 0} = \frac{4}{2 - 1} = 4$$

Let $\theta = \pi$.

$$r = \frac{4}{2 - \cos\pi} = \frac{4}{2 + 1} = \frac{4}{3}$$

Vertices on major axis are at $(4, 0)$ and $\left(\frac{4}{3}, \pi\right)$.

$\theta = \frac{\pi}{2}$.

$$r = \frac{4}{2 - \cos\frac{\pi}{2}} = \frac{4}{2 - 0} = 2$$

Let $\theta = \frac{3\pi}{2}$.

$$r = \frac{4}{2 - \cos\frac{3\pi}{2}} = \frac{4}{2 - 0} = 2$$

Vertices on minor axis of $\frac{2}{1 - \frac{1}{2}\cos\theta}$ are at

$\left(2, \frac{\pi}{2}\right)$ and $\left(2, \frac{3\pi}{2}\right)$.

Thus, the equation $r = \frac{4\sec\theta}{2\sec\theta - 1}$ will have holes at

$\left(2, \frac{\pi}{2}\right)$ and $\left(2, \frac{3\pi}{2}\right)$

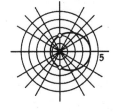

11.

$$r = \frac{12\csc\theta}{6\csc\theta - 2}$$

$$= \frac{\dfrac{12}{\sin\theta}}{\dfrac{6}{\sin\theta} - 2} = \frac{12}{6 - 2\sin\theta}$$

$$= \frac{2}{1 - \frac{1}{3}\sin\theta}$$

$e = \frac{1}{3}$ The graph is an ellipse.

The major axis is on $\theta = \frac{\pi}{2}$.

Let $\theta = \frac{\pi}{2}$.

$$r = \frac{12}{6 - 2\sin\frac{\pi}{2}} = \frac{12}{6 - 2} = 3$$

Let $\theta = \frac{3\pi}{2}$.

$$r = \frac{12}{6 - 2\sin\frac{3\pi}{2}} = \frac{12}{6 + 2} = \frac{3}{2}$$

Vertices on major axis are at $\left(3, \frac{\pi}{2}\right)$ and $\left(\frac{3}{2}, \frac{3\pi}{2}\right)$.

The equation $r = \frac{12\csc\theta}{6\csc\theta - 2}$ has holes at $(2, 0)$ and $(2, \pi)$.

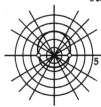

Copyright © Houghton Mifflin Company. All rights reserved.

13.

$$r = \frac{3}{\cos\theta - 1}$$

$$= \frac{-3}{1 - \cos\theta}$$

$e = 1$ The graph is a parabola.
The axis of symmetry is the polar axis.

Let $\theta = \pi$.

$$r = \frac{-3}{1 - \cos\pi} = \frac{-3}{1 - (-1)} = \frac{-3}{1+1} = -\frac{3}{2}$$

Vertex is at $\left(-\frac{3}{2}, \pi\right)$.

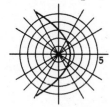

15.

$$r = \frac{12}{3 - 6\cos\theta}$$

$$r(3 - 6\cos\theta) = 12$$

$$3r - 6r\cos\theta = 12$$

$$3\sqrt{x^2 + y^2} - 6x = 12$$

$$3\sqrt{x^2 + y^2} = 6x + 12$$

$$\sqrt{x^2 + y^2} = 2x + 4$$

$$x^2 + y^2 = 4x^2 + 16x + 16$$

$$3x^2 - y^2 + 16x + 16 = 0$$

17.

$$r = \frac{8}{4 + 3\sin\theta}$$

$$r(4 + 3\sin\theta) = 8$$

$$4r + 3r\sin\theta = 8$$

$$4\sqrt{x^2 + y^2} + 3y = 8$$

$$4\sqrt{x^2 + y^2} = -3y + 8$$

$$16x^2 + 16y^2 = 9y^2 - 48y + 64$$

$$16x^2 + 7y^2 + 48y - 64 = 0$$

19.

$$r = \frac{9}{3 - 3\sin\theta}$$

$$r(3 - 3\sin\theta) = 9$$

$$3r - 3r\sin\theta = 9$$

$$3\sqrt{x^2 + y^2} - 3y = 9$$

$$3\sqrt{x^2 + y^2} = 3y + 9$$

$$3\sqrt{x^2 + y^2} = 3(y + 3)$$

$$\sqrt{x^2 + y^2} = y + 3$$

$$x^2 + y^2 = y^2 + 6y + 9$$

$$x^2 - 6y - 9 = 0$$

21. $e = 2$, $r\cos\theta = -1$,

$$d = |-1| = 1$$

$$r = \frac{ed}{1 - e\cos\theta}$$

$$= \frac{(2)(1)}{1 - (2)\cos\theta}$$

$$= \frac{2}{1 - 2\cos\theta}$$

23. $e = 1$, $r\sin\theta = 2$, $d = |2| = 2$

$$r = \frac{ed}{1 + e\sin\theta}$$

$$= \frac{(1)(2)}{1 + (1)\sin\theta}$$

$$= \frac{2}{1 + \sin\theta}$$

25. $e = \frac{2}{3}$, $r\sin\theta = -4$,

$$d = |-4| = 4$$

$$r = \frac{ed}{1 - e\sin\theta}$$

$$= \frac{\left(\frac{2}{3}\right)(4)}{1 - \left(\frac{2}{3}\right)\sin\theta}$$

$$= \frac{\frac{8}{3}}{1 - \frac{2}{3}\sin\theta}$$

$$= \frac{8}{3 - 2\sin\theta}$$

Copyright © Houghton Mifflin Company. All rights reserved.

27.

$$e = \frac{3}{2}, r = 2\sec\theta$$

$$r = \frac{2}{\cos\theta}$$

$$r\cos\theta = 2, \quad d = |2| = 2$$

$$r = \frac{ed}{1 + e\cos\theta}$$

$$= \frac{\left(\frac{3}{2}\right)(2)}{1 + \left(\frac{3}{2}\right)\cos\theta}$$

$$= \frac{3}{1 + \frac{3}{2}\cos\theta}$$

$$= \frac{6}{2 + 3\cos\theta}$$

29. vertex: $(2, \pi)$, curve: parabola

$$r = \frac{ed}{1 - e\cos\theta} \quad e = 1 \text{ (by definition of a parabola)}$$

When $\theta = \pi, r = 2$. Substituting into

$$r = \frac{ed}{1 - e\cos\theta},$$

we have

$$2 = \frac{1 \cdot d}{1 - 1 \cdot \cos(\pi)} = \frac{d}{2}$$

Therefore, $d = 4$. Substituting $e = 1$ and $d = 4$ yields

$$r = \frac{(1)(4)}{1 - (1)\cos\theta} \quad \text{or} \quad r = \frac{4}{1 - \cos\theta}$$

31. vertex: $(1, 3\pi/2)$, $e = 2$

$$r = \frac{ed}{1 - e\sin\theta}$$

When $\theta = \frac{3\pi}{2}, r = 1$. Substituting into

$$r = \frac{ed}{1 - e\sin\theta},$$

we have

$$1 = \frac{2d}{1 - 2\sin\left(\frac{3\pi}{2}\right)} = \frac{2d}{3}$$

Therefore $d = \frac{3}{2}$. Substituting $e = 2$ and $d = \frac{3}{2}$ yields

$$r = \frac{(2)\left(\frac{2}{3}\right)}{1 - (2)\sin\theta}$$

$$= \frac{3}{1 - 2\sin\theta}$$

33.

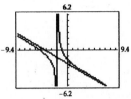

Rotate the graph of Exercise 1 $\frac{\pi}{6}$ radians counterclockwise about the pole.

35.

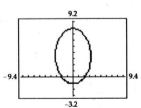

Rotate the graph of Exercise 3 π radians counterclockwise about the pole.

37.

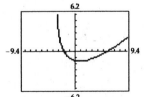

Rotate the graph of Exercise 5 $\frac{\pi}{6}$ radians clockwise about the pole.

39.

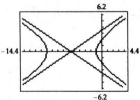

Rotate the graph of Exercise 7 π radians clockwise about the pole.

Copyright © Houghton Mifflin Company. All rights reserved.

●●●

41.

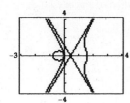

43.

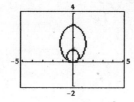

45. $0 \le \theta \le 12\pi$

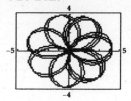

47. Convert the equations for the conic and the directrix to rectangular form.

$$r = \frac{ed}{1 - e\cos\theta} \qquad\qquad d = -r\cos\theta \ \text{(by definition)}$$

$$r(1 - e\cos\theta) = ed \qquad\qquad\qquad = -x$$

$$r - er\cos\theta = ed \qquad\qquad\qquad x = -d$$

$$\sqrt{x^2 + y^2} - ex = ed$$

$$\sqrt{x^2 + y^2} = ex + ed = e(x + d)$$

$$x^2 + y^2 = e^2(x^2 + 2dx + d^2)$$

$$x^2 + y^2 - e^2 x^2 - 2e^2 dx - e^2 d^2 = 0$$

$$(1 - e^2)x^2 + y^2 - (2e^2 d)x - (e^2 d^2) = 0$$

Solving for y^2 yields $y^2 = (e^2 - 1)x^2 + (2ed)x + (e^2 d^2)$.

Now, with $y^2 = (e^2 - 1)x^2 + (2ed)x + (e^2 d^2)$ and a directrix of $x = -d$, let $k = \dfrac{d(P,F)}{d(P,D)}$, where the focus is at the origin (by definition).

$$k = \frac{d(P,F)}{d(P,D)} = \frac{\sqrt{x^2 + y^2}}{|x + d|}$$

We can substitute $y^2 = (e^2 - 1)x^2 + (2ed)x + (e^2 d^2)$ to obtain

$$k = \frac{\sqrt{x^2 + (e^2 - 1)x^2 + (2ed)x + (e^2 d^2)}}{|x + d|} = \frac{\sqrt{e^2 x^2 + 2e^2 dx + e^2 d^2}}{|x + d|}.$$

Solving for k^2 gives us $k^2 = \dfrac{e^2 x^2 + 2e^2 dx + e^2 d^2}{x^2 + 2dx + d^2} = \dfrac{e^2(x^2 + 2dx + d^2)}{x^2 + 2dx + d^2} = e^2$.

Since $k^2 = e^2$, $k = \pm\sqrt{e^2} = \pm e$. But, since $k = \dfrac{d(P,F)}{d(P,D)}$, and the ratio of two distances must be positive, k cannot be negative.

Therefore, $k = e$, or $\dfrac{d(P,F)}{d(P,D)} = e$.

●●●●●●●●●●●●●●●●●●●●●●●●●●●●●●●●●●●●●●●

49. $y^2 + 3y + \left(\dfrac{3}{2}\right)^2 = \left(y + \dfrac{3}{2}\right)^2$

50. $y = x^2 = (2t + 1)^2 = 4t^2 + 4t + 1$

51. ellipse

52. $x^2 + y^2 = (\sin t)^2 + (\cos t)^2 = 1$

53. $y = \ln t$
$e^y = t$

54. Domain: $(-\infty, \infty)$
Range: $[-3, 3]$

Copyright © Houghton Mifflin Company. All rights reserved.

Section 6.7

1. A table of five arbitrarily chosen values of t and the corresponding values of x and y are shown in the table below.

t	$x = 2t$	$y = -t$	(x, y)
-2	-4	2	(-4, 2)
-1	-2	1	(-2, 1)
0	0	0	(0, 0)
1	2	-1	(2, -1)
2	4	-2	(4, -2)

Plotting points for several values of t yields the following graph.

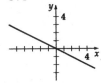

3. A table of five arbitrarily chosen values of t and the corresponding values of x and y are shown in the table below

t	$x = -t$	$y = t^2 - 1$	(x, y)
-2	2	3	(2, 3)
-1	1	0	(1, 0)
0	0	-1	(0, -1)
1	-1	0	(-1, 0)
2	-2	3	(-2, 3)

Plotting points for several values of t yields the following graph.

5. A table of five arbitrarily chosen values of t and the corresponding values of x and y are shown the table below.

t	$x = t^2$	$y = t^3$	(x, y)
-2	4	-8	(4, -8)
-1	1	-1	(1, -1)
0	0	0	(0, 0)
1	1	1	(1, 1)
2	4	8	(4, 8)

Plotting points for several values of t yields the following graph.

7. A table of eight values of t in the specified interval and the corresponding values of x and y are shown the table below.

t	$x = 2\cos t$	$y = 3\sin t$	(x, y)
0	2	0	(2, 0)
$\pi/4$	$\sqrt{2}$	$3\sqrt{2}/2$	$(\sqrt{2}, 3\sqrt{2}/2)$
$\pi/2$	0	3	(0, 3)
$3\pi/4$	$-\sqrt{2}$	$3\sqrt{2}/2$	$(-\sqrt{2}, 3\sqrt{2}/2)$
π	-2	0	(-2, 0)
$5\pi/4$	$-\sqrt{2}$	$-3\sqrt{2}/2$	$(-\sqrt{2}, -3\sqrt{2}/2)$
$3\pi/2$	0	-3	(0, -3)
$7\pi/4$	$\sqrt{2}$	$-3\sqrt{2}/2$	$(\sqrt{2}, -3\sqrt{2}/2)$

Plotting points for several values of t yields the following graph.

Copyright © Houghton Mifflin Company. All rights reserved.

224

Chapter 6: Topics in Analytic Geometry

9. A table of five arbitrarily chosen values of t and the corresponding values of x and y are shown the table below.

t	$x = 2^t$	$y = 2^{t+1}$	(x, y)
-2	1/4	1/2	(1/4, 1/2)
-1	1/2	1	(1/2, 1)
0	1	2	(1, 2)
1	2	4	(2, 4)
2	4	8	(4, 8)

Plotting points for several values of t yields the following graph.

11.
$$x = \sec t \qquad -\frac{\pi}{2} < t < \frac{\pi}{2}$$
$$y = \tan t$$

$$\tan^2 t + 1 = \sec^2 t$$
$$y^2 + 1 = x^2$$
$$x^2 - y^2 - 1 = 0 \qquad x \geq 1, \ y \in R$$

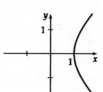

13.
$$x = 2 - t^2 \qquad t \in R$$
$$y = 3 + 2t^2$$

$$x = 2 - t^2 \rightarrow t^2 = 2 - x$$
$$y = 3 + 2(2 - x)$$
$$y = -2x + 7$$

Because $x = 2 - t^2$ and $t^2 \geq 0$ for all real numbers t, $x \leq 2$ for all t. Similarly, $y \geq 3$ for all t.

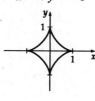

15.
$$x = \cos^3 t \qquad 0 \leq t < 2\pi$$
$$y = \sin^3 t$$

$$\cos^2 t = x^{2/3}$$
$$\sin^2 t = y^{2/3}$$

$$\cos^2 t + \sin^2 t = 1 \qquad -1 \leq x \leq 1$$
$$x^{2/3} + y^{2/3} = 1 \qquad -1 \leq y \leq 1$$

17.
$$x = \sqrt{t+1} \qquad t \geq -1$$
$$y = t$$

$$x = \sqrt{y+1} \qquad x \geq 0$$
$$y = x^2 - 1 \qquad y \geq -1$$

19.
$$x = t^3 \qquad t > 0, \ x > 0$$
$$y = 3 \ln t \qquad y \in R$$

$$x = t^3 \rightarrow t = x^{1/3}$$
$$y = 3 \ln x^{1/3}$$
$$y = \ln x \text{ for } x > 0 \text{ and } y \in R$$

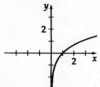

21.
$$C_1: x = 2 + t^2$$
$$y = 1 - 2t^2$$

$$x = 2 + t^2 \rightarrow t^2 = x - 2$$
$$y = 1 - 2(x - 2)$$
$$y = -2x + 5 \qquad x \geq 2, \ y \leq 1$$

$$C_2: x = 2 + t$$
$$y = 1 - 2t$$

$$x = 2 + t \rightarrow t = x - 2$$
$$y = 1 - 2(x - 2)$$
$$y = -2x + 5 \qquad x \in R, \ y \in R$$

The graph of C_1 is a ray beginning at (2, 1) with slope -2.
The graph of C_2 is a line passing through (2, 1) with slope -2.

Copyright © Houghton Mifflin Company. All rights reserved.

23.　$x = \sin t$
　　$y = \csc t$

$$\csc t = \frac{1}{\sin t}$$

$$y = \frac{1}{x}$$

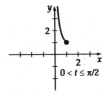

25.

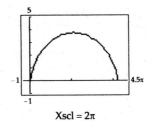

$\text{Xscl} = 2\pi$

Range for graph 1:　$0 < t \le \dfrac{\pi}{2}$

$$0 < x \le 1$$

$$y \ge 1$$

Range for graph 2:　$\pi \le t \le \dfrac{3\pi}{2}$

$$-1 \le x \le 0$$

$$y \le -1$$

27.

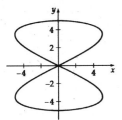

29.

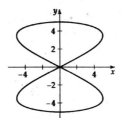

31.

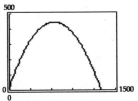

Maximum height (to the nearest foot) of 462 feet is attained when $t \approx 5.38$ seconds.

The projectile has a range (to the nearest foot) of 1295 feet and hits the ground in about 10.75 seconds.

33.

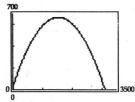

Maximum height (to the nearest foot) of 694 feet is attained when $t \approx 6.59$ seconds.

The projectile has a range (to the nearest foot) of 3084 feet and hits the ground in about 13.17 seconds.

Copyright © Houghton Mifflin Company. All rights reserved.

●●●

35. Let $P_1(x_1, y_1)$ and $P_2(x_2, y_2)$ be two distinct points on a line.

If $P(x, y)$ is any other point on the line, then

$$\frac{y - y_1}{x - x_1} = \frac{y_2 - y_1}{x_2 - x_1}$$ (Slope is constant along entire line.)This

equation can be rewritten as

$$\frac{y - y_1}{y_2 - y_1} = \frac{x - x_1}{x_2 - x_1}$$ Let this value equal t.

Thus, $\dfrac{x - x_1}{x_2 - x_1} = t$ and $\dfrac{y - y_1}{y_2 - y_1} = t$.

Solving for x and y, respectively, we have
$$x = x_1 + t(x_2 - x_1) \quad \text{and} \quad y = y_1 + t(y_2 - y_1)$$

39. Because the circle moves without slipping, $b\theta = a\alpha$.

Therefore, $\alpha = \dfrac{b\theta}{a}$. Let $P(x, y)$ be the coordinates of the moving point.

Angle $\phi = \dfrac{\pi}{2} - \left(\dfrac{b - a}{a}\right)\theta$

Thus, $x = (b - a)\cos\theta + a\sin\left[\dfrac{\pi}{2} - \left(\dfrac{b - a}{a}\right)\theta\right]$

$\qquad y = (b - a)\sin\theta - a\cos\left[\dfrac{\pi}{2} - \left(\dfrac{b - a}{a}\right)\theta\right]$

Simplifying, we have

$x = (b - a)\cos\theta + a\cos\left(\dfrac{b - a}{a}\theta\right)$

$y = (b - a)\sin\theta - a\sin\left(\dfrac{b - a}{a}\theta\right)$

37. radius $= a$, $\theta = \angle TOR$

The x-coordinate of $P(x, y)$ is given by $x = OR + QP$.
The y-coordinate is given by $y = TR - TQ$.

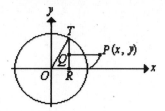

From the figure,

$OR = a\cos\theta$ and $QP = a\theta\sin\theta$. Thus,

$\quad x = a\cos\theta + a\theta\sin\theta$

$TR = a\sin\theta$ and $TQ = a\theta\cos\theta$. Thus,

$\quad y = a\sin\theta - a\theta\cos\theta$

The parametric equations are

$x = a\cos\theta + a\theta\sin\theta$
$y = a\sin\theta - a\theta\cos\theta$

●●●●●●●●●●●●●●●●●●●●●●●●●●●●●●●●●●●●   **C**hapter **6 T**rue/**F**alse **E**xercises

1. False, a parabola has no asymptotes.

2. True

3. False, $x^2 - y^2 = 1$ has a transverse axis of 2 and a conjugate axis of 2. By keeping the foci fixed and varying the asymptotes, we can make the conjugate axis any size needed.

4. False, it also depends on the eccentricity. $\dfrac{x^2}{25} + \dfrac{y^2}{9} = 1$ and $\dfrac{x^2}{36} + \dfrac{y^2}{20} = 1$ have the same c's but different a's.

5. False, parabolas have no asymptotes.

6. True

Copyright © Houghton Mifflin Company. All rights reserved.

7. False, a parabola can be a function. The graph of the function $f(x) = x^2$ is a parabola.

8. False, $x = \cos t$, $y = \sin t$ graphs to be a circle.

9. True

10. True

●●●

1. $x^2 - y^2 = 4$ [6.3]

$\dfrac{x^2}{4} - \dfrac{y^2}{4} = 1$

hyperbola
center: (0, 0)
vertices: (± 2, 0)
foci: ($\pm 2\sqrt{2}$, 0)
asymptotes: $y = \pm x$

2. $y^2 = 16x$ [6.1]

parabola
vertex: (0, 0)
focus: (4, 0)
directrix: $x = -4$

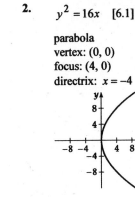

3.
$$x^2 + 4y^2 - 6x + 8y - 3 = 0 \qquad [6.2]$$
$$x^2 - 6x + 4(y^2 + 2y) = 3$$
$$(x^2 - 6x + 9) + 4(y^2 + 2y + 1) = 3 + 9 + 4$$
$$(x - 3)^2 + 4(y + 1)^2 = 16$$
$$\frac{(x-3)^2}{16} + \frac{(y+1)^2}{4} = 1$$

ellipse
center: (3, −1)
vertices: $(3 \pm 4, -1) = (7, -1), (-1, -1)$
foci: $(3 \pm 2\sqrt{3}, -1) = (3 + 2\sqrt{3}, -1), (3 - 2\sqrt{3}, -1)$

4.
$$3x^2 - 4y^2 + 12x - 24y - 36 = 0 \qquad [6.3]$$
$$3(x^2 + 4x) - 4(y^2 + 6y) = 36$$
$$3(x^2 + 4x + 4) - 4(y^2 + 6y + 9) = 36 + 12 - 36$$
$$3(x + 2)^2 - 4(y + 3)^2 = 12$$
$$\frac{(x+2)^2}{4} - \frac{(y+3)^2}{3} = 1$$

hyperbola
center: (−2, −3)
vertices: $(-2 \pm 2, -3) = (0, -3), (-4, -3)$
foci: $(-2 \pm \sqrt{7}, -3) = (-2 + \sqrt{7}, -3), (-2 - \sqrt{7}, -3)$
asymptotes: $y + 3 = \pm \dfrac{\sqrt{3}}{2}(x + 2)$

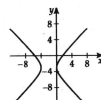

Copyright © Houghton Mifflin Company. All rights reserved.

5.
$$3x - 4y^2 + 8y + 2 = 0 \qquad [6.1]$$
$$-4(y^2 - 2y) = -3x - 2$$
$$-4(y^2 - 2y + 1) = -3x - 2 - 4$$
$$-4(y - 1)^2 = -3(x + 2)$$
$$(y - 1)^2 = \frac{3}{4}(x + 2)$$

parabola
vertex: $(-2, 1)$

focus: $\left(-2 + \frac{3}{16}, 1\right) = \left(-\frac{29}{16}, 1\right)$

directrix: $x = -2 - \frac{3}{16}$, or $x = -\frac{35}{16}$

6.
$$3x + 2y^2 - 4y - 7 = 0 \qquad [6.1]$$
$$2(y^2 - 2y) = -3x + 7$$
$$2(y^2 - 2y + 1) = -3x + 7 + 2$$
$$2(y - 1)^2 = -3(x - 3)$$
$$(y - 1)^2 = -\frac{3}{2}(x - 3)$$

parabola
vertex: $(3, 1)$

focus: $\left(3 - \frac{3}{8}, 1\right) = \left(\frac{21}{8}, 1\right)$

directrix: $x = 3 + \frac{3}{8}$, or $x = \frac{27}{8}$

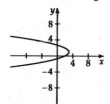

7.
$$9x^2 + 4y^2 + 36x - 8y + 4 = 0 \qquad [6.2]$$
$$9(x^2 + 4x) + 4(y^2 - 2y) = -4$$
$$9(x^2 + 4x + 4) + 4(y^2 - 2y + 1) = -4 + 36 + 4$$
$$9(x + 2)^2 + 4(y - 1)^2 = 36$$
$$\frac{(x + 2)^2}{4} + \frac{(y - 1)^2}{9} = 1$$

ellipse
center: $(-2, 1)$
vertices: $(-2, 1 \pm 3) = (-2, 4), (-2, -2)$
foci: $(-2, 1 \pm \sqrt{5}) = (-2, 1 + \sqrt{5}), (-2, 1 - \sqrt{5})$

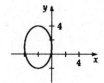

8.
$$11x^2 - 25y^2 - 44x - 50y - 256 = 0 \qquad [6.3]$$
$$11(x^2 - 4x) - 25(y^2 + 2y) = 256$$
$$11(x^2 - 4x + 4) - 25(y^2 + 2y + 1) = 256 + 44 - 25$$
$$11(x - 2)^2 - 25(y + 1)^2 = 275$$
$$\frac{(x - 2)^2}{25} - \frac{(y + 1)^2}{11} = 1$$

hyperbola
center: $(2, -1)$
vertices: $(2 \pm 5, -1) = (7, -1), (-3, -1)$
foci: $(2 \pm 6, -1) = (8, -1), (-4, -1)$
asymptotes: $y + 1 = \pm \frac{\sqrt{11}}{5}(x - 2)$

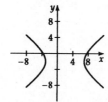

Copyright © Houghton Mifflin Company. All rights reserved.

9.
$$4x^2 - 9y^2 - 8x + 12y - 144 = 0 \qquad [6.3]$$

$$4(x^2 - 2x) - 9\left(y^2 - \frac{4}{3}y\right) = 144$$

$$4(x^2 - 2x + 1) - \left(9y^2 - \frac{4}{3}y + \frac{4}{9}\right) = 144 + 4 - 4$$

$$4(x-1)^2 - 9\left(y - \frac{2}{3}\right)^2 = 144$$

$$\frac{(x-1)^2}{36} - \frac{\left(y - \frac{2}{3}\right)^2}{16} = 1$$

hyperbola

center: $\left(1, \frac{2}{3}\right)$

vertices: $\left(1 \pm 6, \frac{2}{3}\right) = \left(7, \frac{2}{3}\right), \left(-5, \frac{2}{3}\right)$

foci: $\left(1 \pm 2\sqrt{13}, \frac{2}{3}\right) = \left(1 + 2\sqrt{13}, \frac{2}{3}\right), \left(1 - 2\sqrt{13}, \frac{2}{3}\right)$

asymptotes: $y - \frac{2}{3} = \pm \frac{2}{3}(x-1)$

10.
$$9x^2 + 16y^2 + 36x - 16y - 104 = 0 \qquad [6.2]$$

$$9(x^2 + 4x) + 16(y^2 - y) = 104$$

$$9(x^2 + 4x + 4) + 16\left(y^2 - y + \frac{1}{4}\right) = 104 + 36 + 4$$

$$9(x+2)^2 + 16\left(y - \frac{1}{2}\right)^2 = 144$$

$$\frac{(x+2)^2}{16} + \frac{\left(y - \frac{1}{2}\right)^2}{9} = 1$$

ellipse

center: $\left(-2, \frac{1}{2}\right)$

vertices: $\left(-2 \pm 4, \frac{1}{2}\right) = \left(2, \frac{1}{2}\right), \left(-6, \frac{1}{2}\right)$

foci: $\left(-2 \pm \sqrt{7}, \frac{1}{2}\right) = \left(-2 + \sqrt{7}, \frac{1}{2}\right), \left(-2 - \sqrt{7}, \frac{1}{2}\right)$

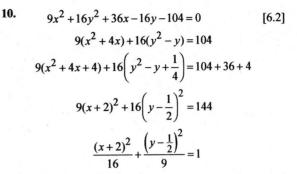

11.
$$4x^2 + 28x + 32y + 81 = 0 \qquad [6.1]$$

$$4(x^2 + 7x) = -32y - 81$$

$$4\left(x^2 + 7x + \frac{49}{4}\right) = -32y - 81 + 49$$

$$4\left(x + \frac{7}{2}\right)^2 = -32(y+1)$$

$$\left(x + \frac{7}{2}\right)^2 = -8(y+1)$$

parabola

$4p = -8 \Rightarrow p = -2$

vertex: $\left(-\frac{7}{2}, -1\right)$

focus: $\left(-\frac{7}{2}, -1-2\right) = \left(-\frac{7}{2}, -3\right)$

directrix: $y = 1$

12.
$$x^2 - 6x - 9y + 27 = 0 \qquad [6.1]$$

$$x^2 - 6x = 9y - 27$$

$$x^2 - 6x + 9 = 9y - 27 + 9$$

$$(x-3)^2 = 9(y-2)$$

parabola

$4p = 9 \Rightarrow p = \frac{9}{4}$

vertex: (3, 2)

focus: $\left(3, 2 + \frac{9}{4}\right) = \left(3, \frac{17}{4}\right)$

directrix: $y = 3 - \frac{9}{4}$, or $y = -\frac{1}{4}$

Copyright © Houghton Mifflin Company. All rights reserved.

13. $2a = |7 - (-3)| = 10$ [6.2]

$a = 5$

$a^2 = 25$

$2b = 8$

$b = 4$

$b^2 = 16$

center $(2, 3)$

$$\frac{(x-2)^2}{25} + \frac{(y-3)^2}{16} = 1$$

14. $2a = |4 - (-2)| = 6$ [6.3]

$a = 3$

$a^2 = 9$

$e = \dfrac{c}{a} = \dfrac{4}{3}$

$\dfrac{c}{3} = \dfrac{4}{3}$

$c = 4$

$c^2 = a^2 + b^2$

$16 = 9 + b^2$

$b^2 = 7$

center $(1, 1)$

$$\frac{(x-1)^2}{9} - \frac{(y-1)^2}{7} = 1$$

15. center $(-2, 2)$, $c = 3$ [6.3]

$2a = 4$

$a = 2$

$a^2 = 4$

$c^2 = a^2 + b^2$

$9 = 4 + b^2$

$b^2 = 5$

$$\frac{(x+2)^2}{4} - \frac{(y-2)^2}{5} = 1$$

16. $(h, k) = \left(\dfrac{6+2}{2}, \dfrac{-3-3}{2}\right) = (4, -3)$ [6.1]

$p = 2 - 4$

$p = -2$

$4p = -8$

$(y+3)^2 = -8(x-4)$

17. $(x-h)^2 = 4p(y-k)$ or $(y-k)^2 = 4p(x-h)$

$(3-0)^2 = 4p(4+2)$ $(4+2)^2 = 4p(3-0)$

$9 = 4p(6)$ $36 = 4p(3)$

$p = \dfrac{3}{8}$ $p = 3$

Thus, there are two parabolas that satisfy the given conditions:

$$x^2 = \frac{3}{2}(y+2) \quad \text{or} \quad (y+2)^2 = 12x \quad [6.1]$$

18. center $(-2, -1)$

$c = 2$

$e = \dfrac{c}{a} = \dfrac{2}{3}$

$\dfrac{2}{a} = \dfrac{2}{3}$

$a = 3$

$c^2 = a^2 - b^2$

$4 = 9 - b^2$

$b^2 = 5$

$$\frac{(x+2)^2}{9} + \frac{(y+1)^2}{5} = 1 \quad [6.2]$$

19. $a = 6$ and the transverse axis is on the x-axis.

$\pm\dfrac{b}{a} = \pm\dfrac{1}{9}$

$\dfrac{b}{6} = \dfrac{1}{9}$

$b = \dfrac{2}{3}$

$$\frac{x^2}{36} - \frac{y^2}{4/9} = 1 \quad [6.3]$$

20. $(x-h)^2 = 4p(y-k)$

$(1-h)^2 = 4p(0-k)$

$(2-h)^2 = 4p(1-k)$

$(0-h)^2 = 4p(1-k)$

In the last two equations, by substitution:

$(2-h)^2 = (0-h)^2$

$4 - 4h + h^2 = h^2$

$4 - 4h = 0$

$4h = 4$

$h = 1$

Thus:

$(1-1)^2 = 4p(0-k)$

$0 = -4pk$

$k = 0$

$(2-1)^2 = 4p(1-k)$

$1 = 4p(1)$

$p = \dfrac{1}{4}$

The equation is $y = (x-1)^2$ [6.1]

Copyright © Houghton Mifflin Company. All rights reserved.

21. $A = 11, \ B = -6, \ C = 19, \ D = 0, \ E = 0, \ F = -40$

$$\cot 2\alpha = \frac{11-19}{-6} = \frac{-8}{-6} = \frac{4}{3} \qquad (2\alpha \text{ is in quadrant I.})$$

$$\csc^2 2\alpha = 1 + \cot^2 2\alpha = 1 + \frac{16}{9} = \frac{25}{9}$$

$$\csc 2\alpha = \frac{5}{3}$$

Thus, $\sin 2\alpha = \frac{3}{5}$ and $\cos 2\alpha = \frac{4}{5}$.

$$\sin \alpha = \sqrt{\frac{1-\frac{4}{5}}{2}} = \frac{\sqrt{10}}{10} \qquad \cos \alpha = \frac{3\sqrt{10}}{10}$$

$$A' = 11\left(\frac{3\sqrt{10}}{10}\right)^2 - 6\left(\frac{\sqrt{10}}{10}\right)\left(\frac{3\sqrt{10}}{10}\right) + 19\left(\frac{\sqrt{10}}{10}\right)^2$$

$$= \frac{99}{10} - \frac{18}{10} + \frac{19}{10} = \frac{100}{10} = 10$$

$$B' = 0$$

$$C' = 11\left(\frac{\sqrt{10}}{10}\right)^2 + 6\left(\frac{\sqrt{10}}{10}\right)\left(\frac{3\sqrt{10}}{10}\right) + 19\left(\frac{3\sqrt{10}}{10}\right)^2$$

$$= \frac{11}{10} + \frac{18}{10} + \frac{171}{10} = \frac{200}{10} = 20$$

$$F' = F$$

$$10(x')^2 + 20(y')^2 - 40 = 0 \text{ or } (x')^2 + 2(y')^2 - 4 = 0$$

The graph is an ellipse. [8.4]

22. $A = 3, B = 6, C = 3, D = -4, E = 5, F = -12$

$$\cot 2\alpha = \frac{3-3}{6} = 0$$

$$1 + \cot^2 2\alpha = \csc^2 2\alpha$$
$$1 = \csc^2 2\alpha$$
$$\csc 2\alpha = 1$$

Thus, $\sin 2\alpha = 1$ so $2\alpha = 90°$ or $\alpha = 45°$.

Therefore, $\cos \alpha = \frac{\sqrt{2}}{2}$, $\sin \alpha = \frac{\sqrt{2}}{2}$.

$$A' = 3\left(\frac{\sqrt{2}}{2}\right)^2 + 6\left(\frac{\sqrt{2}}{2}\right)\left(\frac{\sqrt{2}}{2}\right) + 3\left(\frac{\sqrt{2}}{2}\right)^2 = \frac{3}{2} + \frac{6}{2} + \frac{3}{2} = 6$$

$$B' = 0$$

$$C' = 3\left(\frac{\sqrt{2}}{2}\right)^2 - 6\left(\frac{\sqrt{2}}{2}\right)\left(\frac{\sqrt{2}}{2}\right) + 3\left(\frac{\sqrt{2}}{2}\right)^2 = \frac{3}{2} - \frac{6}{2} + \frac{3}{2} = 0$$

$$D' = -4\left(\frac{\sqrt{2}}{2}\right) + 5\left(\frac{\sqrt{2}}{2}\right) = \frac{\sqrt{2}}{2}$$

$$E' = 4\left(\frac{\sqrt{2}}{2}\right) + 5\left(\frac{\sqrt{2}}{2}\right) = \frac{9\sqrt{2}}{2}$$

$$F' = -12$$

$$6(x')^2 + \frac{\sqrt{2}}{2}x' + \frac{9\sqrt{2}}{2}y' - 12 = 0$$

The graph is a parabola. [8.4]

23. $A = 1, B = 2\sqrt{3}, C = 3, D = 8\sqrt{3}, E = -8, F = 32$

$$\cot 2\alpha = \frac{1-3}{2\sqrt{3}} = -\frac{1}{\sqrt{3}}. \text{ Thus } 90° < 2\alpha < 180°.$$

$$\cot^2 2\alpha + 1 = \csc^2 2\alpha$$
$$\frac{1}{3} + 1 = \csc^2 2\alpha$$
$$\frac{4}{3} = \csc^2 2\alpha, \text{ or } \csc 2\alpha = \frac{2}{\sqrt{3}}$$

Therefore, $\sin 2\alpha = \frac{\sqrt{3}}{2}$ and $\cos 2\alpha = -\frac{1}{2}$.

Since 2α is in quadrant II,

$$\cos \alpha = \sqrt{\frac{1+(-1/2)}{2}} = \frac{1}{2} \qquad \sin \alpha = \sqrt{\frac{1-(-1/2)}{2}} = \frac{\sqrt{3}}{2}$$

$$A' = 1\left(\frac{1}{2}\right)^2 + 2\sqrt{3}\left(\frac{1}{2}\right)\left(\frac{\sqrt{3}}{2}\right) + 3\left(\frac{\sqrt{3}}{2}\right)^2 = \frac{1}{4} + \frac{6}{4} + \frac{9}{4} = 4$$

$$B' = 0$$

$$C' = 1\left(\frac{\sqrt{3}}{2}\right)^2 - 2\sqrt{3}\left(\frac{1}{2}\right)\left(\frac{\sqrt{3}}{2}\right) + 3\left(\frac{1}{2}\right)^2 = \frac{3}{4} - \frac{6}{4} + \frac{3}{4} = 0$$

$$D' = 8\sqrt{3}\left(\frac{1}{2}\right) - 8\left(\frac{\sqrt{3}}{2}\right) = 0$$

$$E' = 8\sqrt{3}\left(\frac{\sqrt{3}}{2}\right) - 8\left(\frac{1}{2}\right) = -12 - 4 = -16$$

$$F' = 32$$

$$4(x')^2 - 16y' + 32 = 0 \text{ or } (x')^2 - 4y' + 8 = 0$$

The graph is a parabola. [8.4]

24. $A = 0, B = 1, C = 0, D = -1, E = -1, F = -1$

$$\cot 2\alpha = \frac{0-0}{1} = 0. \text{ Thus } \alpha = 45°.$$

Therefore, $\sin \alpha = \frac{\sqrt{2}}{2}$ and $\cos \alpha = \frac{\sqrt{2}}{2}$.

$$A' = 0 + 1\left(\frac{\sqrt{2}}{2}\right)\left(\frac{\sqrt{2}}{2}\right) + 0 = \frac{1}{2}$$

$$B' = 0$$

$$C' = 0 - \frac{\sqrt{2}}{2}\frac{\sqrt{2}}{2} + 0 = -\frac{1}{2}$$

$$D' = -1\left(\frac{\sqrt{2}}{2}\right) - 1\left(\frac{\sqrt{2}}{2}\right) = -\sqrt{2}$$

$$E' = \frac{\sqrt{2}}{2} + (-1)\frac{\sqrt{2}}{2} = 0$$

$$F' = -1$$

$$\frac{1}{2}(x')^2 - \frac{1}{2}(y')^2 - \sqrt{2}x' - 1 = 0$$

The equation is a hyperbola. [8.4]

Copyright © Houghton Mifflin Company. All rights reserved.

25.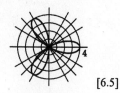

[6.5]

26.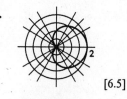

[6.5]

27.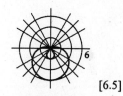

[6.5]

28.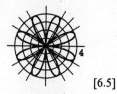

[6.5]

29.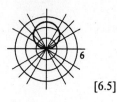

[6.5]

30. vertical line through (3, 0).

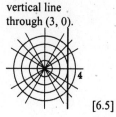

[6.5]

31. horizontal line through (0, 4)

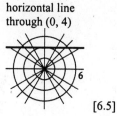

[6.5]

32.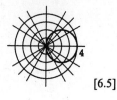

[6.5]

33.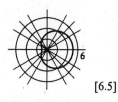

[6.5]

34.

[6.5]

35.
$$y^2 = 16x \quad [6.5]$$
$$(r\sin\ \theta)^2 = 16(r\cos\theta)$$
$$r^2\sin^2\theta = 16r\cos\theta$$
$$r\sin^2\theta = 16\cos\theta$$

36.
$$x^2 + y^2 + 4x + 3y = 0 \quad [6.5]$$
$$(r\cos\theta)^2 + (r\sin\ \theta)^2 + 4(r\cos\theta) + 3(r\sin\ \theta) = 0$$
$$r^2(\cos^2\ \theta + \sin^2\ \theta) + 4r\cos\theta + 3r\sin\ \theta = 0$$
$$r + 4\cos\theta + 3\sin\ \theta = 0$$

37.
$$3x - 2y = 6 \quad [6.6]$$
$$3r\cos\theta - 2r\sin\theta = 6$$

38.
$$xy = 4 \quad [6.5]$$
$$(r\cos\theta)(r\sin\theta) = 4$$
$$r^2\cos\theta\ \sin\theta = 4$$
$$r^2(2)\cos\theta\ \sin\theta = 4(2)$$
$$r^2\sin 2\theta = 8$$

39.
$$r = \frac{4}{1 - \cos\theta} \quad [6.5]$$
$$r - r\cos\ \theta = 4$$
$$\sqrt{x^2 + y^2} - x = 4$$
$$\sqrt{x^2 + y^2} = x + 4$$
$$x^2 + y^2 = x^2 + 8x + 16$$
$$y^2 = 8x + 16$$

40.
$$r^2 = 3r\cos\theta - 4r\sin\theta \quad [6.5]$$
$$x^2 + y^2 = 3x - 4y$$
$$x^2 - 3x + y^2 + 4y = 0$$

41.
$$r^2 = \cos 2\theta \quad [6.5]$$
$$r^2 = \cos^2\theta - \sin^2\theta$$
$$r^4 = r^2\cos^2\theta - r^2\sin^2\theta$$
$$(r^2)^2 = r^2\cos^2\theta - r^2\sin^2\theta$$
$$(x^2 + y^2)^2 = x^2 - y^2$$
$$x^4 + 2x^2y^2 + y^4 = x^2 - y^2$$
$$x^4 + y^4 + 2x^2y^2 - x^2 + y^2 = 0$$

42.
$$\theta = 1 \quad [6.5]$$
$$\tan\theta = \tan 1$$
$$\frac{y}{x} \approx 1.5574$$
$$y = 1.5574x$$

43.
$$r = \frac{4}{3 - 6\sin\theta}$$

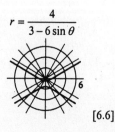

[6.6]

44.
$$r = \frac{2}{1 + \cos\theta}$$

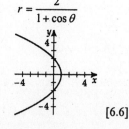

[6.6]

45.
$$r = \frac{2}{2 - \cos\theta}$$

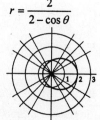

[6.6]

Copyright © Houghton Mifflin Company. All rights reserved.

46. $r = \dfrac{6}{4 + 3\sin\theta}$

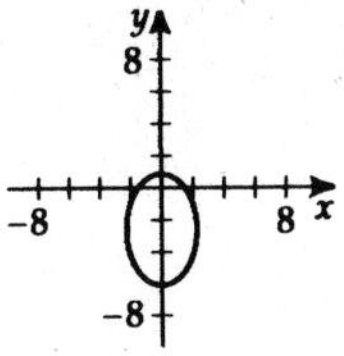

[6.6]

47. $x = 4t - 2,\ y = 3t + 1,\ t \in R$

$4t = x + 2$

$t = \dfrac{x+2}{4}$

$y = 3t + 1$

$y = 3\left(\dfrac{x+2}{4}\right) + 1$

$y = \dfrac{3}{4}x + \dfrac{5}{2}$

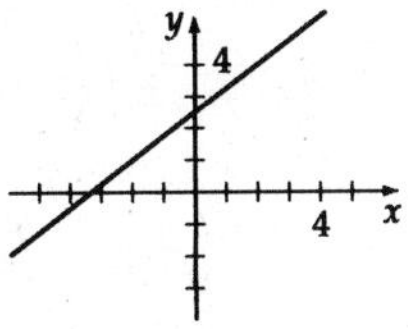

[6.7]

48. $x = 1 - t^2,\ y = 3 - 2t^2,\ t \in R$

$t^2 = -x + 1$

$y = 3 - 2t^2$

$y = 3 - 2(-x + 1)$

$y = 3 + 2x - 2$

$y = 2x + 1,\ x \le 1$

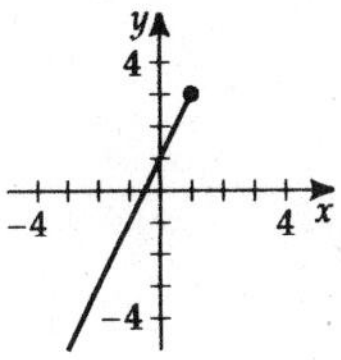

[6.7]

49. $x = 4\sin t \qquad y = 3\cos t \qquad 0 \le t < 2\pi$

$\dfrac{1}{4}x = \sin t \qquad \dfrac{1}{3}y = \cos t$

$\dfrac{1}{16}x^2 = \sin^2 t \qquad \dfrac{1}{9}y^2 = \cos t^2$

Using the trignometric identity $\sin^2 t + \cos^2 t = 1$, we have

$\dfrac{1}{16}x^2 + \dfrac{1}{9}y^2 = 1$

$\dfrac{x^2}{19} + \dfrac{y^2}{9} = 1$

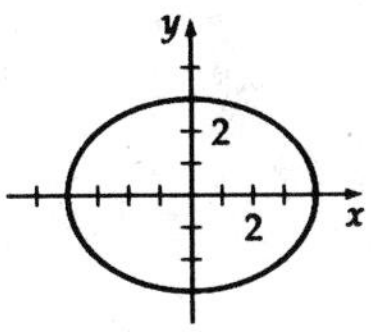

[6.7]

50. $x = \sec t \qquad y = 4\tan t \qquad -\dfrac{\pi}{2} < t < \dfrac{\pi}{2}$

$x^2 = \sec^2 t \qquad \dfrac{1}{4}y = \tan t$

$\left(\dfrac{1}{4}y\right)^2 = \tan^2 t$

$\dfrac{1}{16}y^2 = \tan^2 t$

Using the trignometric identity $1 + \tan^2 t = \sec^2 t$, we have

$1 + \dfrac{y^2}{16} = x^2$

$\dfrac{x^2}{1} - \dfrac{y^2}{16} = 1$

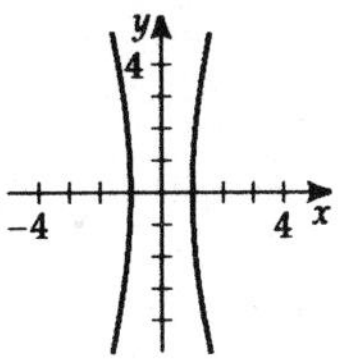

[6.7]

51. $x = \dfrac{1}{t} \qquad y = -\dfrac{2}{t} \qquad t > 0$

$y = -2\left(\dfrac{1}{t}\right)$

$y = -2x,\ x > 0$

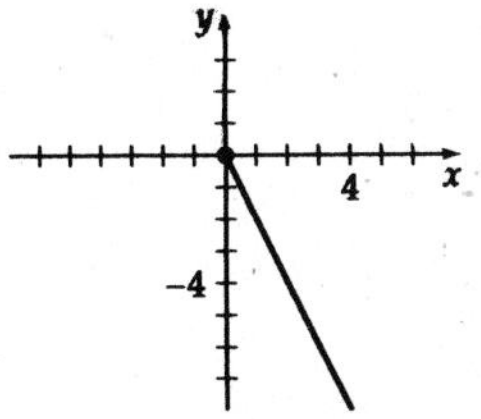

[6.7]

52. $x = 1 + \cos t \qquad y = 2 - \sin t \qquad 0 \le t < 2\pi$

$x - 1 = \cos t \qquad 2 - y = \sin t$

$-(y - 2) = \sin t$

$(x-1)^2 = \cos^2 t \qquad (-(y-2))^2 = \sin^2 t$

$(y-2)^2 = \sin^2 t$

Using the trigonometric identity $\cos^2 t + \sin^2 t = 1$ we have

$(x-1)^2 + (y-2)^2 = 1$

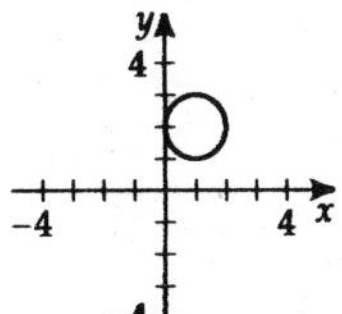

[6.7]

Copyright © Houghton Mifflin Company. All rights reserved.

53. $x = \sqrt{t}, \; y = 2^{-t}, t \geq 0$

$t = x^2$

$y = 2^{-x^2}, \; x \geq 0$

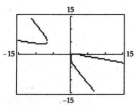

[6.7]

54.

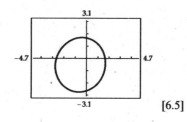

[6.7]

55. Graph $y = \dfrac{-(4x+5) \pm \sqrt{(4x+5)^2 - 8(x^2 - 2x + 1)}}{4}$.

[6.4]

56.

[6.5]

57.

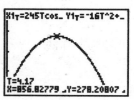

Graph in parametric mode. Use the TRACE feature to determine that the maximum height (to the nearest foot) of 278 feet is attained when $t \approx 4.17$ seconds.   [6.7]

• •

1. $y = \dfrac{1}{8}x^2$ vertex: $(0, 0)$ [6.1]

$x^2 = 8y$ focus: $(0, 2)$

$4p = 8$ directrix: $y = -2$

$p = 2$

2.

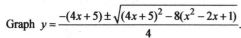

[6.2]

3. $25x^2 - 150x + 9y^2 + 18y + 9 = 0$ [6.2]

$25(x^2 - 6x + 9) + 9(y^2 + 2y + 1) = -9 + 255 + 9$

$25(x-3)^2 + 9(y+1)^2 = 225$

$\dfrac{(x-3)^2}{9} + \dfrac{(y+1)^2}{25} = 1$

$a = 5 \quad b = 3 \quad c = 4$

vertices: $(3, 4), (3, -6)$
foci: $(3, 3), (3, -5)$

4. $2b = 6 \qquad c = 6$ [6.2]

$\quad b = 3$

$a^2 = 9 + 36 = 45$

center $= (0, -3)$

$\dfrac{x^2}{45} + \dfrac{(y+3)^2}{9} = 1$

Copyright © Houghton Mifflin Company. All rights reserved.

5.

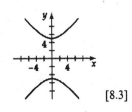

[8.3]

6. $\dfrac{x^2}{36} - \dfrac{y^2}{64} = 1$

vertices: $(6, 0)$, $(-6, 0)$

foci: $(10, 0)$, $(-10, 0)$

asymptotes: $y = \pm\dfrac{4}{3}x$ [8.3]

7. $\dfrac{(y+1)^2}{4} - \dfrac{(x+3)^2}{16} = 1$

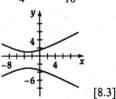

[8.3]

8. $x^2 - 4xy - 5y^2 + 3x - 5y - 20 = 0$ [8.4]

$A = 1$ $B = -4$ $C = -5$ $D = 3$ $E = -5$ $F = -20$

$\cot 2\alpha = \dfrac{A - C}{B} = \dfrac{1 - (-5)}{-4} = -\dfrac{3}{2}$ 2α is in quadrant II.

$\tan 2\alpha = -\dfrac{2}{3}$

$\quad 2\alpha = \tan^{-1}\left(-\dfrac{2}{3}\right)$

$\quad 2\alpha \approx (-33.69° + 180°)$

$\quad \alpha \approx 73.15°$

9. $A = 8$ $B = 5$ $C = 2$ $D = -10$ $E = 5$ $F = 4$

Since $B^2 - 4AC = (5)^2 - 4(8)(2) = -39 < 0$, the graph is an ellipse.
[8.4]

10.

$P(1, -\sqrt{3})$

$r = \sqrt{x^2 + y^2}$

$r = \sqrt{1^2 + (-\sqrt{3})^2}$

$r = 2$

$r\cos\theta = x$ $r\sin\theta = y$ θ is in quadrant IV.

$2\cos\theta = 1$ $2\sin\theta = -\sqrt{3}$ $\theta = 300°$

$\cos\theta = \dfrac{1}{2}$ $\sin\theta = -\dfrac{\sqrt{3}}{2}$

$P(1, -\sqrt{3}) = P(2, 300°)$ [8.5]

11. $r = 4\cos\theta$

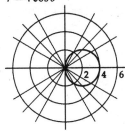

[8.5]

12. $r = 3(1 - \sin\theta)$

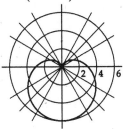

[8.5]

13. $r = 2\sin 4\theta$

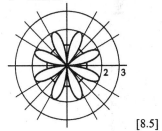

[8.5]

14. $x = r\cos\theta$ $y = r\sin\theta$

$x = 5\cos\dfrac{7\pi}{3}$ $y = 5\sin\dfrac{7\pi}{3}$

$x = \dfrac{5}{2}$ $y = \dfrac{5\sqrt{3}}{2}$

The rectangular coordinates of the point are $(5/2, 5\sqrt{3}/2)$.   [8.5]

15. $r - r\cos x = 4$ [8.5]

$\sqrt{x^2 + y^2} - x = 4$

$\sqrt{x^2 + y^2} = x + 4$

$x^2 + y^2 = x^2 + 8x + 16$

$y^2 - 8x - 16 = 0$

16. $r = \dfrac{4}{1 + \sin\theta}$ [8.6]

$r + r\sin\theta = 4$

$\sqrt{x^2 + y^2} + y = 4$

$x^2 + y^2 = 16 - 8y + y^2$

$x^2 + 8y - 16 = 0$

17. $x = t - 3$

$x + 3 = t$

$(x + 3)^2 = t^2$

$2(x + 3)^2 = 2t^2$

$2(x + 3)^2 = y$

$(x + 3)^2 = \dfrac{1}{2}y$

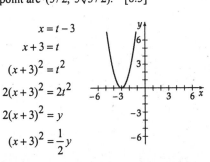

[8.7]

18. $x = 4\sin\theta$ $y = \cos\theta + 2$

$\sin\theta = x/4$ $\cos\theta = y - 2$

$\sin^2\theta + \cos^2\theta = 1$

$\left(\dfrac{x}{4}\right)^2 + (y - 2)^2 = 1$

$\dfrac{x^2}{16} + \dfrac{(y - 2)^2}{1} = 1$

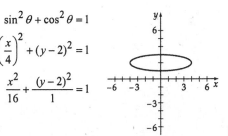

[8.7]

Copyright © Houghton Mifflin Company. All rights reserved.

19.

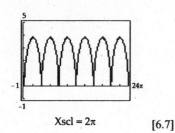

$$Xscl = 2\pi \qquad [6.7]$$

20.

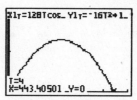

The projectile will travel $256\sqrt{3}$ feet ≈ 443 feet.   [6.7]

●●●●●●●●●●●●●●●●●●●●●●●●●●●●●●●●●●●●●●● $\mathbf{C}$umulative $\mathbf{R}$eview

1.
$$x^2 + 4x - 6 = 0 \quad [1.1]$$
$$x = \frac{-4 \pm \sqrt{4^2 - 4(1)(-6)}}{2(1)} = \frac{-4 \pm \sqrt{40}}{2}$$
$$= \frac{-4 \pm 2\sqrt{10}}{2} = -2 \pm \sqrt{10}$$
The solutions are $2 \pm \sqrt{10}$.

2.
$$y = x^3 - 4x \quad [1.4]$$

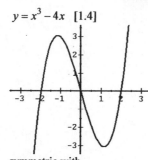

symmetric with
respect to the origin

3.
$$(g \circ f)(x) = g[f(x)] \qquad [1.5]$$
$$= g[\sin x]$$
$$= 3\sin x - 2$$

4.
$$240° = 240°\left(\frac{\pi}{180°}\right) = \frac{4\pi}{3} \quad [2.1]$$

5.
$$v = \omega r \qquad\qquad [2.1]$$
$$= \frac{3 \cdot 2\pi \cdot 60 \cdot 60 \cdot 10}{12 \cdot 5280}$$
$$\approx 11 \text{ mph}$$

6.
$$\sin t = -\frac{\sqrt{3}}{2} = \frac{\text{opp}}{\text{hyp}} \qquad [2.4]$$
$$\text{adj} = \sqrt{2^2 - \left(\sqrt{-3}\right)^2} = \sqrt{1} = 1$$
$$\tan t = \frac{\text{opp}}{\text{adj}} = \frac{-\sqrt{3}}{1} = -\sqrt{3}$$

7.
$$\tan 43° = \frac{a}{20} \qquad [2.4]$$
$$a = 20\tan 43°$$
$$a = 19 \text{ cm}$$

8.
$$y = \frac{1}{2}\cos\left(\frac{\pi x}{3}\right) \quad [2.5]$$
period: 6, amplitude: $\frac{1}{2}$

9.
$$y = 2\tan\left(\frac{\pi x}{3}\right) \quad [2.6]$$
period: 3

10.
$$\frac{\sin x}{1 - \cos x} = \frac{\sin x}{1 - \cos x} \cdot \frac{1 + \cos x}{1 + \cos x} \quad [3.1]$$
$$= \frac{\sin x(1 + \cos x)}{1 - \cos^2 x}$$
$$= \frac{\sin x(1 + \cos x)}{\sin^2 x}$$
$$= \frac{1}{\sin x} + \frac{\cos x}{\sin x}$$
$$= \csc x + \cot x$$

11.
$$\sin \alpha = \frac{3}{5}, \ \cos \alpha = -\frac{4}{5} \qquad [3.2]$$
$$\cos \beta = -\frac{5}{13}, \ \sin \beta = \frac{12}{13}$$
$$\sin(\alpha + \beta) = \sin \alpha \cos \beta + \cos \alpha \sin \beta$$
$$= \left(\frac{3}{5}\right)\left(-\frac{5}{13}\right) + \left(-\frac{4}{5}\right)\left(\frac{12}{13}\right)$$
$$= -\frac{15}{65} - \frac{48}{65} = -\frac{63}{65}$$

12.
$$y = \sin\left[\cos^{-1}\left(\frac{1}{5}\right)\right] \quad [3.5]$$
Let $\alpha = \cos^{-1}\frac{1}{5}, \ \cos \alpha = \frac{1}{5} = \frac{\text{adj}}{\text{hyp}}$
$$\sin \alpha = \frac{\text{opp}}{\text{hyp}} = \frac{\sqrt{5^2 - (1)^2}}{5} = \frac{2\sqrt{6}}{5}$$
$$y = \sin\left[\cos^{-1}\left(\frac{1}{5}\right)\right] = \frac{2\sqrt{6}}{5}$$

Copyright © Houghton Mifflin Company. All rights reserved.

13.
$$\cos^{-1} x = \sin^{-1}\left(\frac{12}{13}\right) \qquad [3.6]$$
$$x = \cos\left[\sin^{-1}\left(\frac{12}{13}\right)\right]$$

Let $\alpha = \sin^{-1}\frac{12}{13}$, $\sin\alpha = \frac{12}{13} = \frac{\text{opp}}{\text{hyp}}$

$$\cos\alpha = \frac{\text{adj}}{\text{hyp}} = \frac{\sqrt{13^2 - 12^2}}{13} = \frac{5}{13}$$

$$x = \cos\left[\sin^{-1}\left(\frac{12}{13}\right)\right] = \frac{5}{13}$$

14.

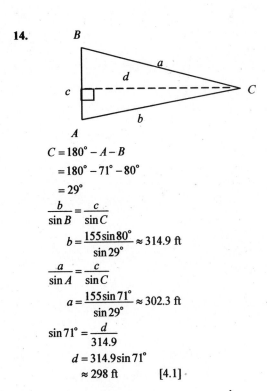

$$C = 180° - A - B$$
$$= 180° - 71° - 80°$$
$$= 29°$$

$$\frac{b}{\sin B} = \frac{c}{\sin C}$$
$$b = \frac{155\sin 80°}{\sin 29°} \approx 314.9 \text{ ft}$$

$$\frac{a}{\sin A} = \frac{c}{\sin C}$$
$$a = \frac{155\sin 71°}{\sin 29°} \approx 302.3 \text{ ft}$$

$$\sin 71° = \frac{d}{314.9}$$
$$d = 314.9\sin 71°$$
$$\approx 298 \text{ ft} \qquad [4.1]$$

15.
$$b^2 = a^2 + c^2 - 2ac\cos B \qquad [4.2]$$
$$\cos B = \frac{a^2 + c^2 - b^2}{2ac} = \frac{4^2 + 3.6^2 - 2.5^2}{2(4)(3.6)} = \frac{22.7}{28.8}$$
$$B \approx 38°$$

16.
$$a_1 = 30\cos 145° \approx -24.6 \quad [4.3]$$
$$a_2 = 30\sin 145° \approx 17.2$$
$$\mathbf{v} = -24.6\mathbf{i} + 17.2\mathbf{j}$$

17.
$$\mathbf{v}\cdot\mathbf{w} = (3\mathbf{i} + 2\mathbf{j})\cdot(5\mathbf{i} - 7\mathbf{j}) \quad [4.3]$$
$$= 3(5) + (2)(-7)$$
$$= 1 \neq 0$$
No. the vectors are not orthogonal.

18.
$$z = -2 + 2i\sqrt{3} \quad [5.2]$$
$$r = \sqrt{(-2)^2 + (2\sqrt{3})^2} = \sqrt{16} = 4$$
$$\alpha = \tan^{-1}\left|\frac{2\sqrt{3}}{-2}\right| = \tan^{-1}\sqrt{3} = \frac{\pi}{3}$$
$$\theta = \pi - \frac{\pi}{3} = \frac{2\pi}{3}$$
$$z = 4\operatorname{cis}\left(\frac{2\pi}{3}\right)$$

19. $r = 3\sin(2\theta)$ [6.5]

20.
$$x = 2t - 1 \quad [6.7]$$
$$x + 1 = 2t$$
$$t = \frac{x+1}{2}$$
$$y = 4t^2 + 1$$
$$y = 4\left(\frac{x+1}{2}\right)^2 + 1 = 4\left(\frac{x^2 + 2x + 1}{4}\right) + 1$$
$$y = x^2 + 2x + 2$$

Copyright © Houghton Mifflin Company. All rights reserved.

Chapter 7
Exponential and Logarithmic Functions

Section 7.1

1.
$f(0)=3^0=1$

$f(4)=3^4=81$

3.
$g(-2)=10^{-2}=\frac{1}{100}$

$g(3)=10^3=1000$

5.
$h(2)=\left(\frac{3}{2}\right)^2=\frac{9}{4}$

$h(-3)=\left(\frac{3}{2}\right)^{-3}=\frac{8}{27}$

7.
$j(-2)=\left(\frac{1}{2}\right)^{-2}=4$

$j(4)=\left(\frac{1}{2}\right)^4=\frac{1}{16}$

9.
$f(3.2)=2^{3.2}\approx9.19$

11.
$g(2.2)=e^{2.2}\approx9.03$

13.
$h(\sqrt{2})=5^{\sqrt{2}}\approx9.74$

15.
$f(x)=5^x$ is a basic exponential graph.

$g(x)=1+5^{-x}$ is the graph of $f(x)$ reflected across the y-axis and moved up 1 unit.

$h(x)=5^{x+3}$ is the graph of $f(x)$ moved to the left 3 units.

$k(x)=5^x+3$ is the graph of $f(x)$ moved up 3 units.

a. $k(x)$ **b.** $g(x)$ **c.** $h(x)$ **d.** $f(x)$

17.
$f(x)=3^x$

19.
$f(x)=10^x$

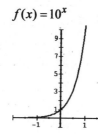

21.
$f(x)=\left(\frac{3}{2}\right)^x$

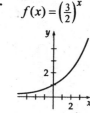

23.
$f(x)=\left(\frac{1}{3}\right)^x$

25. Shift the graph of f vertically upward 2 units.

27. Shift the graph of f horizontally to the right 2 units.

29. Reflect the graph of f across the y-axis.

31. Stretch the graph of f vertically away from the x-axis by a factor of 2.

33. Reflect the graph of f across the y-axis, and then shift this graph vertically upward 2 units.

Copyright © Houghton Mifflin Company. All rights reserved.

35. $f(x) = \dfrac{3^x + 3^{-x}}{2}$

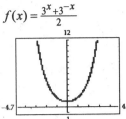

No horizontal asymptote

37. $f(x) = \dfrac{e^x - e^{-x}}{2}$

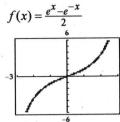

No horizontal asymptote.

39. $f(x) = -e^{(x-4)}$

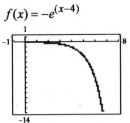

Horizontal asymptote: $y = 0$

41. $f(x) = \dfrac{10}{1 + 0.4e^{-0.5x}}$, with $x \geq 0$

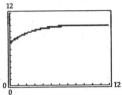

Horizontal asymptote: $y = 10$

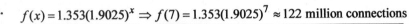

43. **a.** $f(x) = 1.353(1.9025)^x \Rightarrow f(7) = 1.353(1.9025)^7 \approx 122$ million connections

b.

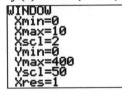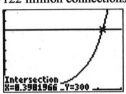

8 years after January 1, 1998 is in 2006.

45. **a.** $d(p) = 25 + 880e^{-0.18p} \Rightarrow d(8) = 25 + 880e^{-0.18(8)} \approx 233$ items per month

$d(p) = 25 + 880e^{-0.18p} \Rightarrow d(18) = 25 + 880e^{-0.18(18)} \approx 59$ items per month

b. As $p \to \infty$, $d(p) \to 25$. The demand will approach 25 items per month.

47. **a.** $P(t) = 1 - e^{-0.75t} \Rightarrow P(1) = 1 - e^{-0.75(1)} \approx 0.53$

b. $P(t) = 1 - e^{-0.75t} \Rightarrow P(3) = 1 - e^{-0.75(3)} \approx 0.89$

c.

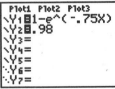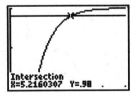

5.2 minutes

d. There is a 98% probability that at least one customer will arrive during the period between 10:00 AM and 10:05.2 AM.

49. **a.** $P(x) = \dfrac{12^x e^{-12}}{x!} \Rightarrow P(9) = \dfrac{12^9 e^{-12}}{9!} \approx 0.087 = 8.7\%$

b. $P(x) = \dfrac{12^x e^{-12}}{x!} \Rightarrow P(18) = \dfrac{12^{18} e^{-12}}{18!} \approx 0.026 = 2.6\%$

Copyright © Houghton Mifflin Company. All rights reserved.

51. **a.** $P(t) = 100 \cdot 2^{2t} \Rightarrow P(3) = 100 \cdot 2^{2(3)} = 100 \cdot 2^6 = 100 \cdot 64 = 6,400$ bacteria

$P(t) = 100 \cdot 2^{2t} \Rightarrow P(6) = 100 \cdot 2^{2(6)} = 100 \cdot 2^{12} = 100 \cdot 4096 = 409,600$ bacteria

b.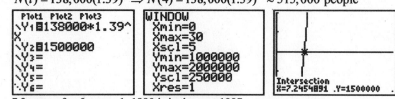

11.6 hours

53. **a.** $N(t) = 138,000(1.39)^t \Rightarrow N(4) = 138,000(1.39)^4 \approx 515,000$ people

b.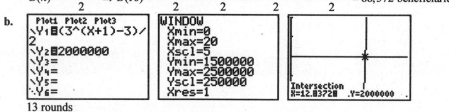

7.2 years after January 1, 1990 is in the year 1997.

55. **a.** $B(n) = \dfrac{3^{n+1} - 3}{2} \Rightarrow B(5) = \dfrac{3^{5+1} - 3}{2} = \dfrac{3^6 - 3}{2} = \dfrac{729 - 3}{2} = \dfrac{726}{2} = 363$ beneficiaries

$B(n) = \dfrac{3^{n+1} - 3}{2} \Rightarrow B(10) = \dfrac{3^{10+1} - 3}{2} = \dfrac{3^{11} - 3}{2} = \dfrac{177147 - 3}{2} = \dfrac{177144}{2} = 88,572$ beneficiaries

b.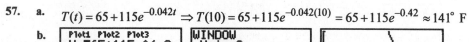

13 rounds

57. **a.** $T(t) = 65 + 115e^{-0.042t} \Rightarrow T(10) = 65 + 115e^{-0.042(10)} = 65 + 115e^{-0.42} \approx 141°$ F

b.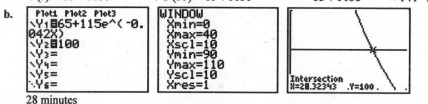

28 minutes

59. **a.** $f(n) = (27.5)2^{(n-1)/12} \Rightarrow f(40) = (27.5)2^{(40-1)/12} = (27.5)2^{39/12} = (27.5)2^{3.25} \approx 261.63$ vibrations per second

b.   No. The function $f(n)$ is not a linear function. Therefore, the graph of $f(n)$ does not increase at a constant rate.

Copyright © Houghton Mifflin Company. All rights reserved.

61. $\sinh(x)=\dfrac{e^x-e^{-x}}{2}$ is an odd function. That is, prove

$\sinh(-x)=-\sinh(x).$

Proof: $\sinh(x)=\dfrac{e^x-e^{-x}}{2}$

$\sinh(-x)=\dfrac{e^{-x}-e^x}{2}$

$\sinh(-x)=\dfrac{-e^{-x}+e^x}{2}$

$\sinh(-x)=\dfrac{\left(e^x-e^{-x}\right)}{2}$

$\sinh(-x)=-F(x)$

63.

65.

domain: $(-\infty,\infty)$

67.

domain: $[0,\infty)$

68. $2^x=16$

$2^x=2^4$

$x=4$

69. $3^{-x}=\dfrac{1}{3^x}=\dfrac{1}{27}$

$\dfrac{1}{3^x}=\dfrac{1}{3^3}$

$x=3$

70. $x^4=625$

$x^4=5^4$

$x=5$

71. $f(x)=\dfrac{2x}{x+3}$

$x=\dfrac{2y}{y+3}$

$xy+3x=2y$

$3x=2y-xy=y(2-x)$

$\dfrac{3x}{2-x}=y$

$f^{-1}(x)=\dfrac{3x}{2-x}$

72. $g(x)=\sqrt{x-2}$

$x-2\geq0$

$x\geq2$

The domain is $\{x\mid x\geq2\}$.

73. The domain is the set of all positive real numbers.

Section 7.2

1. $\log 10=1\Rightarrow10^1=10$

3. $\log_8 64=2\Rightarrow8^2=64$

5. $\log_7 x=0\Rightarrow7^0=x$

7. $\ln x=4\Rightarrow e^4=x$

9. $\ln 1=0\Rightarrow e^0=1$

11. $3^2=9\Rightarrow\log_3 9=2$

13. $4^{-2}=\dfrac{1}{16}\Rightarrow\log_4\dfrac{1}{16}=-2$

15. $b^x=y\Rightarrow\log_b y=x$

17. $y=e^x\Rightarrow\ln y=x$

Copyright © Houghton Mifflin Company. All rights reserved.

19. $100 = 10^2 \Rightarrow \log 100 = 2$

21. $\log_4 16 = 2$ because $4^2 = 16$

23. $\log_3 \dfrac{1}{243} = -5$ because $3^{-5} = \left(\dfrac{1}{3}\right)^5 = \dfrac{1}{243}$

25. $\ln e^3 = 3$ because $e^3 = e^3$

27. $\log \dfrac{1}{100} = -2$ because $10^{-2} = \dfrac{1}{10^2} = \dfrac{1}{100}$

29. $\log_{0.5} 16 = \log_{1/2} 16 = -4$ because $\left(\dfrac{1}{2}\right)^{-4} = 2^4 = 16$

31. $y = \log_4 x$

$x = 4^y$

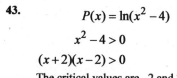

33. $y = \log_{12} x$

$x = 12^y$

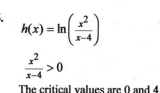

35. $y = \log_{1/2} x$

$x = (1/2)^y$

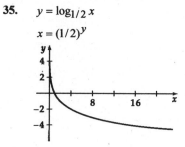

37. $y = \log_{5/2} x$

$x = (5/2)^y$

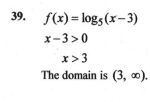

39. $f(x) = \log_5(x - 3)$

$x - 3 > 0$

$x > 3$

The domain is $(3, \infty)$.

41. $k(x) = \log_{2/3}(11 - x)$

$11 - x > 0$

$-x > -11$

$x < 11$

The domain is $(-\infty, 11)$.

43. $P(x) = \ln(x^2 - 4)$

$x^2 - 4 > 0$

$(x + 2)(x - 2) > 0$

The critical values are -2 and 2.
The product is positive.
The domain is $(-\infty, -2) \cup (2, \infty)$.

45. $h(x) = \ln\left(\dfrac{x^2}{x-4}\right)$

$\dfrac{x^2}{x-4} > 0$

The critical values are 0 and 4.
The quotient is positive.
$x > 4$
The domain is $(4, \infty)$.

47. $x^3 - x > 0$

$x(x^2 - 1) > 0$

$x(x + 1)(x - 1) > 0$

Critical values are 0, -1 and 1.
Product is positive.
$-1 < x < 0$ or $x > 1$
$(-1, 0) \cup (1, \infty)$

49. Shift 3 units to the right.

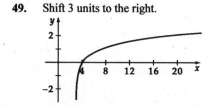

51. Shift 2 units up.

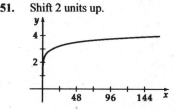

53. Shift 3 units up.

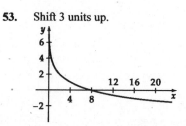

55. Shift 4 units to the right and 1 unit up

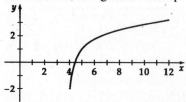

Copyright © Houghton Mifflin Company. All rights reserved.

57. The graph of $f(x) = \log_5(x-2)$ is the graph of $y = \log_5 x$ shifted 2 units to the right.

The graph of $g(x) = 2 + \log_5 x$ is the graph of $y = \log_5 x$ shifted 2 units up.

The graph of $h(x) = \log_5(-x)$ is the graph of $y = \log_5 x$ reflected across the y-axis.

The graph of $k(x) = -\log_5(x+3)$ is the graph of $y = \log_5 x$ reflected across the x-axis and shifted left 3 units.

    a. $k(x)$ **b.** $f(x)$ **c.** $g(x)$ **d.** $h(x)$

59. 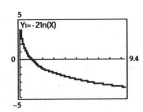   **61.** 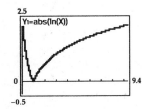   **63.**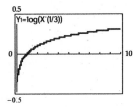

65. 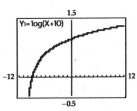   **67.**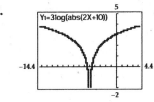

69.   **a.** $r(t) = 0.69607 + 0.60781\ln t \Rightarrow r(9) = 0.69607 + 0.60781\ln 9 \approx 2.0\%$

    b.

                45 months

71.
$$N(x) = 2750 + 180\ln\left(\frac{x}{1000} + 1\right)$$

    a.
$$N(20{,}000) = 2750 + 180\ln\left(\frac{20{,}000}{1000} + 1\right) = 2750 + 180\ln(21) \approx 3298 \text{ units}$$

$$N(40{,}000) = 2750 + 180\ln\left(\frac{40{,}000}{1000} + 1\right) = 2750 + 180\ln(41) \approx 3418 \text{ units}$$

$$N(60{,}000) = 2750 + 180\ln\left(\frac{60{,}000}{1000} + 1\right) = 2750 + 180\ln(61) \approx 3490 \text{ units}$$

    b.
$$N(0) = 2750 + 180\ln\left(\frac{0}{1000} + 1\right) = 2750 + 180\ln(1) = 2750 + 180(0) = 2750 + 0 = 2750 \text{ units}$$

73. $BSA = 0.0003207 \cdot H^{0.3} \cdot W^{(0.7285 - 0.0188\log W)}$

    $BSA = 0.0003207 \cdot (185.42)^{0.3} \cdot (81{,}646.6)^{(0.7285 - 0.0188\log 81{,}646.6)} \approx 2.05$ square meters

75. $N = \text{int}(x\log b) + 1$

    a. $N = \text{int}(10\log 2) + 1 = 3 + 1 = 4$ digits

    b. $N = \text{int}(200\log 3) + 1 = 95 + 1 = 96$ digits

    c. $N = \text{int}(4005\log 7) + 1 = 3384 + 1 = 3385$ digits

    d. $N = \text{int}(13466917\log 2) + 1 = 4{,}053{,}945 + 1 = 4{,}053{,}946$ digits

Copyright © Houghton Mifflin Company. All rights reserved.

77. $f(x)$ and $g(x)$ are inverse functions

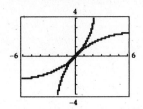

79. The domain of the inverse is the range of the function.
Range of f: $\{y \mid -1 < y \le 1\}$.
The domain of the function is the range of the inverse.
Range of g: all real numbers.

81. $\log 3 + \log 2 \approx 0.77815$
$\log 6 \approx 0.77815$

82. $\ln 8 - \ln 3 \approx 0.98083$
$\ln\left(\frac{8}{3}\right) \approx 0.98083$

83. $3\log 4 \approx 1.80618$
$\log(4^3) \approx 1.80618$

84. $2\ln 5 \approx 3.21888$
$\ln(5^2) \approx 3.21888$

85. $\ln 5 \approx 1.60944$
$\dfrac{\log 5}{\log e} \approx 1.60944$

86. $\log 8 \approx 0.90309$
$\dfrac{\ln 8}{\ln 10} \approx 0.90309$

Section 7.3

1. $\log_b(xyz) = \log_b x + \log_b y + \log_b z$

3.
$$\ln \frac{x}{z^4} = \ln x - \ln z^4$$
$$= \ln x - 4\ln z$$

5.
$$\log_2 \frac{\sqrt{x}}{y^3} = \log_2 \sqrt{x} - \log_2 y^3$$
$$= \log_2 x^{1/2} - \log_2 y^3$$
$$= \tfrac{1}{2}\log_2 x - 3\log_2 y$$

7.
$$\log_7 \frac{\sqrt{xz}}{y^2} = \log_7 \frac{(xz)^{1/2}}{y^2} = \log_7 \frac{x^{1/2}z^{1/2}}{y^2} = \log_7 x^{1/2} + \log_7 z^{1/2} - \log_7 y^2 = \tfrac{1}{2}\log_7 x + \tfrac{1}{2}\log_7 z - 2\log_7 y$$

9.
$$\log(x+5) + 2\log x = \log(x+5) + \log x^2 = \log[x^2(x+5)]$$

11.
$$\ln(x^2 - y^2) - \ln(x - y) = \ln \frac{x^2 - y^2}{x - y} = \ln \frac{(x+y)(x-y)}{x-y} = \ln(x+y)$$

13.
$$3\log x + \tfrac{1}{3}\log y + \log(x+1) = \log x^3 + \log y^{1/3} + \log(x+1) = \log x^3 + \log \sqrt[3]{y} + \log(x+1) = \log\left[x^3 \cdot \sqrt[3]{y}(x+1)\right]$$

15.
$$\log_7 20 = \frac{\log 20}{\log 7} \approx 1.5395$$

17.
$$\log_{11} 8 = \frac{\log 8}{\log 11} \approx 0.8672$$

19.
$$\log_6 \tfrac{1}{3} = \frac{\log \tfrac{1}{3}}{\log 6} \approx -0.6131$$

Copyright © Houghton Mifflin Company. All rights reserved.

21.
$$\log_9 \sqrt{17} = \frac{\log \sqrt{17}}{\log 9} \approx 0.6447$$

23.
$$f(x) = \log_4 x = \frac{\log x}{\log 4}$$

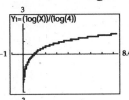

25.
$$g(x) = \log_8 (x-3) = \frac{\log(x-3)}{\log 8}$$

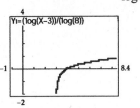

27.
$$h(x) = \log_3 (x-3)^2$$
$$= \frac{\log(x-3)^2}{\log 3}$$
$$= \frac{2\log(x-3)}{\log 3}$$

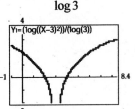

29.
$$F(x) = -\log_5 |x-2| = -\frac{\log|x-2|}{\log 5}$$

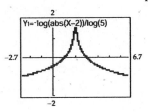

31. False. $\log 10 + \log 10 = 1+1 = 2$
but $\log(10+10) = \log 20 \neq 2$.

33. True.

35. False.
$\log 100 - \log 10 = 2-1 = 1$
but $\log(100-10) = \log 90 \neq 1$

37. False. $\dfrac{\log 100}{\log 10} = \dfrac{2}{1} = 2$
but $\log 100 - \log 10 = 2-1$
$= 1$

39. False. $(\log 10)^2 = 1^2 = 1$ but $2\log 10 = 2(1) = 2$

41.
$$\log_3 5 \cdot \log_5 7 \cdot \log_7 9 = \frac{\log 5}{\log 3} \cdot \frac{\log 7}{\log 5} \cdot \frac{\log 9}{\log 7} = \frac{\cancel{\log 5}}{\log 3} \cdot \frac{\cancel{\log 7}}{\cancel{\log 5}} \cdot \frac{\log 9}{\cancel{\log 7}} = \frac{\log 9}{\log 3} = \frac{\log 3^2}{\log 3} = \frac{2\log 3}{\log 3} = \frac{2\cancel{\log 3}}{\cancel{\log 3}} = 2$$

43.
$\ln 500^{501} = 501\ln 500 \approx 3113.52$
$\ln 506^{500} = 500\ln 506 \approx 3113.27$
$\ln 500^{501}$ is larger.

45.
$$S_n = S_0 \cdot 10^{\frac{n}{N}(\log S_f - \log S_0)}$$
$$S_1 = 1,000,000 \cdot 10^{\frac{1}{5}(\log 500,000 - \log 1,000,000)}$$
$$= 870,551$$
$$S_2 = 1,000,000 \cdot 10^{\frac{2}{5}(\log 500,000 - \log 1,000,000)}$$
$$= 757,858$$
$$S_3 = 1,000,000 \cdot 10^{\frac{3}{5}(\log 500,000 - \log 1,000,000)}$$
$$= 659,754$$
$$S_4 = 1,000,000 \cdot 10^{\frac{4}{5}(\log 500,000 - \log 1,000,000)}$$
$$= 574,349$$
$$S_5 = 1,000,000 \cdot 10^{\frac{5}{5}(\log 500,000 - \log 1,000,000)}$$
$$= 500,000$$
The scales are 1:870,551; 1:757,858; 1:659,754; 1:574,349; 1:500,000.

Copyright © Houghton Mifflin Company. All rights reserved.

47.
$$pH = -\log[H^+]$$
$$pH = -\log[3.97 \times 10^{-11}]$$
$$pH = 10.4$$
$10.4 > 7$; milk of magnesia is a base

49.
$$pH = -\log[H^+]$$
$$9.5 = -\log[H^+]$$
$$-9.5 = \log[H^+]$$
$$10^{-9.5} = 10^{\log[H^+]}$$
$$[H^+] = 3.16 \times 10^{-10} \text{ mole per liter}$$

51.
$$dB(I) = 10\log\left(\frac{I}{I_0}\right)$$

a.
$$dB(1.58 \times 10^8 \cdot I_0) = 10\log\left(\frac{1.58 \times 10^8 \cdot I_0}{I_0}\right)$$
$$= 10\log(1.58 \times 10^8)$$
$$\approx 82.0 \text{ decibels}$$

b.
$$dB(10,800 \cdot I_0) = 10\log\left(\frac{10,800 \cdot I_0}{I_0}\right)$$
$$= 10\log(10,800)$$
$$\approx 40.3 \text{ decibels}$$

c.
$$dB(3.16 \times 10^{11} \cdot I_0) = 10\log\left(\frac{3.16 \times 10^{11} \cdot I_0}{I_0}\right)$$
$$= 10\log(3.16 \times 10^{11})$$
$$\approx 115.0 \text{ decibels}$$

d.
$$dB(1.58 \times 10^{15} \cdot I_0) = 10\log\left(\frac{1.58 \times 10^{15} \cdot I_0}{I_0}\right)$$
$$= 10\log(1.58 \times 10^{15})$$
$$\approx 152.0 \text{ decibels}$$

53.
$$dB(I) = 10\log\left(\frac{I}{I_0}\right)$$
$$120 = 10\log\left(\frac{I_{120}}{I_0}\right)$$
$$12 = \log\left(\frac{I_{120}}{I_0}\right)$$
$$10^{12} = \frac{I_{120}}{I_0}$$
$$10^{12} \cdot I_0 = I_{120}$$

$$110 = 10\log\left(\frac{I_{110}}{I_0}\right)$$
$$11 = \log\left(\frac{I_{110}}{I_0}\right)$$
$$10^{11} = \frac{I_{110}}{I_0}$$
$$10^{11} \cdot I_0 = I_{110}$$

$$\frac{I_{120}}{I_{110}} = \frac{10^{12} \cdot I_0}{10^{11} \cdot I_0}$$
$$= \frac{10^{12}}{10^{11}} = 10^{12-11} = 10^1$$
$$= 10 \text{ times more intense}$$

55.
$$M = \log\left(\frac{100,000 I_0}{I_0}\right) = \log 100,000 = \log 10^5 = 5$$

57.
$$\log\left(\frac{I}{I_0}\right) = M$$
$$\log\left(\frac{I}{I_0}\right) = 6.5$$
$$\frac{I}{I_0} = 10^{6.5}$$
$$I = 10^{6.5} I_0$$
$$I \approx 3,162,277.7 I_0$$

Copyright © Houghton Mifflin Company. All rights reserved.

59.

$$M = \log\left(\frac{I}{I_0}\right)$$

$$M_5 = \log\left(\frac{I_5}{I_0}\right) \Rightarrow 5 = \log\left(\frac{I_5}{I_0}\right) \Rightarrow 10^5 = \frac{I_5}{I_0} \Rightarrow 10^5 I_0 = I_5$$

$$M_3 = \log\left(\frac{I_3}{I_0}\right) \Rightarrow 3 = \log\left(\frac{I_3}{I_0}\right) \Rightarrow 10^3 = \frac{I_3}{I_0} \Rightarrow 10^3 I_0 = I_3$$

$$\frac{I_5}{I_3} = \frac{10^5 I_0}{10^3 I_0} = \frac{10^5}{10^3} = 10^{5-3} = 10^2 = 100 \text{ to } 1 \qquad \bullet \text{ short cut: begin with this line}$$

61. $\dfrac{10^{8.9}}{10^{7.1}} = \dfrac{10^{8.9-7.1}}{1} = 10^{1.8} \text{ to } 1 \approx 63 \text{ to } 1$

63. $M = \log A + 3\log 8t - 2.92$
$= \log 18 + 3\log[8(31)] - 2.92 \approx 5.5$

65. a. $M \approx 6$ **b.** $M \approx 4$

c. When $t = 40$, $M = \log A + 3\log 8t - 2.92 = \log 50 + 3\log[8(40)] - 2.92 \approx 6.3$

When $t = 30$, $M = \log A + 3\log 8t - 2.92 = \log 1 + 3\log[8(30)] - 2.92 \approx 4.2$

The results from parts **a.** and **b.** are close to the magnitudes of 6.3 and 4.2 produced by the Amplitude-Time-Difference Formula.

●●

Prepare for Section 7.4

66. $3^6 = 729 \Rightarrow \log_3 729 = 6$

67. $\log_5 625 = 4 \Rightarrow 5^4 = 625$

68. $a^{x+2} = b \Rightarrow \log_a b = x + 2$

69.
$$4a = 7bx + 2cx$$
$$7bx + 2cx = 4a$$
$$x(7b + 2c) = 4a$$
$$x = \frac{4a}{7b + 2c}$$

70.
$$165 = \frac{300}{1 + 12x}$$
$$165(1 + 12x) = 300$$
$$165 + 1980x = 300$$
$$1980x = 135$$
$$x = \frac{135}{1980} = \frac{3}{44}$$

71.
$$A = \frac{100 + x}{100 - x}$$
$$A(100 - x) = 100 + x$$
$$100A - Ax = 100 + x$$
$$100A - 100 = Ax + x$$
$$100(A - 1) = x(A + 1)$$
$$x = \frac{100(A - 1)}{A + 1}$$

Section 7.4

1.
$$2^x = 64$$
$$2^x = 2^6$$
$$x = 6$$

3.
$$49^x = \frac{1}{343}$$
$$7^{2x} = 7^{-3}$$
$$2x = -3$$
$$x = -\frac{3}{2}$$

5.
$$2^{5x+3} = \frac{1}{8}$$
$$2^{5x+3} = 2^{-3}$$
$$5x + 3 = -3$$
$$5x = -6$$
$$x = -\frac{6}{5}$$

7.
$$\left(\frac{2}{5}\right)^x = \frac{8}{125}$$
$$\left(\frac{2}{5}\right)^x = \left(\frac{2}{5}\right)^3$$
$$x = 3$$

Copyright © Houghton Mifflin Company. All rights reserved.

9.
$$5^x = 70$$
$$\log(5^x) = \log 70$$
$$x \log 5 = \log 70$$
$$x = \frac{\log 70}{\log 5}$$

11.
$$3^{-x} = 120$$
$$\log(3^{-x}) = \log 120$$
$$-x \log 3 = \log 120$$
$$-x = \frac{\log 120}{\log 3}$$
$$x = -\frac{\log 120}{\log 3}$$

13.
$$10^{2x+3} = 315$$
$$\log 10^{2x+3} = \log 315$$
$$(2x+3) \log 10 = \log 315$$
$$2x + 3 = \log 315$$
$$x = \frac{\log 315 - 3}{2}$$

15.
$$e^x = 10$$
$$\ln e^x = \ln 10$$
$$x = \ln 10$$

17.
$$2^{1-x} = 3^{x+1}$$
$$\log 2^{1-x} = \log 3^{x+1}$$
$$(1-x) \log 2 = (x+1) \log 3$$
$$\log 2 - x \log 2 = x \log 3 + \log 3$$
$$\log 2 - x \log 2 - x \log 3 = \log 3$$
$$-x \log 2 - x \log 3 = \log 3 - \log 2$$
$$-x(\log 2 + \log 3) = \log 3 - \log 2$$
$$x = -\frac{(\log 3 - \log 2)}{(\log 2 + \log 3)}$$
$$x = \frac{\log 2 - \log 3}{\log 2 + \log 3} \text{ or } \frac{\log 2 - \log 3}{\log 6}$$

19.
$$2^{2x-3} = 5^{-x-1}$$
$$\log 2^{2x-3} = \log 5^{-x-1}$$
$$(2x-3) \log 2 = (-x-1) \log 5$$
$$2x \log 2 - 3 \log 2 = -x \log 5 - \log 5$$
$$2x \log 2 + x \log 5 - 3 \log 2 = -\log 5$$
$$2x \log 2 + x \log 5 = 3 \log 2 - \log 5$$
$$x(2 \log 2 + \log 5) = 3 \log 2 - \log 5$$
$$x = \frac{3 \log 2 - \log 5}{2 \log 2 + \log 5}$$

21.
$$\log(4x - 18) = 1$$
$$4x - 18 = 10^1$$
$$4x - 18 = 10$$
$$4x = 28$$
$$x = 7$$

23.
$$\ln(x^2 - 12) = \ln x$$
$$x^2 - 12 = x$$
$$x^2 - x - 12 = 0$$
$$(x - 4)(x + 3) = 0$$
$$x = 4 \quad \text{or} \quad x = -3 \text{ (No; not in domain.)}$$
$$x = 4$$

25.
$$\log_2 + \log_2(x-4) = 2$$
$$\log_2 x(x-4) = 2$$
$$\log_2(x^2 - 4x) = 2$$
$$2^2 = x^2 - 4x$$
$$0 = x^2 - 4x - 4$$
$$x = \frac{4 \pm \sqrt{16 - 4(1)(-4)}}{2}$$
$$x = \frac{4 \pm 4\sqrt{2}^2}{2}$$
$$x = 2 \pm 2\sqrt{2}$$

$2 - 2\sqrt{2}$ is not a solution because the logarithm of a negative number is not defined. The solution is $x = 2 + 2\sqrt{2}$.

Copyright © Houghton Mifflin Company. All rights reserved.

27.
$$\log_3 x + \log 3(x+6) = 3$$
$$\log_3 x(x+6) = 3$$
$$3^3 = x(x+6)$$
$$27 = x^2 + 6x$$
$$0 = x^2 + 6x - 27$$
$$0 = (x+9)(x-3)$$
$$x = 3$$
$$x = -9$$

$\log_3(-9)$ is not defined. The solution is $x = 3$.

29.
$$\ln(1-x) + \ln(3-x) = \ln 8$$
$$\ln[(1-x)(3-x)] = \ln 8$$
$$(1-x)(3-x) = 8$$
$$3 - 4x + x^2 = 8$$
$$x^2 - 4x - 5 = 0$$
$$(x+1)(x-5) = 0$$
$$x = -1 \quad \text{or} \quad x = 5 \text{ (No; not in domain.)}$$

The solution is $x = -1$.

31.
$$\log \sqrt{x^3 - 17} = \frac{1}{2}$$
$$\frac{1}{2}\log\left(x^3 - 17\right) = \frac{1}{2}$$
$$10^1 = x^3 - 17$$
$$27 = x^3$$
$$\sqrt[3]{27} = \sqrt[3]{x^3}$$
$$3 = x$$

The solution is $x = 3$.

33.
$$\log(\log x) = 1$$
$$10^1 = \log x$$
$$10^{10} = x$$

35.
$$\ln(e^{3x}) = 6$$
$$3x \ln e = 6$$
$$3x(1) = 6$$
$$3x = 6$$
$$x = 2$$

37.
$$e^{\ln(x-1)} = 4$$
$$\ln e^{\ln(x-1)} = \ln 4$$
$$\ln(x-1) \ln e = \ln 4$$
$$\ln(x-1)(1) = \ln 4$$
$$(x-1) = 4$$
$$x = 5$$

39.
$$\frac{10^x - 10^{-x}}{2} = 20$$
$$10^x\left(10^x - 10^{-x}\right) = 40\left(10^x\right)$$
$$10^{2x} - 1 = 40\left(10^x\right)$$
$$10^{2x} - 40(10)^x - 1 = 0$$

Let $u = 10^x$.
$$u^2 - 40u - 1 = 0$$
$$u = \frac{40 \pm \sqrt{40^2 - 4(1)(-1)}}{2}$$
$$= \frac{40 \pm \sqrt{1600 + 4}}{2}$$
$$= \frac{40 \pm \sqrt{1604}}{2}$$
$$= \frac{40 \pm 2\sqrt{401}}{2}$$
$$= 20 \pm \sqrt{401}$$
$$10^x = 20 + \sqrt{401}$$
$$\log 10^x = \log\left(20 + \sqrt{401}\right)$$
$$x = \log\left(20 + \sqrt{401}\right)$$

41.
$$\frac{10^x + 10^{-x}}{10^x - 10^{-x}} = 5$$
$$10^x + 10^{-x} = 5\left(10^x - 10^{-x}\right)$$
$$10^x\left(10^x + 10^{-x}\right) = 5\left(10^x - 10^{-x}\right)10^x$$
$$10^{2x} + 1 = 5\left(10^{2x} - 1\right)$$
$$4\left(10^{2x}\right) = 6$$
$$2\left(10^{2x}\right) = 3$$
$$\left(10^x\right)^2 = \frac{3}{2}$$
$$10^x = \sqrt{\frac{3}{2}}$$
$$x \log 10 = \log \sqrt{\frac{3}{2}}$$
$$x = \log \sqrt{\frac{3}{2}}$$
$$x = \frac{1}{2}\log\left(\frac{3}{2}\right)$$

Copyright © Houghton Mifflin Company. All rights reserved.

43.
$$\frac{e^x + e^{-x}}{2} = 15$$
$$e^x\left(e^x + e^{-x}\right) = (30)e^x$$
$$e^{2x} + 1 = e^x(30)$$
$$e^{2x} - 30e^x + 1 = 0$$
Let $u = e^x$.
$$u^2 - 30u + 1 = 0$$
$$u = \frac{30 \pm \sqrt{900 - 4}}{2}$$
$$u = \frac{30 \pm \sqrt{896}}{2}$$
$$u = \frac{30 \pm 8\sqrt{14}}{2}$$
$$u = 15 \pm 4\sqrt{14}$$
$$e^x = 15 \pm 4\sqrt{14}$$
$$x \ln e = \ln(15 \pm 4\sqrt{14})$$
$$x = \ln(15 \pm 4\sqrt{14})$$

45.
$$\frac{1}{e^x - e^{-x}} = 4$$
$$1 = 4(e^x - e^{-x})$$
$$1(e^x) = 4(e^x)(e^x - e^{-x})$$
$$e^x = 4(e^{2x} - 1)$$
$$e^x = 4e^{2x} - 4$$
$$0 = 4e^{2x} - e^x - 4$$
Let $u = e^x$.
$$0 = 4u^2 - u - 4$$
$$u = \frac{1 \pm \sqrt{1 - 4(4)(-4)}}{8}$$
$$u = \frac{1 \pm \sqrt{65}}{8}$$
$$e^x = \frac{1 + \sqrt{65}}{8}$$
$$x \ln e = \ln\left(\frac{1 + \sqrt{65}}{8}\right)$$
$$x = \ln(1 + \sqrt{65}) - \ln 8$$

47.
$$2^{-x+3} = x + 1$$
Graph $f = 2^{-x+3} - (x+1)$.
Its x-intercept is the solution.
$x \approx 1.61$

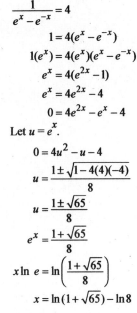

Xmin = −4, Xmax = 4, Xscl = 1,
Ymin = −4, Ymax = 4, Yscl = 1

49.
$$e^{3-2x} - 2x = 1$$
Graph $f = e^{3-2x} - 2x - 1$.
Its x-intercept is the solution.
$x \approx 0.96$

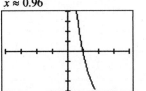

Xmin = −4, Xmax = 4, Xscl = 1,
Ymin = −4, Ymax = 4, Yscl = 1

51.
$$3 \log_2(x-1) = -x + 3$$
Graph $f = \frac{3 \log(x-1)}{\log 2} + x - 3$.
Its x-intercept is the solution.
$x \approx 2.20$

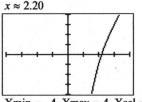

Xmin = −4, Xmax = 4, Xscl = 1,
Ymin = −4, Ymax = 4, Yscl = 1

53.
$$\ln(2x+4) + \frac{1}{2}x = -3$$
Graph $f = \ln(2x+4) + \frac{1}{2}x + 3$.
Its x-intercept is the solution.
$x \approx -1.93$

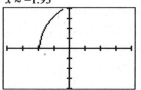

Xmin = −4, Xmax = 4, Xscl = 1,
Ymin = −4, Ymax = 4, Yscl = 1

55.
$$2^{x+1} = x^2 - 1$$
Graph $f = 2^{x+1} - x^2 + 1$.
Its x-intercept is the solution.
$x \approx -1.34$

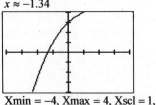

Xmin = −4, Xmax = 4, Xscl = 1,
Ymin = −4, Ymax = 4, Yscl = 1

Copyright © Houghton Mifflin Company. All rights reserved.

57.   **a.**

$$P(0) = 8500(1.1)^0 = 8500(1) = 8500$$

$$P(2) = 8500(1.1)^2 = 10,285$$

b.

$$15,000 = 8500(1.1)^t$$

$$\ln 15,000 = 8500(1.1)^t$$

$$\ln 51,000 = \ln 8500 + t \ln(1.1)$$

$$\frac{\ln 15,000 - \ln 8500}{\ln(1.1)} = t$$

$$6 \approx t$$

The population will reach 15,000 in 6 years.

59.   **a.**

$$T(10) = 36 + 43e^{-0.058(10)} = 36 + 43e^{-0.58}$$

$$T \approx 60°F$$

b.

$$45 = 36 + 43e^{-0.058t}$$

$$\ln(45 - 36) = \ln 43 - 0.058t \ln e$$

$$\frac{\ln(45 - 36) - \ln 43}{-0.058} = t$$

$$t \approx 27 \text{ minutes}$$

61.   **a.**

   b.   48 hours
   c. $P = 100$
   d.   As the number of hours of training increases, the test scores approach 100%.

63.   **a.**

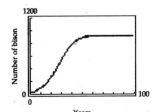

   b.   in 27 years or 2026
   c. $B = 1000$
   d.   As the number of years increases, the bison population approaches but never exceeds 1000.

65.   **a.**

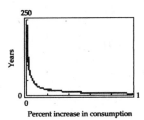

   b.   When $r = 3\%$, or 0.03, $T \approx 78$ years

   c.   When $T = 100$, $r \approx 0.019$, or 1.9%

67.   **a.**

$$t = 2.43 \ln \frac{150 + v}{150 - v}$$

$$5 = 2.43 \ln \frac{150 + v}{150 - v}$$

$$\frac{5}{2.43} = \ln \frac{150 + v}{150 - v}$$

$$e^{5/2.43} = \frac{150 + v}{150 - v}$$

$$150e^{5/2.43} - ve^{5/2.43} = 150 + v$$

$$150e^{5/2.43} - 150 = ve^{5/2.43} + v$$

$$150(e^{5/2.43} - 1) = v(e^{5/2.43} + 1)$$

$$v = \frac{150(e^{5/2.43} - 1)}{e^{5/2.43} + 1}$$

$$v \approx 116 \text{ ft/sec}$$

   b.   The vertical asymptote occurs when the denominator of $\frac{150 + v}{150 - v}$ is zero, or when $v = 150$.

   c.   The velocity of the package approaches, but never reaches or exceeds 150 feet per second.

Copyright © Houghton Mifflin Company. All rights reserved.

69. a.

$$v = 100\left(\frac{e^{0.64t} - 1}{e^{0.64t} + 1}\right)$$

$$50 = 100\left(\frac{e^{0.64t} - 1}{e^{0.64t} + 1}\right)$$

$$\frac{50}{100} = \frac{e^{0.64t} - 1}{e^{0.64t} + 1}$$

$$0.5 = \frac{e^{0.64t} - 1}{e^{0.64t} + 1}$$

$$0.5(e^{0.64t} + 1) = e^{0.64t} - 1$$

$$0.5e^{1.64t} + 0.5 = e^{0.64t} - 1$$

$$0.5e^{0.64t} - e^{0.64t} = -1.5$$

$$-0.5e^{0.64t} = -1.5$$

$$e^{0.64t} = 3$$

$$0.64t = \ln 3$$

$$t = \frac{\ln 3}{0.64}$$

$$t \approx 1.72$$

In approximately 1.72 seconds, the velocity will be 50 feet per second.

b. The horizontal asymptote is the value of

$$100\left[\frac{e^{0.64t} - 1}{e^{0.64t} + 1}\right] \text{ as } t \to \infty. \text{ Therefore, the horizontal}$$

asymptote is $v = 100$ feet per second.

c. The object cannot fall faster than 100 feet per second.

73.

Graph $V = 400{,}000 - 150{,}000(1.005)^x$
   and $V = 100{,}000$.

They intersect when $x \approx 138.97$.

After 138 withdrawals, the account has $101,456.39.

After 139 withdrawals, the account has $99,963.67.

The designer can make at most 138 withdrawals and still have $100,000.

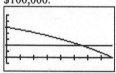

Xmin = 0,Xmax = 200,Xscl = 25

Ymin = −50000,Ymax = 350000,Yscl = 50000

71. a.

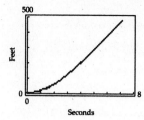

b. When $s = 100$, $t \approx 2.6$ seconds.

Copyright © Houghton Mifflin Company. All rights reserved.

75. The second step because log $0.5 < 0$. Thus the inequality sign must be reversed.

77.
$$\log(x+y) = \log x + \log y$$
$$\log(x+y) = \log xy$$
Therefore $x + y = xy$
$$x - xy = -y$$
$$x(1-y) = -y$$
$$x = \frac{-y}{1-y}$$
$$x = \frac{y}{y-1}$$

79. Since $e^{0.336} \approx 1.4$,
$$F(x) = (1.4)^x \approx (e^{0.336})^x = e^{0.336x} = G(x)$$

81.
$$A = 1000\left(1 + \frac{0.1}{12}\right)^{12(2)} = 1220.39$$

82.
$$A = 600\left(1 + \frac{0.04}{4}\right)^{4(8)} = 824.96$$

83.
$$0.5 = e^{14k}$$
$$\ln 0.5 = \ln e^{14k}$$
$$\ln 0.5 = 14k$$
$$\frac{\ln 0.5}{14} = k$$
$$-0.0495 \approx k$$

84.
$$0.85 = 0.5^{t/5730}$$
$$\ln 0.85 = \ln 0.5^{t/5730}$$
$$\ln 0.85 = \frac{t}{5730}\ln 0.5$$
$$\frac{5730\ln 0.85}{\ln 0.5} = t$$
$$1340 \approx t$$

85.
$$6 = \frac{70}{5 + 9e^{-12k}}$$
$$6(5 + 9e^{-12k}) = 70$$
$$30 + 54e^{-12k} = 70$$
$$54e^{-12k} = 40$$
$$e^{-12k} = \frac{20}{27}$$
$$\ln e^{-12k} = \ln\frac{20}{27}$$
$$-12k = \ln\frac{20}{27}$$
$$k = -\frac{1}{12}\ln\frac{20}{27}$$
$$k \approx 0.025$$

86.
$$2{,}000{,}000 = \frac{3^{n+1} - 3}{2}$$
$$4{,}000{,}000 = 3^{n+1} - 3$$
$$3{,}999{,}997 = 3^{n+1}$$
$$\ln 3{,}999{,}997 = \ln 3^{n+1}$$
$$\ln 3{,}999{,}997 = (n+1)\ln 3$$
$$\frac{\ln 3{,}999{,}997}{\ln 3} = n+1$$
$$\frac{\ln 3{,}999{,}997}{\ln 3} - 1 = n$$
$$12.8 \approx n$$

Copyright © Houghton Mifflin Company. All rights reserved.

Section 7.5

1. a. $P = 8000,\ r = 0.05,\ t = 4,\ n = 1$

$$B = 8000\left(1 + \frac{0.05}{1}\right)^4 \approx \$9724.05$$

b.

$$t = 7,\ B = 8000\left(1 + \frac{0.05}{1}\right)^7 \approx \$11,256.80$$

3. a. $P = 38,000,\ r = 0.065,\ t = 4,\ n = 1$

$$B = 38,000\left(1 + \frac{0.065}{1}\right)^4 \approx \$48,885.72$$

b.

$$n = 365,\ B = 38,000\left(1 + \frac{0.065}{365}\right)^{4(365)} \approx \$49,282.20$$

c.

$$n = 8760,\ B = 38,000\left(1 + \frac{0.065}{8760}\right)^{4(8760)} \approx \$49,283.30$$

5. $P = 15,000,\ r = 0.1, t = 5$

$$B = 15,000e^{5(0.1)} \approx \$24,730.82$$

7. $t = \dfrac{\ln 2}{r}$ $\qquad r = 0.0784$

$t = \dfrac{\ln 2}{0.0784}$

$t \approx 8.8$ years

9. $B = Pe^{rt}$ $\quad$ Let $B = 3P$

$3P = Pe^{rt}$

$3 = e^{rt}$

$\ln 3 = rt \ln e$

$t = \dfrac{\ln 3}{r}$

11. $t = \dfrac{\ln 3}{r}$ $\qquad r = 0.076$

$t = \dfrac{\ln 3}{0.076}$

$t \approx 14$ years

13. a. $t = 0$ hours, $N(0) = 2200(2)^0 = 2200$ bacteria

15. a.

$$N(t) = N_0 e^{kt} \text{ where } N_0 = 24600$$

$$N(5) = 22,600e^{k(5)}$$

$$24,200 = 22,600e^{5k}$$

$$\frac{24,200}{22,600} = e^{5k}$$

$$\ln\left(\frac{24,200}{22,600}\right) = \ln\left(e^{5k}\right)$$

$$\ln\left(\frac{24,200}{22,600}\right) = 5k$$

$$\frac{1}{5}\left[\ln\frac{24,200}{22,600}\right] = k$$

$$0.01368 \approx k$$

$$N(t) = 22,600e^{0.01368t}$$

b. $t = 15$

$$N(15) = 22,600e^{0.01368(15)}$$

$$= 22,600e^{0.2052}$$

$$\approx 27,700$$

b. $t = 3$ hours, $N(3) = 2200(2)^3 = 17,600$ bacteria

17. a. $P = 10,130(1.005)^t$ where $t = 12$

$= 10.130(1.005)^{12}$

$= 110.130(1.061677812)$

$= 10,754.79623$ thousand

$P \approx 10,775,000$

b.

$$13,000 = 10,130(1.005)^t$$

$$\frac{13,000}{10,130} = 1.005^t$$

$$\log\left(\frac{13,000}{10,130}\right) = \log 1.005^t$$

$$\log\left(\frac{13,000}{10,130}\right) = t \log 1.005$$

$$\frac{\log\left(\frac{13,000}{10,130}\right)}{\log 1.005} = t \approx 50$$

in 50 years, or in 2042

Copyright © Houghton Mifflin Company. All rights reserved.

19. **a.**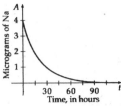

b. $A(5) = 4e^{-0.23} \approx 3.18$ micrograms

c. Since $A = 4$ micrograms are present when $t = 0$, find the time t at which half remains—that is when $A = 2$.

$$2 = 4e^{-0.046t}$$
$$\frac{1}{2} = e^{-0.046t}$$
$$\ln\left(\frac{1}{2}\right) = -0.046t$$
$$\frac{\ln\left(\frac{1}{2}\right)}{-0.046} = t$$
$$15.07 \approx t$$

The half-life of sodium-24 is about 15.07 hours.

d.
$$1 = 4e^{-0.046t}$$
$$\frac{1}{4} = e^{-0.046t}$$
$$\ln\left(\frac{1}{4}\right) = -0.046t$$
$$\frac{\ln\left(\frac{1}{4}\right)}{-0.046t} = t$$
$$30.14 \approx t$$

The amount of sodium-24 will be 1 microgram after 30.14 hours.

21.
$$N(t) = N_0 (0.5)^{t/5730}$$
$$N(t) = 0.45N_0$$
$$0.45N_0 = N_0 (0.5)^{t/5730}$$
$$\ln(0.45) = \frac{t}{5730} \ln 0.5$$
$$5730 \frac{\ln 0.45}{\ln 0.5} = t$$
$$6601 \approx t$$

The bone is about 6601 years old.

23.
$$N(t) = N_0 (0.5)^{t/5730}$$
$$N(t) = 0.75N_0$$
$$0.75N_0 = N_0 (0.5)^{t/5730}$$
$$\ln 0.75 = \frac{t}{5730} \ln 0.5$$
$$5730 \frac{\ln 0.75}{\ln 0.5} = t$$
$$2378 \approx t$$

The Rhind papyrus is about 2378 years old.

25. **a.** $A = 34°\text{F}, T_0 = 75°\text{F}, T_t = 65°\text{F}, t = 5.$ Find k.

$$65 = 34 + (75 - 34)e^{-5k}$$
$$31 = 41e^{-5k}$$
$$\frac{31}{41} = e^{-5k}$$
$$\ln\left(\frac{31}{41}\right) = -5k$$
$$k = -\frac{1}{5}\ln\left(\frac{31}{41}\right)$$
$$k \approx 0.056$$

b. $A = 34°\text{F}, k = 0.056, T_0 = 75°\text{F}, t = 30$

$$T_t = 34 + (75 - 34)e^{-30(0.056)}$$
$$T_t = 34 + (41)e^{-1.68}$$
$$T_t \approx 42°\text{F}$$

c. $T_t = 36°\text{F}, k = 0.056, T_t = 75°\text{F}, A = 34°\text{F}$

$$36 = 34 + (75 - 34)e^{-0.056t}$$
$$2 = 41e^{-0.056t}$$
$$t \approx 54 \text{ minutes}$$

27. **a.** 10% of 80,000 is 8000.

$$8000 = 80,000\left(1 - e^{-0.0005t}\right)$$
$$0.1 = 1 - e^{-0.0005t}$$
$$-0.9 = -e^{-0.0005t}$$
$$0.9 = e^{-0.0005t}$$
$$\ln 0.9 = -0.0005t \ln e$$
$$\ln 0.9 = -0.0005t$$
$$\frac{\ln 0.9}{-0.0005} = t$$
$$211 \approx t$$

b. 50% of 80,000 is 40,000.

$$40,000 = 80,000\left(1 - e^{-0.0005t}\right)$$
$$0.5 = 1 - e^{-0.0005t}$$
$$-0.5 = -e^{-0.0005t}$$
$$0.5 = e^{-0.0005t}$$
$$\ln(0.5) = \ln\left(e^{-0.0005t}\right)$$
$$\ln(0.5) = -0.0005t$$
$$\frac{\ln(0.5)}{-0.0005} = t$$
$$1386 \approx t$$

Copyright © Houghton Mifflin Company. All rights reserved.

29.

$$V(t) = V_0(1-r)^t$$

$$0.5V_0 = V_0(1-0.20)^t$$

$$0.5 = (1-0.20)^t$$

$$0.5 = 0.8^t$$

$$\ln 0.5 = \ln 0.8^t$$

$$\ln 0.5 = t \ln 0.8$$

$$\frac{\ln 0.5}{\ln 0.8} = t$$

$$3.1 \text{ years} \approx t$$

31. a.

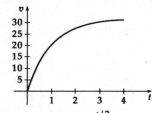

b.

$$20 = 64(1 - e^{-t/2})$$

$$0.625 = 1 - e^{-t/2}$$

$$e^{-t/2} = 0.375$$

$$-t/2 = \ln 0.375$$

$$t \approx 0.98 \text{ seconds}$$

c. The horizontal asymptote is $v = 32$.

d. As time increases, the object's velocity approaches but never exceeds 32 ft/sec.

33. a.

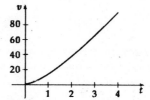

b. The graphs of $s = 32t + 32(e^{-t} - 1)$ and $s = 50$ intersect when $t \approx 2.5$ seconds.

c. The slope m of the secant line containing $(1, s(1))$ and $(2, s(2))$ is $m = \dfrac{s(2)-s(1)}{2-1} \approx 24.56$ ft/sec

d. The average speed of the object was 24.56 feet per second between $t = 1$ and $t = 2$.

35. a. 1900

  b. 0.16

  c. $$P(0) = \frac{1900}{1 + 8.5e^{-0.16(0)}} = 200$$

37. a. 157,500

  b. 0.04

  c. $$P(0) = \frac{157,500}{1 + 2.5e^{-0.04(0)}} = 45,000$$

39. a. 2400

  b. 0.12

  c. $$P(0) = \frac{2400}{1 + 7e^{-0.12(0)}} = 300$$

Copyright © Houghton Mifflin Company. All rights reserved.

41.
$$a=\frac{c-P_0}{P_0}=\frac{5500-400}{400}=12.75$$

$$P(t)=\frac{c}{1+ae^{-bt}}$$

$$P(2)=\frac{5500}{1+12.75e^{-b(2)}}$$

$$780=\frac{5500}{1+12.75e^{-2b}}$$

$$780(1+12.75e^{-2b})=5500$$

$$780+9945e^{-2b}=5500$$

$$9945e^{-2b}=4720$$

$$e^{-2b}=\frac{4720}{9945}$$

$$\ln e^{-2b}=\ln\frac{4720}{9945}$$

$$-2b=\ln\frac{4720}{9945}$$

$$b=-\frac{1}{2}\ln\frac{4720}{9945}$$

$$b\approx0.37263$$

$$P(t)=\frac{5500}{1+12.75e^{-0.37263t}}$$

43.
$$a=\frac{c-P_0}{P_0}=\frac{100-18}{18}=4.55556$$

$$P(t)=\frac{c}{1+ae^{-bt}}$$

$$P(3)=\frac{100}{1+4.55556e^{-b(3)}}$$

$$30=\frac{100}{1+4.55556e^{-3b}}$$

$$30(1+4.55556e^{-3b})=100$$

$$30+136.67e^{-3b}=100$$

$$136.67e^{-3b}=70$$

$$e^{-3b}=\frac{70}{136.67}$$

$$\ln e^{-3b}=\ln\frac{70}{136.67}$$

$$-3b=\ln\frac{70}{136.67}$$

$$b=-\frac{1}{3}\ln\frac{70}{136.67}$$

$$b\approx0.22302$$

$$P(t)=\frac{100}{1+4.55556e^{-0.22302t}}$$

45. **a.**
$$R(t)=\frac{625,000}{1+3.1e^{-0.045t}}$$

$$R(1)=\frac{625,000}{1+3.1e^{-0.045(1)}}\approx\$158,000$$

$$R(2)=\frac{625,000}{1+3.1e^{-0.045(2)}}\approx\$163,000$$

b.
$$R(t)=\frac{625,000}{1+3.1e^{-0.045t}}\text{, as } t\rightarrow\infty,\ R(t)\rightarrow\$625,000$$

47. **a.**
$$a=\frac{c-P_0}{P_0}=\frac{1600-312}{312}=4.12821$$

$$P(t)=\frac{c}{1+ae^{-bt}}$$

$$P(6)=\frac{1600}{1+4.12821e^{-b(6)}}$$

$$416=\frac{1600}{1+4.12821e^{-6b}}$$

$$416(1+4.12821e^{-6b})=1600$$

$$416+1717.34e^{-6b}=1600$$

$$1717.34e^{-6b}=1184$$

$$e^{-6b}=\frac{1184}{1717.34}$$

$$\ln e^{-6b}=\ln\frac{1184}{1717.34}$$

$$-6b=\ln\frac{1184}{1717.34}$$

$$b=-\frac{1}{6}\ln\frac{1184}{1717.34}$$

$$b\approx0.06198$$

$$P(t)=\frac{1600}{1+4.12821e^{-0.06198t}}$$

b.
$$P(10)=\frac{1600}{1+4.12821e^{-0.06198(10)}}\approx497\text{ wolves}$$

Copyright © Houghton Mifflin Company. All rights reserved.

49. a.

$$a = \frac{c - P_0}{P_0} = \frac{8500 - 1500}{1500} = 4.66667$$

$$P(t) = \frac{c}{1 + ae^{-bt}}$$

$$P(2) = \frac{8500}{1 + 4.66667e^{-b(2)}}$$

$$1900 = \frac{8500}{1 + 4.66667e^{-2b}}$$

$$1900(1 + 4.66667e^{-2b}) = 8500$$

$$1900 + 8866.673e^{-2b} = 8500$$

$$8866.673e^{-2b} = 6600$$

$$e^{-2b} = \frac{6600}{8866.673}$$

$$\ln e^{-2b} = \ln \frac{6600}{8866.673}$$

$$-2b = \ln \frac{6600}{8866.673}$$

$$b = -\frac{1}{2}\ln \frac{6600}{8866.673}$$

$$b \approx 0.14761$$

$$P(t) = \frac{8500}{1 + 4.66667e^{-0.14761t}}$$

b.

$$4000 = \frac{8500}{1 + 4.66667e^{-0.14761t}}$$

$$4000(1 + 4.66667e^{-0.14761t}) = 8500$$

$$1 + 4.66667e^{-0.14761t} = 2.125$$

$$4.66667e^{-0.14761t} = 1.125$$

$$e^{-0.14761t} = \frac{1.125}{4.66667}$$

$$\ln e^{-0.14761t} = \ln \frac{1.125}{4.66667}$$

$$-0.14761t = \ln \frac{1.125}{4.66667}$$

$$t = -\frac{1}{0.14761}\ln \frac{1.125}{4.66667}$$

$$t \approx 9.6$$

The population will exceed 4000 in 2010.

51.

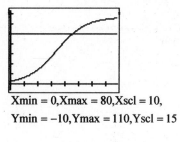

Xmin = 0, Xmax = 80, Xscl = 10,
Ymin = −10, Ymax = 110, Yscl = 15

When $P = 75\%$, $t \approx 45$ hours.

Copyright © Houghton Mifflin Company. All rights reserved.

53.

a. $A(1) = 0.5^{1/2}$

≈ 0.71 gram

b. $A(4) = 0.5^{4/2} + 0.5^{(4-3)/2}$

$= 0.5^2 + 0.5^{1/2}$

≈ 0.96 gram

c. $A(9) = 0.5^{9/2} + 0.5^{(9-3)/2} + 0.5^{(9-6)/2}$

$= 0.5^{4.5} + 0.5^3 + 0.5^{1.5}$

≈ 0.52 gram

55.

a. $P(0) = \dfrac{4.1^0 e^{-4.1}}{0!} = \dfrac{1 \cdot e^{-4.1}}{1} \approx 0.017 = 1.7\%$

b. $P(2) = \dfrac{4.1^2 e^{-4.1}}{2!} = \dfrac{16.81 e^{-4.1}}{2} \approx 0.139 = 13.9\%$

c. $P(3) = \dfrac{4.1^3 e^{-4.1}}{3!} = \dfrac{68.921 e^{-4.1}}{6} \approx 0.190 = 19.0\%$

d. $P(4) = \dfrac{4.1^4 e^{-4.1}}{4!} = \dfrac{282.5761 e^{-4.1}}{24} \approx 0.195 = 19.5\%$

e. $P(9) = \dfrac{4.1^9 e^{-4.1}}{9!} = \dfrac{327381.9344 e^{-4.1}}{362880} \approx 0.015 = 1.5\%$

As $x \to \infty$, $P \to 0$.

57. $\dfrac{3^8 2^{12} 8! 12!}{2 \cdot 3 \cdot 2} = 3^7 2^{10} 8! 12!$

$\text{time} = \left(3^7 2^{10} 8! 12! \text{ arrangements}\right)\left(\dfrac{1 \text{ second}}{\text{arrangement}}\right)\left(\dfrac{1 \text{ minute}}{60 \text{ seconds}}\right)\left(\dfrac{1 \text{ hour}}{60 \text{ minutes}}\right)\left(\dfrac{1 \text{ day}}{24 \text{ hours}}\right)\left(\dfrac{1 \text{ year}}{365 \text{ days}}\right)\left(\dfrac{1 \text{ century}}{100 \text{ years}}\right)$

$= 3^7 2^{10} 8! 12! \left(\dfrac{1}{60^2 \cdot 24 \cdot 365 \cdot 100}\right)$

$= \dfrac{(2187)(1024)(40320)(479001600)}{(3600)(24)(365)(100)}$

$= 13,715,120,270$ centuries

59. decreasing

60. decreasing

61. $P(0) = \dfrac{108}{1 + 2e^{-0.1(0)}} = \dfrac{108}{1+2} = 36$

62. $N(0) = 840 e^{1.05(0)} = 840$

63.
$$10 = \dfrac{20}{1 + 2.2e^{-0.05t}}$$
$$10(1 + 2.2e^{-0.05t}) = 20$$
$$10 + 22e^{-0.05t} = 20$$
$$e^{-0.05t} = \dfrac{10}{22}$$
$$\ln e^{-0.05t} = \ln \dfrac{10}{22}$$
$$-0.05t = \ln \dfrac{10}{22}$$
$$t = -20 \ln \dfrac{10}{22}$$
$$t \approx 15.8$$

64. $P(t) = \dfrac{55}{1 + 3e^{-0.08t}}$

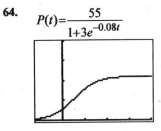

There is a horizontal asymptote at $P = 0$ and $P = 55$.

Copyright © Houghton Mifflin Company. All rights reserved.

Section 7.6

1.

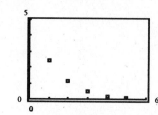

increasing exponential function

3.

decreasing exponential function;
decreasing logarithmic function

5.

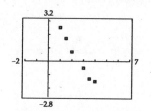

decreasing logarithmic function;
decreasing exponential function

7.

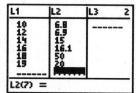

$y \approx 0.99628(1.20052)^x$; $r \approx 0.85705$

9.

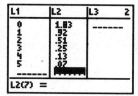

$y \approx 1.81505(0.51979)^x$; $r \approx -0.99978$

11.

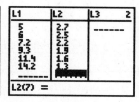

$y \approx 4.89060 - 1.35073 \ln x$; $r \approx -0.99921$

13.

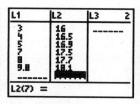

$y \approx 14.05858 + 1.76393 \ln x$; $r \approx 0.99983$

Copyright © Houghton Mifflin Company. All rights reserved.

15.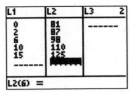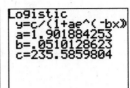

$$y \approx \frac{235.58598}{1+1.90188e^{-0.05101x}}$$

17.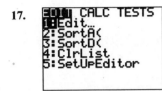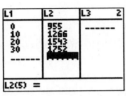

$$y \approx \frac{2098.68307}{1+1.19794e^{-0.06004x}}$$

19.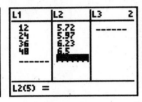

a. $y = 5.48184(1.00356)^x$; $y = 5.48184(1.00356)^{60} \approx 6.78\%$

b.
$$7 = 5.48184(1.00356)^x \Rightarrow \frac{7}{5.48184} = (1.00356)^x \Rightarrow \ln\frac{7}{5.48184} = \ln(1.00356)^x$$

$$\Rightarrow \ln 7 - \ln 5.48184 = x \ln 1.00356 \Rightarrow \frac{\ln 7 - \ln 5.48184}{\ln 1.00356} = x$$

$$\Rightarrow x \approx 69 \text{ months}$$

21.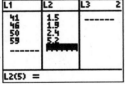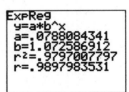

a. Exponential: $T \approx 0.06273(1.07078)^F$

b. $T \approx 0.06273(1.07078)^{65} \approx 5.3$ hours

23.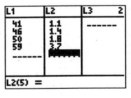

a. $T \approx 0.7881(1.07259)^F$

b. $T \approx 0.07881(1.07259)^{65} \approx 7.5$ hours

   $7.5 - 5.3 = 2.2$ hours.

Copyright © Houghton Mifflin Company. All rights reserved.

25.

a. $N(t) \approx 1500.093(1.940)^t$

b.
$$1,000,000 \approx 1500.093(1.940)^t \Rightarrow \frac{1,000,000}{1500.093} \approx (1.940)^t \Rightarrow \ln\frac{1,000,000}{1500.093} \approx \ln(1.940)^t$$

$$\Rightarrow \ln 1,000,000 - \ln 1500.093 \approx x \ln 1.940 \Rightarrow x \approx \frac{\ln 1,000,000 - \ln 1500.093}{\ln 1.940}$$

$$\Rightarrow x \approx 9.8 \text{ years after } 1996 \Rightarrow \text{ in } 2005$$

27.

a. $p \approx 7.862(1.026)^y$

b. $p \approx 7.862(1.026)^{60} \approx 36 \text{ cm}$

29.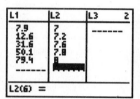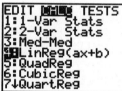

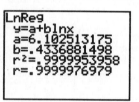

a.   Linear: $\text{pH} \approx 0.01353q + 7.02852$; $r \approx 0.956627$.

Logarithmic: $\text{pH} \approx 6.10251 + 0.43369 \ln q$; $r \approx 0.999998$. The logarithmic model provides the better fit.

b.
$$8.2 \approx 6.10251 + 0.43369 \ln q \Rightarrow 2.09749 \approx 0.43369 \ln q \Rightarrow \frac{2.09749}{0.43369} \approx \ln q \Rightarrow q \approx e^{\frac{2.09749}{0.43369}} \approx 126.0$$

31.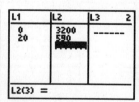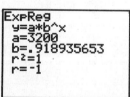

a. $p \approx 3200(0.91894)^t$; $200 \approx 3200(0.91894)^t \Rightarrow \frac{1}{16} \approx (0.91894)^t \Rightarrow \ln\frac{1}{16} \approx \ln(0.91894)^t$

$$\Rightarrow \ln 1 - \ln 16 \approx t \ln 0.91894 \Rightarrow t \approx \frac{-\ln 16}{\ln 0.91894}$$

$$\Rightarrow t \approx 32.8 \text{ years after } 1980 \Rightarrow \text{ in } 2012$$

b.   No. The model fits the data perfectly because there are only two data points.

Copyright © Houghton Mifflin Company. All rights reserved.

33.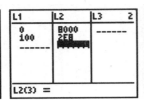

a. $a \approx 8000(1.10657)^t$; $a \approx 8000(1.10657)^{110} \approx 550{,}500{,}000$ automobiles

b. $300{,}000{,}000 \approx 8000(1.10657)^t \Rightarrow 37500 \approx 1.10657^t \Rightarrow \ln 37500 \approx \ln 1.10657^t$

$\Rightarrow \ln 37500 \approx t \ln 1.10657 \Rightarrow t \approx \dfrac{\ln 37500}{\ln 1.10657}$

$\Rightarrow t \approx 104$ years after $1900 \Rightarrow$ in 2004.

35.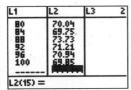

a. Logistic: $\qquad y \approx \dfrac{71.84128}{1 + 13.7774 e^{-0.07915x}}$

Logarithmic: $\qquad y \approx -22.5660 + 20.91276 \ln x$

b. Logistic: $\qquad\qquad\qquad$ Logarithmic:

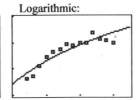

The logistical growth model fits the data better than the logarithmic model.

c. $y \approx \dfrac{71.84128}{1 + 13.7774 e^{-0.07915(108)}} \approx 71.65$ ft

37.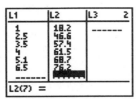

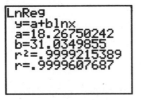

a. Linear: $w \approx 10.17227t + 16.45111$; $r \approx 0.95601$. Logarithmic: $w \approx 18.26750 + 31.03499 \ln t$; $r \approx 0.99996$.

b. The logarithmic model provides the better fit.

c. $w \approx 18.26750 + 31.03499 \ln 10 \approx 89.7$ cubic yards

Copyright © Houghton Mifflin Company. All rights reserved.

39.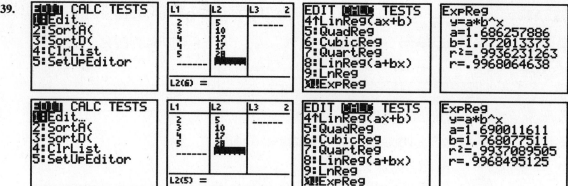

A and *B* have different exponential regression functions.

41.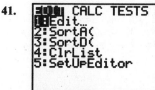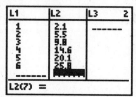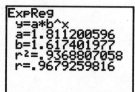

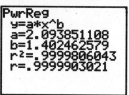

a. Exponential: $y \approx 1.81120(1.61740)^x$; $r \approx 0.96793$. Power: $y \approx 2.09385(x^{1.40246})$; $r \approx 0.99999$.

b. The power regression provides the better fit.

●●●

Chapter 7 True/False Exercises

1. True

2. True

3. True

4. False, because f is not defined for negative values of x, and thus $g[f(x)]$ is undefined for negative values of x.

5. False; $h(x)$ is not an increasing function for $0 < b < 1$.

6. False; $j(x)$ is not an increasing function for $0 < b < 1$.

7. True

8. True

9. True, because $f(-x) = \dfrac{2^{-x} + 2^{-(-x)}}{2} = \dfrac{2^{-x} + 2^x}{2} = f(x)$

10. True

11. False,
$\log x + \log y = \log xy$.

12. True

13. True

14. True

Copyright © Houghton Mifflin Company. All rights reserved.

●●●●●●●●●●●●●●●●●●●●●●●●●●●●●●●●●

1. $\log_5 25 = x$ [7.2]

$5^x = 25$

$5^x = 5^2$

$x = 2$

2. $\log_3 81 = x$ [7.2]

$3^x = 81$

$3^x = 3^4$

$x = 4$

3. $\ln e^3 = x$ [7.2]

$e^x = e^3$

$x = 3$

4. $\ln e^\pi = x$ [7.2]

$e^x = e^\pi$

$x = \pi$

5. $3^{2x+7} = 27$ [7.4]

$3^{2x+7} = 3^3$

$2x + 7 = 3$

$2x = -4$

$x = -2$

6. $5^{x-4} = 625$ [7.4]

$5^{x-4} = 5^4$

$x - 4 = 4$

$x = 8$

7. $2^x = \dfrac{1}{8}$ [7.4]

$2^x = 2^{-3}$

$x = -3$

8. $27(3^x) = 3^{-1}$ [7.4]

$27(3^x) = \dfrac{1}{3}$

$3^x = \dfrac{1}{81}$

$3^x = 3^{-4}$

$x = -4$

9. $\log x^2 = 6$ [7.4]

$10^6 = x^2$

$1,000,000 = x^2$

$\pm\sqrt{1,000,000} = x$

$\pm 1000 = x$

10. $\dfrac{1}{2}\log|x| = 5$ [7.4]

$\log|x| = 10$

$10^{10} = |x|$

$x = \pm 10^{10}$

11. $10^{\log 2x} = 14$ [7.4]

$2x = 14$

$x = 7$

12. $e^{\ln x^2} = 64$ [7.4]

$x^2 = 64$

$x = \pm 8$

13.

14.

15.

16.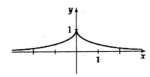

17.

18.

19.

20.

21.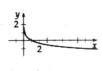

22.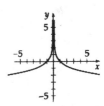

23.

24.

25. $\log_4 64 = 3$ [7.2]

$4^3 = 64$

26. $\log_{1/2} 8 = -3$ [7.2]

$\left(\dfrac{1}{2}\right)^{-3} = 8$

27. $\log_{\sqrt{2}} 4 = 4$ [7.2]

$\left(\sqrt{2}\right)^4 = 4$

28. $\ln 1 = 0$ [7.2]

$e^0 = 1$

Copyright © Houghton Mifflin Company. All rights reserved.

29.
$$5^3 = 125 \quad [7.2]$$
$$\log_5 125 = 3$$

30.
$$2^{10} = 1024 \quad [7.2]$$
$$\log_2 1024 = 10$$

31.
$$10^0 = 1 \quad [7.2]$$
$$\log_{10} 1 = 0$$

32.
$$8^{1/2} = 2\sqrt{2} \quad [7.2]$$
$$\log_8 2\sqrt{2} = \frac{1}{2}$$

33.
$$\log_b \frac{x^2 y^3}{z} = 2\log_b x + 3\log_b y - \log_b z \quad [7.3]$$

34.
$$\log_b \frac{\sqrt{x}}{y^2 z} = \frac{1}{2}\log_b x - \left(2\log_b y + \log_b z\right) \quad [7.3]$$
$$= \frac{1}{2}\log_b x - 2\log_b y - \log_b z$$

35.
$$\ln xy^3 = \ln x + 3\ln y \quad [7.3]$$

36.
$$\ln \frac{\sqrt{xy}}{z^4} = \frac{1}{2}\left(\ln x + \ln y\right) - 4\ln z \quad [7.3]$$
$$= \frac{1}{2}\ln x + \frac{1}{2}\ln y - 4\ln z$$

37.
$$2\log x + \frac{1}{3}\log(x+1) = \log\left(x^2 \sqrt[3]{x+1}\right) \quad [7.3]$$

38.
$$5\log x - 2\log(x+5) = \log \frac{x^5}{(x+5)^2} \quad [7.3]$$

39.
$$\frac{1}{2}\ln 2xy - 3\ln z = \ln \frac{\sqrt{2xy}}{z^3} \quad [7.3]$$

40.
$$\ln x - (\ln y - \ln z) = \ln \frac{x}{y/z} = \ln \frac{xz}{y} \quad [7.3]$$

41.
$$\log_5 101 = \frac{\log 101}{\log 5} \approx 2.86754 \quad [7.3]$$

42.
$$\log_3 40 = \frac{\log 40}{\log 3} \approx 3.35776 \quad [7.3]$$

43.
$$\log_4 0.85 = \frac{\log 0.85}{\log 4} \approx -0.117233 \quad [7.3]$$

44.
$$\log_8 0.3 = \frac{\log 0.3}{\log 8} \approx -0.578989 \quad [7.3]$$

45.
$$4^x = 30 \quad [7.4]$$
$$\log 4^x = \log 30$$
$$x\log 4 = \log 30$$
$$x = \frac{\log 30}{\log 4}$$

46.
$$5^{x+1} = 41 \quad [7.4]$$
$$(x+1)\log 5 = \log 41$$
$$x+1 = \frac{\log 41}{\log 5}$$
$$x = \frac{\log 41}{\log 5} - 1$$

47.
$$\ln(3x) - \ln(x-1) = \ln 4 \quad [7.4]$$
$$\ln \frac{3x}{x-1} = \ln 4$$
$$\frac{3x}{x-1} = 4$$
$$3x = 4(x-1)$$
$$3x = 4x - 4$$
$$4 = x$$

48.
$$\ln(3x) + \ln 2 = \ln 1 \quad [7.4]$$
$$\ln(3x \cdot 2) = 1$$
$$\ln(6x) = 1$$
$$e^1 = 6x$$
$$\frac{e}{6} = x$$

49.
$$e^{\ln(x+2)} = 6 \quad [7.4]$$
$$(x+2) = 6$$
$$x+2 = 6$$
$$x = 4$$

50.
$$10^{\log(2x+1)} = 31 \quad [7.4]$$
$$2x+1 = 31$$
$$2x = 30$$
$$x = 15$$

Copyright © Houghton Mifflin Company. All rights reserved.

51.
$$\frac{4^x+4^{-x}}{4^x-4^{-x}}=2$$
$$4^x\left(4^x+4^{-x}\right)=2\left(4^x-4^{-x}\right)4^x$$
$$4^{2x}+1=2\left(4^{2x}-1\right)$$
$$4^{2x}+1=2\left(4^{2x}-1\right)$$
$$4^{2x}-2\cdot4^{2x}+3=0$$
$$4^{2x}=3$$
$$2^x\ln 4=\ln 3$$
$$x=\frac{\ln 3}{2\ln 4}\qquad\text{[7.4]}$$

52.
$$\frac{5^x+5^{-x}}{2}=8$$
$$5^x\left(5^x+5^{-x}\right)=16\left(5^x\right)$$
$$5^{2x}+1=16\left(5^x\right)$$
$$5^{2x}-16\left(5^x\right)+1=0$$
Let $5^x=u$
$$u^2-16u+1=0$$
$$u=\frac{16\pm\sqrt{16^2-4(1)(1)}}{2}$$
$$u=\frac{16\pm\sqrt{252}}{2}$$
$$u=\frac{16\pm6\sqrt{7}}{2}$$
$$u=8\pm3\sqrt{7}$$
$$5^x=8\pm3\sqrt{7}$$
$$x=\frac{\ln\left(8\pm3\sqrt{7}\right)}{\ln 5}\quad\text{[7.4]}$$

53.
$$\log\left(\log x\right)=3\qquad\text{[7.4]}$$
$$10^3=\log x$$
$$10^{\left(10^3\right)}=x$$
$$10^{1000}=x$$

54.
$$\ln\left(\ln x\right)=2\qquad\text{[7.4]}$$
$$e^2=\ln x$$
$$e^{\left(e^2\right)}=x$$

55.
$$\log\sqrt{x-5}=3\qquad\text{[7.4]}$$
$$10^3=\sqrt{x-5}$$
$$10^6=x-5$$
$$10^6+5=x$$
$$x=1,000,005$$

56.
$$\log x+\log\left(x-15\right)=1$$
$$\log x\left(x-15\right)=1$$
$$10=x^2-15x$$
$$0=x^2-15x-10$$
$$x=\frac{15\pm\sqrt{15^2-4(1)(-10)}}{2}$$
$$x=\frac{15\pm\sqrt{265}}{2}$$
$$x=\frac{15+\sqrt{265}}{2}\qquad\text{[7.4]}$$

57.
$$\log_4\left(\log_3 x\right)=1$$
$$4=\log_3 x$$
$$3^4=x$$
$$81=x\qquad\text{[7.4]}$$

58.
$$\log_7\left(\log_5 x^2\right)=0$$
$$7^0=\log_5 x^2$$
$$1=\log_5 x^2$$
$$5=x^2$$
$$\pm\sqrt{5}=x\qquad\text{[7.4]}$$

59.
$$\log_5 x^3=\log_5 16x\quad\text{[7.4]}$$
$$x^3=16x$$
$$x^2=16$$
$$x=4$$

60.
$$25=16^{\log_4 x}\quad\text{[7.4]}$$
$$25=4^{2\log_4 x}$$
$$25=4^{\log_4 x^2}$$
$$25=x^2$$
$$\pm5=x$$
$$5=x$$

61.
$$m=\log\left(\frac{I}{I_0}\right)\qquad\text{[7.3]}$$
$$=\log\left(\frac{51,782,000 I_0}{I_0}\right)$$
$$=\log 51,782,000$$
$$\approx 7.7$$

Copyright © Houghton Mifflin Company. All rights reserved.

62. $M = \log A + 3\log 8t - 2.92$ [7.3]

$\quad = \log 18 + 3\log 8(21) - 2.92$

$\quad = \log 18 + 3\log 168 - 2.92$

$\quad \approx 5.0$

63.

$\log\left(\dfrac{I_1}{I_0}\right) = 7.2 \quad$ and $\quad \log\left(\dfrac{I_2}{I_0}\right) = 3.7 \quad$ [7.3]

$\quad\quad \dfrac{I_1}{I_0} = 10^{7.2} \quad\quad\quad\quad\quad \dfrac{I_2}{I_0} = 10^{3.7}$

$\quad\quad I_1 = 10^{7.2} I_0 \quad\quad\quad\quad\quad I_2 = 10^{3.7} I_0$

$\dfrac{I_1}{I_2} = \dfrac{10^{7.2} I_0}{10^{3.7} I_0} = \dfrac{10^{3.5}}{1} \approx \dfrac{3162}{1}$

3162 to 1

64. $\dfrac{I_1}{I_2} = 600 = 10^x$ [7.3]

$\log 600 = \log 10^x$

$\log 600 = x$

$\quad 2.8 \approx x$

65. $pH = -\log\left[H_3O^+\right]$ [7.3]

$\quad = -\log\left[6.28 \times 10^{-5}\right]$

$\quad \approx 4.2$

66.

$5.4 = -\log\left[H_3O^+\right]$ [7.3]

$-5.4 = \log\left[H_3O^+\right]$

$10^{-5.4} = H_3O^+$

$0.00000398 \approx H_3O^+$

$H_3O^+ \approx 3.98 \times 10^{-6}$

67. $P = 16,000, r = 0.08, t = 3$ [7.5]

a.

$B = 16,000\left(1 + \dfrac{0.08}{12}\right)^{36} \approx \$20,323.79$

b.

$B = 16,000e^{0.08(3)}$

$B = 16,000e^{0.24} \approx \$20,339.99$

68. $P = 19,000, r = 0.06, t = 5$ [7.5]

a.

$B = 19,000\left(1 + \dfrac{0.06}{365}\right)^{1825} \approx \$25,646.69$

b.

$B = 19,000e^{0.3} \approx \$25,647.32$

69. $S(n) = P(1-r)^n, P = 12,400, r = 0.29, t = 3$ [7.5]

$S(n) = 12,400(1 - 0.29)^3 \approx \$4438.10$

70. a.

$N(t) = N_0 e^{-0.12t}$ [7.5]

$N(10) = N_0 e^{-0.12(10)}$

$\dfrac{N(10)}{N_0} = e^{-1.2}$

$\quad\quad = .301$

$\dfrac{N(10)}{N_0} = 30.1\%$ healed

$100\% - 30.1\% = 69.9\%$ healed

b.

$\dfrac{N(t)}{N_0} = 0.5$

$0.5 = e^{-0.12t}$

$\ln 0.5 = -0.12t$

$\dfrac{\ln 0.5}{-0.12} = t$

$\quad t \approx 6$ days

c.

$\dfrac{N(t)}{N_0} = 0.1$

$0.1 = e^{-0.12t}$

$\ln 0.1 = -0.12t$

$\dfrac{\ln 0.1}{-0.12} = t$

$\quad t \approx 19$ days

Copyright © Houghton Mifflin Company. All rights reserved.

71. $N(0)=1$

$1=N_0e^{k(0)}$

$1=N_0$

$N(2)=5$

$5=e^{2k}$

$\ln 5=2k$

$k=\dfrac{\ln 5}{2}\approx 0.8047$

Thus $N(t)=e^{0.8047t}$ [7.5]

72. $N(0)=N_0=2$ and $N(3)=N_0e^{3k}=2e^{3k}=11$

$e^{3k}=\dfrac{11}{2}$

$e^{3k}=\dfrac{11}{2}$

$3k=\ln\left(\dfrac{11}{2}\right)$

$k=\dfrac{1}{3}\ln\left(\dfrac{11}{2}\right)$

≈ 0.5682

Thus $N(t)=2e^{0.5682t}$ [7.5]

73. $4=N(1)=N_0e^k$ and thus $\dfrac{4}{N_0}=e^k$. Now, we also have $N(5)=5=N_0e^{5k}=N_0\left(\dfrac{4}{N_0}\right)^5=\dfrac{1024}{N_0^4}$.

$N_0=\sqrt[4]{\dfrac{1024}{5}}\approx 3.783$

Thus $4=3.783e^k$.

$\ln\left(\dfrac{4}{3.783}\right)=k$

$k\approx 0.0558$

Thus $N_0=3.783e^{0.0558t}$. [7.5]

74. $1=N(0)=N_0$ and $2=N(-1)=N_0e^{-k}$.

Since $N_0=1$, we have $2=1\cdot e^{-k}$.

$\ln 2=-k$

$k\approx -0.6931$

Thus $N(t)=e^{-0.6931t}$. [7.5]

75. **a.** $N(1)=25{,}200e^{k(1)}=26{,}800$ [7.5]

$e^k=\dfrac{26{,}800}{25{,}200}$

$\ln e^k=\ln\left(\dfrac{26{,}800}{25{,}200}\right)$

$k\approx 0.061557893$

$N(t)=25{,}200e^{0.061557893t}$

b. $N(7)=25{,}200e^{0.061557893(7)}$

$=25{,}200e^{0.430905251}$

$\approx 38{,}800$

76. $P(t)=0.5^{t/5730}=0.96$ [7.5]

$\log\left(0.5^{t/5730}\right)=\log 0.96$

$\dfrac{t}{5730}\log 0.5=\log 0.96$

$\dfrac{t}{5730}=\dfrac{\log 0.96}{\log 0.5}$

$t=5730\left(\dfrac{\log 0.96}{\log 0.5}\right)\approx 340$ years

77. **a.**

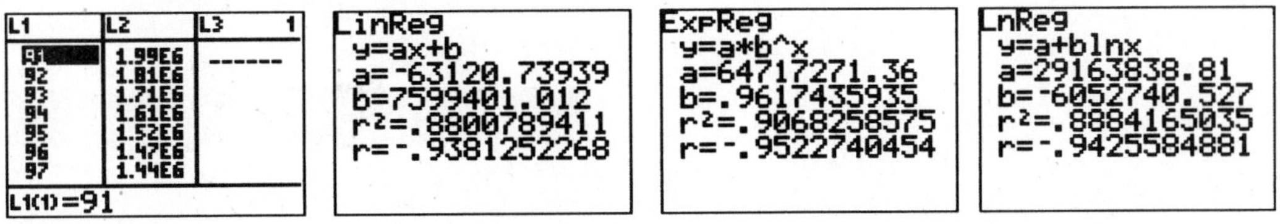
L1 L2 L3 1
91 1.99E6
92 1.81E6
93 1.71E6
94 1.61E6
95 1.52E6
96 1.47E6
97 1.44E6
L1(1)=91
LinReg
y=ax+b
a=⁻63120.73939
b=7599401.012
r²=.8800789411
r=⁻.9381252268
ExpReg
y=a*b^x
a=64717271.36
b=.9617435935
r²=.9068258575
r=⁻.9522740454
LnReg
y=a+blnx
a=29163838.81
b=⁻6052740.527
r²=.8884165035
r=⁻.9425584881

linear: $P=-633121t+7{,}599{,}401,\ \ r\approx -0.93813$

exponential: $P=64{,}717{,}271\left(0.91674359^t\right),\ \ r\approx -0.95228$

logarithmic: $P=29{,}163{,}839-6{,}052{,}740\ln t,\ \ r\approx -0.94256$

b. The exponential equation provides a better fit to the data.

c. $P=64{,}717{,}271\left(0.91674359^{106}\right)=1{,}040{,}000$ [7.6]

Copyright © Houghton Mifflin Company. All rights reserved.

78. a.

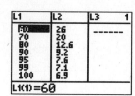

| LinReg
y=ax+b
a=-.4752966016
b=53.1037402
r²=.9627103239
r=-.9811780287 | ExpReg
y=a*b^x
a=207.5443969
b=.966205573
r²=.9932187045
r=-.9966035844 | LnReg
y=a+blnx
a=181.2017885
b=-38.05863628
r²=.9815538066
r=-.9907339737 |

linear: $P = -0.475297t + 53.1037, \ r \approx -0.98118$

exponential: $P = 207.544\left(0.966206^t\right), \ r \approx -0.99660$

logarithmic: $P = 181.202 - 38.0586\ln t, \ r \approx -0.99073$

b. The exponential equation provides a better fit to the data.

c. $P = 207.544\left(0.966206^{108}\right) = 5.1$ per 1000 live births [7.6]

79. a.

$$P(t) = \frac{mP_0}{P_0 + (m - P_0)e^{-kt}}$$

$$P(3) = 360 = \frac{1400(210)}{210 + (1400 - 210)e^{-k(3)}}$$

$$360 = \frac{294000}{210 + 1190e^{-3k}}$$

$$360\left(210 + 1190e^{-3k}\right) = 294000$$

$$210 + 1190e^{-3k} = \frac{294000}{360}$$

$$1190e^{-3k} = \frac{29400}{36} - 210$$

$$e^{-3k} = \frac{29400/36 - 210}{1190}$$

$$\ln e^{-3k} = \ln\left(\frac{29400/36 - 210}{1190}\right)$$

$$-3k = \ln\left(\frac{29400/36 - 210}{1190}\right)$$

$$k = -\frac{1}{3}\ln\left(\frac{29400/36 - 210}{1190}\right)$$

$$k \approx 0.2245763649$$

$$P(t) = \frac{294000}{210 + 1190e^{-0.22458t}} = \frac{1400}{1 + \frac{17}{3}e^{-0.22458t}}$$

b.

$$P(13) = \frac{294000}{210 + 1190e^{-0.22458(13)}}$$

$$= \frac{294000}{210 + 1190e^{-2.919492744}}$$

$$\approx 1070 \qquad\qquad [7.5]$$

80. a.

$$P(0) = \frac{128}{1 + 5e^{-0.27(0)}} = \frac{128}{1 + 5e^0} = \frac{128}{1 + 5} = \frac{128}{6} = 21\frac{1}{3}$$

b. As $t \to \infty, \ e^{-0.27t} \to 0$.

$$P(t) \to \frac{128}{1 + 5(0)} = \frac{128}{1} = 128 \quad [7.5]$$

•••

Chapter **T**est

1. a. $\log_b(5x - 3) = c$ [7.2]

$b^c = 5x - 3$

b. $3^{x/2} = y$

$\log_3 y = \dfrac{x}{2}$

2. $\log_b \dfrac{z^2}{y^3\sqrt{x}} = \log_b z^2 - \log_b y^3 - \log_b x^{1/2}$ [7.3]

$= 2\log_b z - 3\log_b y - \dfrac{1}{2}\log_b x$

3. $\log_{10}(2x+3) - 3\log_{10}(x-2) = \log_{10}(2x+3) - \log_{10}(x-2)^3$

$= \log_{10} \dfrac{2x+3}{(x-2)^3}$ [7.3]

Copyright © Houghton Mifflin Company. All rights reserved.

4.
$$\log_4 12 = \frac{\log 12}{\log 4} \quad [7.3]$$
$$\approx 1.7925$$

5.

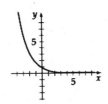

6.

7.
$$5^x = 22 \quad [7.4]$$
$$x \log 5 = \log 22$$
$$x = \frac{\log 22}{\log 5}$$
$$x \approx 1.9206$$

8.
$$4^{5-x} = 7^x \quad [7.4]$$
$$\ln 4^{5-x} = \ln 7^x$$
$$(5-x)\ln 4 = x\ln 7$$
$$5\ln 4 - x\ln 4 = x\ln 7$$
$$5\ln 4 = x\ln 7 + x\ln 4$$
$$5\ln 4 = x(\ln 7 + \ln 4)$$
$$\frac{5\ln 4}{\ln 28} = x$$

9.
$$\log(x+99) - \log(3x-2) = 2 \quad [7.4]$$
$$\log \frac{x+99}{3x-2} = 2$$
$$\frac{x+99}{3x-2} = 10^2$$
$$x+99 = 100(3x-2)$$
$$x+99 = 300x - 200$$
$$-299x = -299$$
$$x = 1$$

10.
$$\ln(2-x) + \ln(5-x) = \ln(37-x)$$
$$\ln(2-x)(5-x) = \ln(37-x)$$
$$(2-x)(5-x) = (37-x)$$
$$10 - 7x + x^2 = 37 - x$$
$$x^2 - 6x - 27 = 0$$
$$(x-9)(x+3) = 0$$
$$x = 9 \text{ (not in domain) or } x = -3$$
$$x = -3 \quad [7.4]$$

11. a.
$$A = P\left(1+\frac{r}{n}\right)^{nt} \quad [7.5]$$
$$= 20{,}000\left(1+\frac{0.078}{12}\right)^{12(5)}$$
$$= 20{,}000(1.0065)^{60}$$
$$= \$29{,}502.36$$

b.
$$A = Pe^{rt}$$
$$= 20{,}000e^{0.078(5)}$$
$$= 20{,}000e^{0.39}$$
$$= \$29{,}539.62$$

12.
$$A = P\left(1+\frac{r}{n}\right)^{nt} \quad [7.5]$$
$$2P = P\left(1+\frac{0.04}{12}\right)^{12t}$$
$$2 = \left(1+\frac{0.04}{12}\right)^{12t}$$
$$\ln 2 = \ln\left(1+\frac{0.04}{12}\right)^{12t}$$
$$\ln 2 = 12t\ln\left(1+\frac{0.04}{12}\right)$$
$$12t = \frac{\ln 2}{\ln\left(1+\frac{0.04}{12}\right)}$$
$$t = \frac{1}{12} \cdot \frac{\ln 2}{\ln\left(1+\frac{0.04}{12}\right)}$$
$$t \approx 17.36 \text{ years}$$

13. a.
$$M = \log\left(\frac{I}{I_0}\right) \quad [7.3]$$
$$= \log\left(\frac{42{,}304{,}000I_0}{I_0}\right)$$
$$= \log 42{,}304{,}000$$
$$\approx 7.6$$

b.
$$\log\left(\frac{I_1}{I_0}\right) = 6.3 \qquad \text{and} \qquad \log\left(\frac{I_2}{I_0}\right) = 4.5$$
$$\frac{I_1}{I_0} = 10^{6.3} \qquad\qquad \frac{I_2}{I_0} = 10^{4.5}$$
$$I_1 = 10^{6.3}I_0 \qquad\qquad I_2 = 10^{4.5}I_0$$
$$\frac{I_1}{I_2} = \frac{10^{6.3}I_0}{10^{4.5}I_0} = \frac{10^{1.8}}{1} \approx \frac{63}{1}$$

Therefore the ratio is 63 to 1.

14. a.
$$N(3) = 34600e^{k(3)} = 39800$$
$$34600e^{3k} = 39800$$
$$e^{3k} = \frac{39800}{34600}$$
$$\ln e^{3k} = \ln\left(\frac{398}{346}\right)$$
$$3k = \ln\left(\frac{398}{346}\right)$$
$$k = \frac{1}{3}\ln\left(\frac{398}{346}\right)$$
$$k \approx 0.0466710767$$
$$N(t) = 34600e^{0.0466710767\,t} \quad [7.5]$$

b.
$$N(10) = 34600e^{0.0466710767(10)}$$
$$= 34600e^{0.466710767}$$
$$\approx 55{,}000$$

Copyright © Houghton Mifflin Company. All rights reserved.

15.

$$P(t) = 0.5^{t/5730} = 0.92 \quad [7.5]$$

$$\log 0.5^{t/5730} = \log 0.92$$

$$\frac{t}{5730}\log 0.5 = \log 0.92$$

$$\frac{t}{5730} = \frac{\log 0.92}{\log 0.5}$$

$$t = 5730\left(\frac{\log 0.92}{\log 0.5}\right)$$

$$t \approx 690 \text{ years}$$

16. a.

 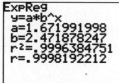

$$y = 1.671991998(2.471878247)^x$$

b.
$$y = 1.671991998(2.471878247)^{7.8} \quad [7.6]$$
$$\approx 1945$$

17. a.

$$R(t) = 1.74830 + 0.78089\ln t$$
$$R(3.5) = 1.74830 + 0.78089\ln 3.5$$
$$\approx 2.73$$

The predicted interest rate for 3.5 years is 2.73%.

b.
$$1.74830 + 0.78089\ln t = 2.5 \qquad [7.6]$$
$$0.78089\ln t = 0.7517$$
$$\ln t = \frac{0.7517}{0.78089}$$
$$e^{\ln t} = e^{0.7517/0.78089}$$
$$t = e^{0.7517/0.78089}$$
$$t \approx 2.6 \text{ years}$$

18. a.

$$a = \frac{c - P_0}{P_0} = \frac{1100 - 160}{160} = 5.875$$

$$P(t) = \frac{c}{1 + ae^{-bt}}$$

$$P(1) = \frac{1100}{1 + 5.875e^{-b(1)}}$$

$$190 = \frac{1100}{1 + 5.875e^{-b}}$$

$$190(1 + 5.875e^{-b}) = 1100$$

$$190 + 1116.25e^{-b} = 1100$$

$$1116.25e^{-b} = 910$$

$$e^{-b} = \frac{910}{1116.25}$$

$$\ln e^{-b} = \ln \frac{910}{1116.25}$$

$$-b = \ln \frac{910}{1116.25}$$

$$b = -\ln \frac{910}{1116.25}$$

$$b \approx 0.20429$$

$$P(t) = \frac{1100}{1 + 5.875e^{-0.20429t}} \quad [7.5]$$

b.

$$P(t) = \frac{1100}{1 + 5.875e^{-0.20429(7)}} \approx 457 \text{ raccoons}$$

Copyright © Houghton Mifflin Company. All rights reserved.

1. $(f \circ g)(x) = f[g(x)]$ [1.5]
$$= f(x^2 + 1)$$
$$= \cos(x^2 + 1)$$

2. $f(x) = 2x + 8$ [1.6]
$$y = 2x + 8$$
$$x = 2y + 8$$
$$x - 8 = 2y$$
$$\frac{x - 8}{2} = y$$
$$f^{-1}(x) = \frac{1}{2}x - 4$$

3. $c = \sqrt{3^2 + 4^2} = \sqrt{25} = 5$ [2.2]
$$\sin\theta = \frac{3}{5}, \;\; \cos\theta = \frac{4}{5}, \;\; \tan\theta = \frac{3}{4}$$

4. $\cos 26° = \dfrac{15}{a}$ [2.2]
$$a = \frac{15}{\cos 26°} \approx 16.7 \text{ cm}$$

5. $y = 3\sin\left(\dfrac{1}{3}x - \dfrac{\pi}{2}\right)$ [2.5]
$$0 \le 2x - \frac{\pi}{2} \le 2\pi$$
$$\frac{\pi}{2} \le 2x \le \frac{5\pi}{2}$$
$$\frac{\pi}{4} \le x \le \frac{5\pi}{4}$$
amplitude = 4, period = π,
phase shift = $\dfrac{\pi}{4}$

6. $y = \sin x + \cos x$ [3.4]
Amplitude: $\sqrt{2}$, period: 2π

7. $f(-x) = \sin(-x) = -\sin x = -f(x)$
[2.4]
odd function

8. $\dfrac{1}{\sin x} - \sin x = \dfrac{1 - \sin^2 x}{\sin x}$ [3.1]
$$= \frac{\cos^2 x}{\sin x}$$
$$= \cos x \frac{\cos x}{\sin x}$$
$$= \cos x \cot x$$

9. $\tan\left(\sin^{-1}\left(\dfrac{12}{13}\right)\right)$ [3.5]

Let $\alpha = \sin^{-1}\dfrac{12}{13}$, $\sin\alpha = \dfrac{12}{13} = \dfrac{\text{opp}}{\text{hyp}}$

$$\tan\alpha = \frac{\text{opp}}{\text{adj}} = \frac{12}{\sqrt{13^2 - 12^2}} = \frac{12}{5}$$

$$\tan\left[\sin^{-1}\left(\frac{12}{13}\right)\right] = \frac{12}{5}$$

10. $2\cos^2 x + \sin x - 1 = 0$ [3.6]
$$2(1 - \sin^2 x) + \sin x - 1 = 0$$
$$2\sin^2 x - \sin x - 1 = 0$$
$$(2\sin x + 1)(\sin x - 1) = 0$$

$2\sin x + 1 = 0$ $\sin x - 1 = 0$

$\sin x = -\dfrac{1}{2}$ $\sin x = 1$

$x = \dfrac{7\pi}{6}, \dfrac{11\pi}{6}$ $x = \dfrac{\pi}{2}$

11. $\|\mathbf{v}\| = \sqrt{(-3)^2 + 4^2}$
$$\|\mathbf{v}\| = \sqrt{9 + 16}$$
$$\|\mathbf{v}\| = 5$$

$\alpha = \tan^{-1}\left|\dfrac{4}{-3}\right| = \tan^{-1}\dfrac{4}{3}$ [4.3]
$$\alpha \approx 53.1°$$
$$\theta = 180° - \alpha$$
$$\theta \approx 180° - 53.1°$$
$$\theta \approx 126.9°$$

12. $\cos\theta = \dfrac{\mathbf{v} \cdot \mathbf{w}}{\|\mathbf{v}\| \, \|\mathbf{w}\|}$ [4.3]

$$\cos\theta = \frac{(2\mathbf{i} - 3\mathbf{j}) \cdot (-3\mathbf{i} + 4\mathbf{j})}{\sqrt{2^2 + (-3)^2}\sqrt{(-3)^2 + 4^2}}$$

$$\cos\theta = \frac{2(-3) + (-3)(4)}{\sqrt{13}\sqrt{25}}$$

$$\cos\theta = \frac{-18}{\sqrt{325}} = -0.9985$$

$$\theta = 176.8°$$

13 $\mathbf{AB} = 400(\cos 42 \mathbf{i} + \sin 42 \mathbf{j}) \approx 297.3\mathbf{i} + 267.7\mathbf{j}$ [4.3]
$$\mathbf{AD} = 55[\cos(-25°)\mathbf{i} + \sin(-25°)\mathbf{j}] \approx 49.8\mathbf{i} - 23.2\mathbf{j}$$
$$\mathbf{AC} = \mathbf{AB} + \mathbf{AD}$$
$$\mathbf{AC} = 297.3\mathbf{i} + 267.7\mathbf{j} + 49.8\mathbf{i} - 23.2\mathbf{j}$$
$$\mathbf{AC} \approx 347.1\mathbf{i} + 244.5\mathbf{j}$$
$$\|\mathbf{AC}\| = \sqrt{347.1^2 + (244.5)^2}$$
$$\|\mathbf{AC}\| \approx 425 \text{ mph}$$

$$\alpha = 90° - \theta = 90° - \tan^{-1}\left(\frac{244.5}{347.1}\right) \approx 55°$$

Copyright © Houghton Mifflin Company. All rights reserved.

14.
$$\frac{a}{\sin A} = \frac{b}{\sin B} \qquad \text{[4.1]}$$
$$\sin A = \frac{a \sin B}{b} = \frac{42 \sin 32°}{50} \approx 0.445$$
$$A \approx 26°$$

15. $z = 1 - i$ [5.3]

$$r = \sqrt{1^2 + (-1)^2} \qquad\qquad \alpha = \tan^{-1}\left|\frac{-1}{1}\right| = 45°$$
$$r = \sqrt{2}$$
$$\theta = 315°$$

$$z = \sqrt{2}(\cos 315° + i \sin 315°)$$
$$(1-i)^8 = [\sqrt{2}(\cos 315° + i \sin 315°)]^8$$
$$= 16[\cos(8 \cdot 315°) + i \sin(8 \cdot 315°)]$$
$$= 16(\cos 2520° + i \sin 2520°)$$
$$= 16(\cos 0° + i \sin 0°)$$
$$= 16 + 0i = 16$$

16. $i = 1(\cos 90° + i \sin 90°)$ [5.3]

$$w_k = 1^{1/2}\left(\cos \frac{90° + 360°k}{2} + i \sin \frac{90° + 360°k}{2}\right) \qquad k = 0, 1$$

$$w_0 = \cos \frac{90°}{2} + i \sin \frac{90°}{2} \qquad\qquad w_1 = \cos \frac{90° + 360°}{2} + i \sin \frac{90° + 360°}{2}$$
$$w_0 = \cos 45° + i \sin 45° \qquad\qquad w_1 = \cos 225° + i \sin 225°$$
$$w_0 = \frac{\sqrt{2}}{2} + \frac{\sqrt{2}}{2}i \qquad\qquad w_1 = -\frac{\sqrt{2}}{2} - \frac{\sqrt{2}}{2}i$$

17.
$$r = \sqrt{x^2 + y^2} \qquad\qquad \theta = \tan^{-1}\frac{y}{x} \quad \text{[6.5]}$$
$$= \sqrt{1^2 + 1^2} \qquad\qquad = \tan^{-1}\left(\frac{1}{1}\right)$$
$$= \sqrt{2} \qquad\qquad\qquad = 45°$$

The polar coordinates of the point are $(\sqrt{2}, 45°)$.

18. $r = 2 - 2\cos\theta$ [6.5]

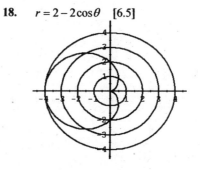

19.
$$5^x = 10 \qquad \text{[7.4]}$$
$$\log 5^x = \log 10$$
$$x \log 5 = 1$$
$$x = \frac{1}{\log 5} \approx 1.43$$

20.
$$N(t) = N_0 e^{kt}$$
$$N(138) = N_0 e^{138k}$$
$$0.5 N_0 = N_0 e^{138k}$$
$$0.5 = e^{138k}$$
$$\ln 0.5 = \ln e^{138k}$$
$$\ln 0.5 = 138k \ln e$$
$$\ln 0.5 = 138k$$
$$\frac{\ln 0.5}{138} = k$$
$$-0.005023 \approx k$$
$$N(t) = N0(0.5)^{t/138} \approx N_0 e^{-0.005023t}$$
$$N(100) = 3(0.5)^{100/138} \approx 1.8 \text{ mg} \quad \text{[7.5]}$$

Copyright © Houghton Mifflin Company. All rights reserved.

Chapter Tests

CHAPTER 1 of *College Trigonometry* Chapter Test

Functions and Graphs

1. Solve $4(3x-2)-3(2x-5)=25$

2. Solve by factoring and apply the zero product property.
$12x^2 -17x+6=0$

3. Use interval notation to express the solution set.
$|x-3| \le 2$

4. Find the midpoint and the length of the line segment with endpoints (–3, 4) and (–7, –5).

5. Determine the x- and y-intercepts, and then graph the equation $x+y^2=6$.

6. Graph the equation $y=-2|x+3|+4$.

7. Find the center and radius of the circle that has the general form $x^2+6x+y^2-8y-11=0$.

8. Determine the domain of the function $f(x)=\sqrt{25-x^2}$.

9. Use the formula $d=\dfrac{|mx_1+b-y_1|}{\sqrt{1+m^2}}$ to find the distance from the point (–2, 5) to the line given by the equation $y=4x-3$.

10. Graph $f(x)=\begin{cases} x^2 & \text{if } x\le-1 \\ |x| & \text{if } -1<x<1 \\ 1 & \text{if } x\ge1 \end{cases}$.

Identify the intervals over which the function is
a. increasing
b. constant
c. decreasing

11. Graph the function $f(x)=(x-2)^2-1$. From the graph, find the domain and range of the function.

12. Use the graph of $f(x)=x^2$ to graph $y=3f(x+1)+2$.

13. Classify each of the following as either an even function, an odd function, or neither an even nor an odd function.
a. $f(x)=x^2+3$
b. $f(x)=x^2+2x$
c. $f(x)=4x$

14. Let $f(x)=4x^2-3$ and $g(x)=x+6$. Find $(f-g)$ and (fg).

15. Find the difference quotient of the function
$f(x)=x^2-x-4$.

16. Evaluate $(f\circ g)$, where $f(x)=\sqrt{2x}$ and $g(x)=\frac{1}{2}x^2$.

17. Find the inverse of $f(x)=\frac{2}{3}x-4$. Graph f and f^{-1} on the same coordinate axes.

18. Find the inverse of $f(x)=\dfrac{x-3}{3x+6}$. State the domain and the range of f^{-1}.

19. The distance traveled by a ball rolling down a ramp is given by $s(t)=4t^2$, where t is the time in seconds after the ball is released and $s(t)$ is measured in feet. Evaluate the average velocity of the ball for each of the following time intervals.
a. [1, 2]   b. [1, 1.5]   c. [1, 1.01]

20. a. Determine the linear regression equation for the given set.

x	2	3	4	6	10
y	3	4	6	10	16

b. Using the linear model from part a, find the expected y-value when the x-value is 8.

Copyright © Houghton Mifflin Company. All rights reserved.

Trigonometric Functions

In Problems 1 and 2, convert the measure of the given angle from degrees to radians.

1. $-135°$

2. $240°$

In Problems 3 and 4, convert the measure of the given angle from radians to degrees.

3. $\dfrac{9\pi}{2}$

4. $\dfrac{16\pi}{3}$

5. Determine the measure in radians and in degrees of the central angle that cuts an arc of length 12π centimeters in a circle with radius 8 centimeters.

In Problems 6-9, determine the given trigonometric function value.

6. $\sin\dfrac{5\pi}{4}$

7. $\cos\left(-\dfrac{4\pi}{3}\right)$

8. $\sec\dfrac{19\pi}{2}$

9. $\tan\left(-\dfrac{17\pi}{3}\right)$

In Problems 10 and 11, the given point is located on the terminal side of angle θ. Find $\sin\theta$ and $\tan\theta$.

10. $(-10, 24)$

11. $(-12, -9)$

In Problems 12 and 13, evaluate each quantity.

12. $\cos\left(-270°\right)$

13. $\sin 180°$

In Problems 14-17, sketch the graph on the interval $[0, 2\pi]$.

14. $y = -2 + 3\cos(-x)$

15. $y = -1 + \sin(2x - \pi)$

16. $y = \sec\dfrac{x}{2}$

17. $y = \cot(2x + \pi)$

18. Determine the amplitude and period and then sketch the graph of the function $y = 3\cos\dfrac{1}{2}x$.

19. Determine the amplitude, period, and phase shift and then sketch the graph of the function $y = 14\sin(2x + \pi)$.

20. Sketch the graph of the function $y = -3\sin\dfrac{1}{2}x + x$.

21. What is the angle of elevation of the sun at the time that a flagpole 68 feet tall casts a shadow 25 feet long?

22. A ladder 25 feet long is leaning against a building. If the foot of the ladder is 6 feet from the base of the building (on ground level), what acute angle does the ladder make with the ground?

23. The equation $x = 4\sin\left(\dfrac{t}{2} - 2\pi\right)$ represents simple harmonic motion. Determine the amplitude, period, and frequency. Sketch the graph.

Copyright © Houghton Mifflin Company. All rights reserved.

Trigonometric Identities and Equations

In Problems 1 and 2, verify the identity.

1. $\dfrac{1-\tan x}{\sec x} + \sin x = \cos x$

2. $\dfrac{\tan^2 x + 1}{\tan^2 x + 2} = \dfrac{1}{1+\cos^2 x}$

3. Evaluate $\cot\dfrac{5\pi}{12}$ without using a calculator.

In Problems 4 and 5, verify the identity.

4. $\dfrac{\sin(a+b)}{\cos a \cos b} = \tan a + \tan b$

5. $\dfrac{\tan a - \tan b}{\tan a + \tan b} = \dfrac{\sin(a-b)}{\sin(a+b)}$

6. Given that $\alpha = 315°$, find $\sin 2\alpha$, $\cos 2\alpha$, and $\tan 2\alpha$, using the double-angle identities.

In Problems 7 and 8, from the given information, determine $\cos\dfrac{\alpha}{2}$.

7. $\sin\alpha = -\dfrac{2}{3}, \alpha$ in Quadrant III

8. $\tan\alpha = -\dfrac{3}{2}, \alpha$ in Quadrant II

In Problems 9 and 10, write the given expression as a single trigonometric function.

9. $\dfrac{\cot\varphi - \tan\varphi}{\cot\varphi + \tan\varphi}$

10. $\dfrac{\sin 2\alpha}{1 - \cos 2\alpha}$

In Problems 11 and 12, verify the identity.

11. $\dfrac{1}{\sin x} - \sin x = \dfrac{\sin x}{\tan^2 x}$

12. $\sin\left(\dfrac{\pi}{4} - \alpha\right) = \cos\left(\alpha + \dfrac{\pi}{4}\right)$

In Problems 13 and 14, solve for x. Express the answer in radians.

13. $2\cos^2 x - \cos x = 1$

14. $\sin^2 x + 3\sin x = 4$

In Problems 15 and 16, given that $\sin 72° = 0.95$ and $\cos 72° = 0.31$, find the function value. Round your answer to two decimal places.

15. $\sec 18°$

16. $\cot 18°$

In Problems 17 to 20, find the exact value of each expression.

17. $\sin^{-1}\left(-\dfrac{\sqrt{2}}{2}\right)$

18. $\cos^{-1}\left(-\dfrac{1}{2}\right)$

19. $\cos^{-1}\left(\cos\left(-\dfrac{\pi}{4}\right)\right)$

20. $\sin^{-1}\left(\cos\dfrac{\pi}{6}\right)$

Copyright © Houghton Mifflin Company. All rights reserved.

Applications of Trigonometry

In Problems 1 - 9, solve triangle *ABC*.

1. $a = 17$, $B = 46°$, $C = 81°$

2. $a = 11$, $b = 16$, $A = 40°$

3. $a = 14$, $A = 38°$, $B = 73°$

4. $a = 18$, $b = 7.0$, $B = 23°$

5. $A = 41°$, $B = 83°$, $c = 44.6$

6. $A = 120°$, $b = 20$, $c = 16$

7. $a = 78$, $B = 128°$, $c = 125$

8. $a = 14$, $b = 19$, $c = 13$

9. $a = 10$, $b = 13$, $C = 100°$

In Problems 10 - 12, find the area of triangle *ABC*.

10. $B = 30°$, $C = 135°$, $b = 4\sqrt{2}$

11. $a = 7.0$, $b = 9.0$, $C = 26°$

12. $a = 12$, $b = 15$, $c = 23$

13. The vector **u** has initial point (0, 0) and terminal point (–3, 3). The vector **v** has initial point (0, 0) and terminal point (1, 1).
   a.   Determine the magnitude and direction of vector **u**.
   b.   Draw **u** – 2**v**.

In Problems 14 and 15, determine the dot product of the given vectors. Find the angle between the vectors to the nearest degree.

14. $\mathbf{u} = \langle 3, -1 \rangle$, $\mathbf{v} = \langle -2, -2 \rangle$

15. $\mathbf{u} = \langle 4, 6 \rangle$, $\mathbf{v} = \langle -3, 2 \rangle$

In Problems 16 and 17, use the vectors $\mathbf{u} = \left\langle -\dfrac{1}{2}, \dfrac{7}{2} \right\rangle$ and $\mathbf{v} = \langle 6, -1 \rangle$ to determine the following.

16. $2\mathbf{u} - \dfrac{1}{2}\mathbf{v}$

17. $\mathbf{u} - 4\mathbf{v}$

18. A plane's air speed is 500 mph and its bearing is N 75° E. An 6-mph wind is blowing in the direction of N 79° W. Determine the ground speed and true bearing of the plane.

19. Find the force required to hold a 120-pound crate stationary on a ramp inclined at 25°.

       Copyright © Houghton Mifflin Company. All rights reserved.

Complex Numbers

In Problems 1 – 14, perform the indicated operation; then write the answer in standard form.

1. i^{51}

2. $(-i)^{35}$

3. $(3i)^4 \cdot i^{44}$

4. $(1-3i)+(-2+i)$

5. $(4-3i)-(6-2i)$

6. $5i(4-2i)$

7. $2i - i(8+3i)$

8. $(1-4i)(3+2i)$

9. $(3+5i)(7-4i)$

10. $(2-9i)(2+9i)$

11. $(1+i\sqrt{5})(1-i\sqrt{5})$

12. $\dfrac{4}{3i}$

13. $\dfrac{1}{3+11i}$

14. $\dfrac{6-2i}{2+i}$

In Problems 15 and 16, express the complex number in trigonometric form.

15. $3 - i\sqrt{3}$

16. $-4 - 4i$

17. Write $z = 4 \text{ cis } 45°$ in standard form.

18. Multiply: $(-2 \text{ cis } 225°)(\sqrt{3} \text{ cis } 75°)$. Write the answer in standard form.

19. Divide: $\dfrac{6\sqrt{3}+6i}{1-i\sqrt{3}}$. Write the answer in trigonometric form.

20. Find $\left(\dfrac{-\sqrt{3}}{2}+\dfrac{1}{2}i\right)^5$. Write the answer in standard form.

21. Find the fourth roots of $-1+i\sqrt{3}$. Write the answer in trigonometric form.

Copyright © Houghton Mifflin Company. All rights reserved.

Topics in Analytic Geometry

In Problems 1 – 4, write the equation of each conic in standard form and state the type of conic, the foci, and the vertices of each.

1. $x^2 + 4y^2 - 36 = 0$
2. $y^2 + 2y - x + 3 = 0$
3. $x^2 - 4y^2 - 36 = 0$
4. $y^2 - 6y + x = 0$

In Problems 5 – 8, find the foci and vertices of each conic. If the conic is a hyperbola, find the asymptotes. Graph each equation.

5. $y^2 - 6y - x + 5 = 0$
6. $\dfrac{(x-5)^2}{25} + \dfrac{y^2}{4} = 1$
7. $\dfrac{(y+2)^2}{24} - \dfrac{(x-1)^2}{4} = 1$
8. $4x^2 - 9y^2 - 8x + 18y = 149$

In Problems 9 and 10, the rectangular coordinates of a point are given. Determine two polar representations (r, θ) for the point with $0 \le \theta < 2\pi$.

9. $(-1, 1)$
10. $(-\sqrt{3}, -1)$

In Problems 11 – 14, convert the given polar equation into rectangular form.

11. $r = 3\sin\theta$
12. $r = 2\cos\theta$
13. $\theta = \pi/6$
14. $r^2 + \sin 2\theta = 0$

In Problems 15 and 16, graph each polar equation.

15. $r = 2 + \sin\theta$
16. $r = 3 - 2\cos\theta$

In Problems 17 and 18, graph the conic given by each polar equation.

17. $r = \dfrac{5}{1 + \sin\theta}$
18. $r = \dfrac{18}{2 + 3\cos\theta}$

19. Write the equation $13x^2 + 6\sqrt{3}xy + 7y^2 - 32 = 0$ without an xy term. Name the graph of the equation.

20. Eliminate the parameter and graph the curve given by the parametric equations $x = 2t - 1$, $y = 2t^2 + t$.

   Copyright © Houghton Mifflin Company. All rights reserved.

Exponential and Logarithmic Functions

1. Graph : $f(x) = 2^{-x}$

2. Graph : $f(x) = e^{2x}$

3. Write $\log_4 (3x + 2) = d$ in exponential form.

4. Write $4^{x+3} = y$ in logarithmic form.

5. Write $\log_b \dfrac{\sqrt{y}}{x^3 z^2}$ in terms of logarithms x, y, and z.

6. Write $\log_b (x-1) + \frac{1}{2}\log_b (y+2) - 4\log_b (z+3)$ as a single logarithm with a coefficient of 1.

7. Use the change-of-base formula and a calculator to approximate $\log_5 15$. Round your result to the nearest 0.0001.

8. Graph: $f(x) = \ln (x+2)$

9. Solve: $6^x = 60$. Round your solution to the nearest 0.0001.

10. Find the *exact* solution(s) of
$$\frac{10^x - 10^{-x}}{10^x + 10^{-x}} = \frac{1}{4}$$

11. Solve: $\log (2x+1) - 2 = \log (3x-1)$

12. Solve: $\ln (x+1) + \ln(x-2) = \ln (4x-8)$

13. Find the balance on $12,000 invested at an annual interest rate of 6.7% for 3 years:
    a. compounded quarterly.
    b. compounded continuously

14. a. What, to the nearest 0.1, will an earthquake measure on the Richter scale if it has an intensity of $I = 35{,}792{,}000I_0$?

    b. Compare the intensity of an earthquake that measures 6.4 on the Richter scale to the intensity of an earthquake that measures 5.1 on the Richter scale by finding the ratio of the larger intensity to the smaller intensity. Round to the nearest whole number.

15. a. Find the exponential growth function for a city whose population was 51,300 in 1980 and 240,500 in 2000. Use $t = 0$ to represent the year 1980.
    b. Use the growth function to predict the population of the city in 2010. Round to the nearest 100.

16. Determine, to the nearest 10 years, the age of a bone if it now contains 89% of its original amount of carbon-14. The half-life of carbon-14 is 5730 years.

17. a. Use a graphing utility to find the exponential regression function for the following data:
    {(1.4, 3), (2.6, 7), (3.9, 15), (4.3, 20), (5.2, 26)}
    b. Use the function to predict, to the nearest whole number, the *y*-value associated with $x = 8.2$.

18. An altimeter is used to determine the height of an airplane above sea level. The following table shows the values for the pressure p and altitude h of an altimeter.

Pressure p in pounds per square inch	13.1	12.7	12.3	11.4	10.9
Altitude h in feet	3000	3900	4800	6800	8000

a. Find the logarithmic regression function for the data and use the function to estimate, to the nearest 100 feet, the height of the airplane when the pressure is 12.0 pounds per square inch.

b. According to your function, what will the pressure be, to the nearest 0.1 per square inch, when the airplane is 6200 feet above sea level?

Copyright © Houghton Mifflin Company. All rights reserved.

Solutions to
Chapter Tests

Functions and Graphs

1. $4(3x-2)-3(2x-5)=25$
$12x-8-6x+15=25$
$6x+7=25$
$6x=18$
$x=3$

2. $12x^2-17x+6=0$
$(3x-2)(4x-3)=0$
$3x-2=0$ or $4x-3=0$
$3x=2$ $\qquad$ $4x=3$
$x=\frac{2}{3}$ $\qquad$ $x=\frac{3}{4}$

3. $|x-3|\le 2$
$-2\le x-3\le 2$
$1\le\ x\ \le 5$

$[1,5]$

4. midpoint $=\left(\dfrac{x_1+x_2}{2},\ \dfrac{y_1+y_2}{2}\right)=\left(\dfrac{-3+(-7)}{2},\ \dfrac{4+(-5)}{2}\right)=\left(\dfrac{-3-7}{2},\ \dfrac{4-5}{2}\right)=\left(\dfrac{-10}{2},\ \dfrac{-1}{2}\right)=\left(-5,\ -\dfrac{1}{2}\right)$

length $=d=\sqrt{(x_1-x_2)^2+(y_1-y_2)^2}=\sqrt{[-3-(-7)]^2+[4-(-5)]^2}=\sqrt{(-3+7)^2+(4+5)^2}=\sqrt{(4)^2+(9)^2}=\sqrt{16+81}=\sqrt{97}$

5. $x+y^2=6$

$y=0\Rightarrow x+(0)^2=6\Rightarrow x=6$
Thus the x-intercept is $(6,0)$.
$x=0\Rightarrow 0+y^2=6$
$\qquad\qquad y^2=6$
$\qquad\qquad y=\pm\sqrt{6}$
Thus the y-intercepts are $\left(0,\ -\sqrt{6}\right)$ and $\left(0,\ \sqrt{6}\right)$.

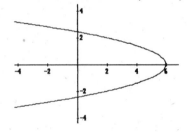

6. $y=-2|x+3|+4$

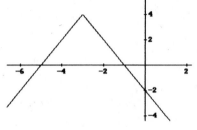

7. $x^2+6x+y^2-8y-11=0.$
$(x^2+6x)+(y^2-8y)=11$
$(x^2+6x+9)+(y^2-8y+16)=11+9+16$
$(x+3)^2+(y-4)^2=36$
$(x+3)^2+(y-4)^2=6^2$
center $(-3,4)$, radius 6

8. $25-x^2\ge 0$
$(5-x)(5+x)\ge 0$
The product is positive or zero.
The critical values are 5 and -5.

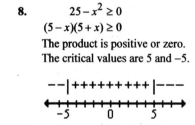

The domain is $\{x|-5\le x\le 5\}$.

9. $y=4x-3,\ (x_1,y_1)=(-2,5)$

$d=\dfrac{|mx_1+b-y_1|}{\sqrt{1+m^2}}=\dfrac{|4(-2)+(-3)-5|}{\sqrt{1+4^2}}=\dfrac{|-8-3-5|}{\sqrt{1+16}}=\dfrac{|-16|}{\sqrt{17}}=\dfrac{16}{\sqrt{17}}=\dfrac{16\sqrt{17}}{17}$

10. $f(x)=\begin{cases}x^2 & \text{if } x\le -1\\ |x| & \text{if } -1<x<1\\ 1 & \text{if } x\ge 1\end{cases}$

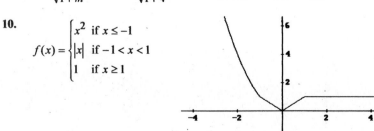

a. increasing on $[0,1]$
b. constant on $[1,\infty)$
c. decreasing on $(-\infty,0]$

Copyright © Houghton Mifflin Company. All rights reserved.

11. $f(x) = (x-2)^2 - 1$

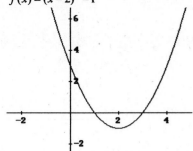

domain: all real numbers
range: $\{y | y \geq -1\}$

12. The graph is shifted 1 unit to the left and 2 units up, and is stretched vertically by multiplying each y-coordinate by 3.

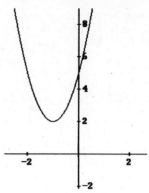

13. a. $f(x) = x^2 + 3$

$f(-x) = (-x)^2 + 3 = x^2 + 3 = f(x)$

$f(x)$ is an even function.

b. $f(x) = x^2 + 2x$

$f(-x) = (-x)^2 + 2(-x) = x^2 - 2x$

$f(-x) = x^2 - 2x \neq f(x)$ not an even function

$f(-x) = x^2 - 2x \neq -f(x)$ not an odd function

$f(x)$ is neither an even function nor an odd function.

c. $f(x) = 4x$

$f(-x) = 4(-x) = -4x = -f(x)$

$f(x)$ is an odd function.

$f(3) = -(3)^2 + 6(3) + 7$

$= -9 + 18 + 7$

$= 16$

The maximum value of the function is 16.

14. $(f - g)(x) = f(x) - g(x)$

$= (4x^2 - 3) - (x + 6)$

$= 4x^2 - 3 - x - 6$

$= 4x^2 - x - 9$

$(fg) = f(x) \cdot g(x)$

$= (4x^2 - 3)(x + 6)$

$= 4x^3 + 24x^2 - 3x - 18$

15. $f(x) = x^2 - x - 4$

$\dfrac{f(x+h) - f(x)}{h} = \dfrac{(x+h)^2 - (x+h) - 4 - (x^2 - x - 4)}{h}$

$= \dfrac{x^2 + 2xh + h^2 - x - h - 4 - x^2 + x + 4}{h}$

$= \dfrac{2xh + h^2 - h}{h}$

$= \dfrac{h(2x + h - 1)}{h}$

$= 2x + h - 1$

16. $f(x) = \sqrt{2x}$ and $g(x) = \frac{1}{2}x^2$

$(f \circ g)(x) = f[g(x)] = f\!\left(\frac{1}{2}x^2\right)$

$= \left(\sqrt{2\left(\frac{1}{2}x^2\right)}\right)$

$= \sqrt{x^2}$

$= x$

17. $y = \frac{2}{3}x - 4$

Interchange x and y, then solve for y.

$x = \frac{2}{3}y - 4$

$3(x) = 3\left(\frac{2}{3}y - 4\right)$

$3x = 2y - 12$

$3x + 12 = 2y$

$\frac{1}{2}(3x + 12) = \frac{1}{2}(2y)$

$\frac{3}{2}x + 6 = y$

$f^{-1}(x) = \frac{3}{2}x + 6$

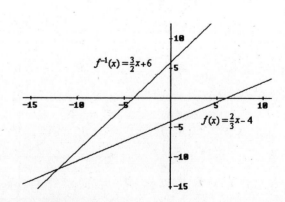

Copyright © Houghton Mifflin Company. All rights reserved.

18. $y = \dfrac{x-3}{3x+6}$

Interchange x and y, then solve for y.

$$x = \frac{y-3}{3y+6}$$

$$x(3y+6) = y-3$$

$$3xy + 6x = y-3$$

$$3xy - y = -6x-3$$

$$y(3x-1) = -6x-3$$

$$y = \frac{-6x-3}{3x-1}$$

$$f^{-1}(x) = \frac{-6x-3}{3x-1}$$

Domain $f^{-1}(x)$: all real numbers except $\frac{1}{3}$

Range $f^{-1}(x)$: all real numbers except -2

19. $s(t) = 4t^2$

a. Average velocity $= \dfrac{4(2)^2 - 4(1)^2}{2-1}$

$$= \frac{4(4) - 4(1)}{1}$$

$$= 16 - 4$$

$$= 12 \text{ ft/sec}$$

b. Average velocity $= \dfrac{4(1.5)^2 - 4(1)^2}{1.5-1}$

$$= \frac{4(2.25) - 4(1)}{1.5-1}$$

$$= \frac{9-4}{0.5}$$

$$= \frac{5}{0.5}$$

$$= 10 \text{ ft/sec}$$

c. Average velocity $= \dfrac{4(1.01)^2 - 4(1)^2}{1.01-1}$

$$= \frac{4(1.0201) - 4(1)}{0.01}$$

$$= \frac{4.0804 - 4}{0.01}$$

$$= \frac{0.0804}{0.01}$$

$$= 8.04 \text{ ft/sec}$$

20. a.   Enter the data on your calculator. The technique for a TI-83 calculator is illustrated here.

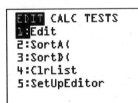

$$y = 1.675x - 0.575$$

b. $y = 1.675(8) - 0.575$

$$= 12.825$$

Copyright © Houghton Mifflin Company. All rights reserved.

Trigonometric Functions

1. $\alpha = \dfrac{d}{180°} \cdot \pi = -\dfrac{135°}{180°} \cdot \pi = -\dfrac{3\pi}{4}$

2. $\alpha = \dfrac{d}{180°} \cdot \pi = \dfrac{240°}{180°} \cdot \pi = \dfrac{4\pi}{3}$

3. $d = \dfrac{\alpha}{\pi} \cdot 180° = \dfrac{\frac{9\pi}{2}}{\pi} \cdot 180° = 810°$

4. $d = \dfrac{\alpha}{\pi} \cdot 180° = \dfrac{\frac{16\pi}{3}}{\pi} \cdot 180° = 960°$

5. $\theta = \dfrac{s}{r} = \dfrac{12\pi}{8} = \dfrac{3\pi}{2}$

   The angle measures $\dfrac{3\pi}{2}$ radians, or $270°$.

6. $\sin\dfrac{5\pi}{4} = -\sin\left(\dfrac{5\pi}{4} - \pi\right) = -\sin\dfrac{\pi}{4} = -\dfrac{\sqrt{2}}{2}$

7. $\cos\left(-\dfrac{4\pi}{3}\right) = \cos\left(2\pi - \dfrac{4\pi}{3}\right) = \cos\left(\dfrac{2\pi}{3}\right) = -\cos\left(\pi - \dfrac{2\pi}{3}\right) = -\cos\dfrac{\pi}{3} = -\dfrac{1}{2}$

8. $\sec\left(\dfrac{19\pi}{2}\right) = \sec\left(5 \cdot 2\pi - \dfrac{19}{2}\pi\right) = \sec\dfrac{\pi}{2}$: undefined

9. $\tan\left(-\dfrac{17\pi}{6}\right) = \tan\left(2 \cdot 2\pi - \dfrac{17\pi}{6}\right) = \tan\dfrac{7\pi}{6} = \tan\left(\dfrac{7\pi}{6} - \pi\right) = \tan\dfrac{\pi}{6} = \dfrac{\sqrt{3}}{3}$

10. $r = \sqrt{x^2 + y^2} = \sqrt{(-10)^2 + 24^2}$
    $= \sqrt{676} = 26$

    $\sin\theta = \dfrac{y}{r} = \dfrac{24}{26} = \dfrac{12}{13}$

    $\tan\theta = \dfrac{y}{x} = \dfrac{24}{-10} = -\dfrac{12}{5}$

11. $r = \sqrt{x^2 + y^2} = \sqrt{(-12)^2 + (-9)^2}$
    $= \sqrt{225} = 15$

    $\sin\theta = \dfrac{y}{r} = \dfrac{-9}{15} = -\dfrac{3}{5}$

    $\tan\theta = \dfrac{y}{x} = \dfrac{-9}{-12} = \dfrac{3}{4}$

12. For the angle $-270°$, use the point $(0, 1)$.

    $\cos(-270°) = \dfrac{x}{r} = \dfrac{0}{1} = 0$

13. For the angle $180°$, use the point $(-1, 0)$.

    $\sin 180° = \dfrac{y}{r} = \dfrac{0}{1} = 0$

14.

15.

16.

17.

Copyright © Houghton Mifflin Company. All rights reserved.

18. Amplitude 3, period 4π

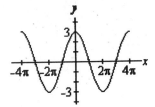

19. Amplitude 4, period π, phase shift $-\frac{\pi}{2}$

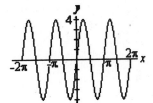

20.

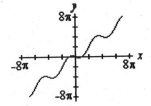

21.

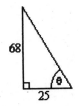

$$\tan\theta = \frac{68}{25}$$
$$\tan\theta = 2.72$$
$$\theta \approx 70°$$

The angle of elevation is approximately $70°$.

22.

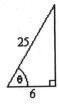

$$\cos\theta = \frac{6}{25}$$
$$\cos\theta = 0.24$$
$$\theta \approx 76°$$

The ladder makes an angle of approximately $76°$ with the ground.

23. Amplitude 4, period 4π, frequency $\frac{1}{4\pi}$

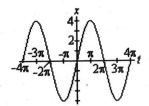

Copyright © Houghton Mifflin Company. All rights reserved.

Trigonometric Identities and Equations

1. $\dfrac{1-\tan x}{\sec x}+\sin x=\dfrac{1-\dfrac{\sin x}{\cos x}}{\dfrac{1}{\cos x}}+\sin x=\cos x-\sin x+\sin x=\cos x$

2. $\dfrac{\tan^2 x+1}{\tan^2 x+2}=\dfrac{\sec^2 x}{\left(\tan^2 x+1\right)+1}=\dfrac{\sec^2 x}{\sec^2 x+1}=\dfrac{\dfrac{1}{\cos^2 x}}{\dfrac{1}{\cos^2 x}+1}\cdot\dfrac{\cos^2 x}{\cos^2 x}=\dfrac{1}{1+\cos^2 x}$

3. $\cot\dfrac{5\pi}{12}=\cot\left(\dfrac{\pi}{4}+\dfrac{\pi}{6}\right)=\dfrac{1}{\tan\left(\dfrac{\pi}{4}+\dfrac{\pi}{6}\right)}=\dfrac{1}{\dfrac{\tan\dfrac{\pi}{4}+\tan\dfrac{\pi}{6}}{1-\tan\dfrac{\pi}{4}\tan\dfrac{\pi}{6}}}=\dfrac{1-\tan\dfrac{\pi}{4}\tan\dfrac{\pi}{6}}{\tan\dfrac{\pi}{4}+\tan\dfrac{\pi}{6}}=\dfrac{1-(1)\left(\dfrac{\sqrt{3}}{3}\right)}{1+\dfrac{\sqrt{3}}{3}}=\dfrac{3-\sqrt{3}}{3+\sqrt{3}}$

4. $\dfrac{\sin(a+b)}{\cos a\cos b}=\dfrac{\sin a\cos b+\cos a\sin b}{\cos a\cos b}=\dfrac{\sin a\cos b}{\cos a\cos b}+\dfrac{\cos a\sin b}{\cos a\cos b}=\dfrac{\sin a}{\cos a}+\dfrac{\sin b}{\cos b}=\tan a+\tan b$

5. $\dfrac{\tan a-\tan b}{\tan a+\tan b}=\dfrac{\dfrac{\sin a}{\cos a}-\dfrac{\sin b}{\cos b}}{\dfrac{\sin a}{\cos a}+\dfrac{\sin b}{\cos b}}\cdot\dfrac{\cos a\cos b}{\cos a\cos b}=\dfrac{\sin a\cos b-\sin b\cos a}{\sin a\cos b+\sin b\cos a}=\dfrac{\sin(a-b)}{\sin(a+b)}$

6. $\sin(2\cdot315°)=2\sin315°\cos315°=2\left(-\dfrac{\sqrt{2}}{2}\right)\left(\dfrac{\sqrt{2}}{2}\right)=-1$

   $\cos(2\cdot315°)=2\cos^2 315°-1=2\left(\dfrac{\sqrt{2}}{2}\right)^2-1=0$

   $\tan(2\cdot315°)=\dfrac{2\tan315°}{1-\tan^2 315°}=\dfrac{2(-1)}{1-1^2}=-\dfrac{2}{0}\ \text{Undefined}$

7. $\sin\alpha=-\dfrac{2}{3}$

   $\dfrac{y}{r}=\dfrac{-2}{3}$

   $x^2=r^2-y^2=3^2-(-2)^2=5$

   $x=-\sqrt{5}$

   $\cos\alpha=\dfrac{x}{r}=-\dfrac{\sqrt{5}}{3}$

   $\cos\dfrac{\alpha}{2}=-\sqrt{\dfrac{1+\cos\alpha}{2}}=-\sqrt{\dfrac{1-\dfrac{\sqrt{5}}{3}}{2}}$

   $=-\sqrt{\dfrac{\dfrac{3-\sqrt{5}}{3}}{2}}=-\sqrt{\dfrac{3-\sqrt{5}}{6}}$

8. $\tan\alpha=-\dfrac{3}{2}$

   $\dfrac{y}{x}=\dfrac{3}{-2}$

   $r=\sqrt{x^2+y^2}=\sqrt{(-2)^2+3^2}=\sqrt{13}$

   $\cos\alpha=\dfrac{x}{r}=\dfrac{-2}{\sqrt{13}}=-\dfrac{2\sqrt{13}}{13}$

   $\cos\dfrac{\alpha}{2}=\sqrt{\dfrac{1+\cos\alpha}{2}}=-\sqrt{\dfrac{1-\dfrac{2\sqrt{13}}{13}}{2}}$

   $=\sqrt{\dfrac{\dfrac{13-2\sqrt{13}}{13}}{2}}=\sqrt{\dfrac{13-2\sqrt{13}}{26}}$

9. $\dfrac{\cot\varphi-\tan\varphi}{\cot\varphi+\tan\varphi}=\dfrac{\dfrac{\cos\varphi}{\sin\varphi}-\dfrac{\sin\varphi}{\cos\varphi}}{\dfrac{\cos\varphi}{\sin\varphi}+\dfrac{\sin\varphi}{\cos\varphi}}\cdot\dfrac{\sin\varphi\cos\varphi}{\sin\varphi\cos\varphi}=\dfrac{\cos^2\varphi-\sin^2\varphi}{\cos^2\varphi+\sin^2\varphi}=\dfrac{\cos^2\varphi}{1}=\cos2\varphi$

Copyright © Houghton Mifflin Company. All rights reserved.

10. $\dfrac{\sin 2\alpha}{1-\cos 2\alpha}=\dfrac{2\sin\alpha\cos\alpha}{1-(1-2\sin^2\alpha)}=\dfrac{2\sin\alpha\cos\alpha}{2\sin^2\alpha}=\dfrac{\cos\alpha}{\sin\alpha}=\cot\alpha$

11. $\dfrac{1}{\sin x}-\sin x=\dfrac{1-\sin^2 x}{\sin x}=\dfrac{\cos^2 x}{\sin x}=\dfrac{\cos^2 x}{\sin x}\cdot\dfrac{\dfrac{\sin x}{\cos^2 x}}{\dfrac{\sin x}{\cos^2 x}}=\dfrac{\sin x}{\dfrac{\sin^2 x}{\cos^2 x}}=\dfrac{\sin x}{\tan^2 x}$

12. $\sin\left(\dfrac{\pi}{4}-\alpha\right)=\sin\dfrac{\pi}{4}\cos\alpha-\cos\dfrac{\pi}{4}\sin\alpha=\dfrac{\sqrt{2}}{2}\cos\alpha-\dfrac{\sqrt{2}}{2}\sin\alpha=\cos\dfrac{\pi}{4}\cos\alpha-\sin\dfrac{\pi}{4}\sin\alpha=\cos\left(\dfrac{\pi}{4}+\alpha\right)$

13.
$$2\cos^2 x-\cos x=1$$
$$2\cos^2 x-\cos x-1=0$$
$$(2\cos x+1)(\cos x-1)=0$$

$2\cos x+1=0$	$\cos x-1=0$
$\cos x=-\dfrac{1}{2}$	$\cos x=1$
$x=\dfrac{2\pi}{3}$ or $\dfrac{4\pi}{3}$	$x=0$

The solutions are $\dfrac{2\pi}{3}+2n\pi,\ \dfrac{4\pi}{3}+2n\pi,$ and $0+2n\pi$.

14.
$$\sin^2 x+3\sin x=4$$
$$\sin^2 x+3\sin x-4=0$$
$$(\sin x-1)(\sin x+4)=0$$

$\sin x-1=0$	$\sin x+4=0$
$\sin x=1$	$\sin x=-4$
$x=\dfrac{\pi}{2}$	no solutions

The solution is $\dfrac{\pi}{2}+2n\pi$.

15. $\sec 18°=\dfrac{1}{\cos 18°}=\dfrac{1}{\sin(90°-18°)}=\dfrac{1}{\sin 72°}=\dfrac{1}{0.95}\approx 1.05$

16. $\cot 18°=\tan(90°-18°)=\tan 72°=\dfrac{\sin 72°}{\cos 72°}=\dfrac{0.95}{0.31}\approx 3.06$

17. $\sin^{-1}\left(-\dfrac{\sqrt{2}}{2}\right)=-\dfrac{\pi}{4}$

18. $\cos^{-1}\left(-\dfrac{1}{2}\right)=\dfrac{2\pi}{3}$

19. $\cos^{-1}\left(\cos\left(-\dfrac{\pi}{4}\right)\right)=\cos^{-1}\left(\dfrac{\sqrt{2}}{2}\right)=\dfrac{\pi}{4}$

20. $\sin^{-1}\left(\cos\dfrac{\pi}{6}\right)=\sin^{-1}\left(\dfrac{\sqrt{3}}{2}\right)=\dfrac{\pi}{3}$

Copyright © Houghton Mifflin Company. All rights reserved.

Applications of Trigonometry

1. $A = 180° - (46° + 81°) = 53°$

$$\frac{17}{\sin 53°} = \frac{b}{\sin 46°} \qquad\qquad \frac{17}{\sin 53°} = \frac{c}{\sin 81°}$$

$$\frac{17}{0.7986} = \frac{b}{0.7193} \qquad\qquad \frac{17}{0.7986} = \frac{c}{0.9877}$$

$$b \approx 15 \qquad\qquad\qquad c \approx 21$$

2. $\dfrac{11}{\sin 40°} = \dfrac{16}{\sin B}$

$\dfrac{11}{0.6428} = \dfrac{16}{\sin B}$

$\sin B \approx 0.9350$

$B \approx 69°$ or $111°$

Case 1:

$B \approx 69°$

$C = 180° - (40° + 69°) = 71°$

$$\frac{11}{\sin 40°} = \frac{c}{\sin 71°}$$

$$\frac{11}{0.6428} = \frac{c}{0.9455}$$

$$c \approx 16$$

Case 2:

$B \approx 111°$

$C = 180° - (40° + 111°) = 29°$

$$\frac{11}{\sin 40°} = \frac{c}{\sin 29°}$$

$$\frac{11}{0.6428} = \frac{c}{0.4848}$$

$$c \approx 8.3$$

Thus, there are two possible triangles: one with $B \approx 69°, C \approx 71°$, and $c \approx 16$, and one with $B \approx 111°, C \approx 29°$, and $c \approx 8.3$.

3. $C = 180° - (38° + 73°) = 69°$

$$\frac{14}{\sin 38°} = \frac{b}{\sin 73°} \qquad\qquad \frac{14}{\sin 38°} = \frac{c}{\sin 69°}$$

$$\frac{14}{0.6157} = \frac{b}{0.9563} \qquad\qquad \frac{14}{0.6157} = \frac{c}{0.9336}$$

$$b \approx 22 \qquad\qquad\qquad c \approx 21$$

4. $\dfrac{18}{\sin A} = \dfrac{7.0}{\sin 23°}$

$\dfrac{18}{\sin A} = \dfrac{7.0}{0.3907}$

$\sin A \approx 1.0047$

Since there is no value A such that $\sin A > 1$, no triangle is possible.

5. $C = 180° - (41° + 83°) = 56°$

$$\frac{a}{\sin 41°} = \frac{44.6}{\sin 56°} \qquad\qquad \frac{b}{\sin 83°} = \frac{44.6}{\sin 56°}$$

$$\frac{a}{0.6561} = \frac{44.6}{0.8290} \qquad\qquad \frac{b}{0.9925} = \frac{44.6}{0.8290}$$

$$a \approx 35 \qquad\qquad\qquad b \approx 53$$

6. $a^2 = b^2 + c^2 - 2bc \cos A$

$a^2 = 20^2 + 16^2 - 2(20)(16)\cos 120°$

$a^2 = 656 - 640(-0.5)$

$a^2 \approx 976$

$a \approx 31$

$b^2 = a^2 + c^2 - 2ac \cos B$

$20^2 = 31^2 + 16^2 - 2(31)(16)\cos B$

$\cos B \approx 0.8236$

$B \approx 35°$

$C \approx 180° - (120° + 35°) = 25°$

Copyright © Houghton Mifflin Company. All rights reserved.

7. $b^2 = a^2 + c^2 - 2ac \cos B$

$b^2 = 78^2 + 125^2 - 2(78)(125)\cos 128°$

$b^2 \approx 21{,}709 - 19{,}500(-0.6157)$

$b^2 \approx 33{,}715.15$

$b \approx 184$

$a^2 = b^2 + c^2 - 2bc \cos A$

$78^2 = 184^2 + 125^2 - 2(184)(125)\cos A$

$\cos A \approx 0.9434$

$A \approx 19°$

$C \approx 180° - (19° + 128°) = 33°$

8. $a^2 = b^2 + c^2 - 2bc \cos A$

$14^2 = 19^2 + 13^2 - 2(19)(13)\cos A$

$\cos A \approx 0.6791$

$A \approx 47°$

$b^2 = a^2 + c^2 - 2ac \cos B$

$19^2 = 14^2 + 13^2 - 2(14)(13)\cos B$

$\cos B \approx 0.0110$

$B \approx 89°$

$C \approx 180° - (47° + 89°) = 44°$

9. $c^2 = a^2 + b^2 - 2ab \cos C$

$c^2 = 10^2 + 13^2 - 2(10)(13)\cos 100°$

$c^2 \approx 269 - 260(-0.1736)$

$c^2 \approx 314.1$

$c \approx 18$

$a^2 = b^2 + c^2 - 2bc \cos A$

$10^2 = 13^2 + 18^2 - 2(13)(18)\cos A$

$\cos A \approx 0.8397$

$A \approx 33°$

$B \approx 180° - (100° + 33°) = 47°$

10. $\dfrac{4\sqrt{2}}{\sin 30°} = \dfrac{c}{\sin 135°}$

$\dfrac{4\sqrt{2}}{\frac{1}{2}} = \dfrac{c}{\frac{\sqrt{2}}{2}}$

$c = 8.0$

$A = 180° - (30° + 135°) = 15°$

$K = \dfrac{1}{2}bc \sin A$

$= \dfrac{1}{2}(4\sqrt{2})(8.0)\sin 15°$

≈ 6 square units

11. $K = \dfrac{1}{2}ab \sin C$

$= \dfrac{1}{2}(7.0)(9.0)\sin 26°$

≈ 14 square units

12. $s = \dfrac{1}{2}(a + b + c)$

$= \dfrac{1}{2}(12 + 15 + 23)$

$= 25$

$K = \sqrt{s(s-a)(s-b)(s-c)}$

$= \sqrt{25(25 - 12)(25 - 15)(25 - 23)}$

$= \sqrt{25(13)(10)(2)}$

$= \sqrt{6500}$

≈ 81 square units

13. a. $d = \sqrt{(-3-0)^2 + (3-0)^2} = \sqrt{18} = 3\sqrt{2}$

$\tan \theta = \dfrac{y}{x}$

$\tan \theta = \dfrac{3}{-3}$

$\tan \theta = -1$

$\theta = 135°$

b.

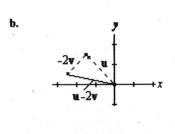

Copyright © Houghton Mifflin Company. All rights reserved.

14. $\langle 3, -1 \rangle \cdot \langle -2, -2 \rangle = 3(-2) + (-1)(-2)$

$$= -4$$

$$\cos\theta = \frac{\mathbf{u} \cdot \mathbf{v}}{\|\mathbf{u}\| \cdot \|\mathbf{v}\|}$$

$$\cos\theta = \frac{-\sqrt{4}}{(\sqrt{3^2 + (-1)^2})(\sqrt{(-2)^2 + (-2)^2})}$$

$$\cos\theta \approx -0.4472$$

$$\theta \approx 117°$$

15. $\langle 4, 6 \rangle \cdot \langle -3, 2 \rangle = 4(-3) + 6(2)$

$$= 0$$

$$\cos\theta = \frac{\mathbf{u} \cdot \mathbf{v}}{\|\mathbf{u}\| \cdot \|\mathbf{v}\|}$$

$$\cos\theta = \frac{0}{(\sqrt{4^2 + 6^2})(\sqrt{(-3)^2 + 2^2})}$$

$$\cos\theta = 0$$

$$\theta = 90°$$

16. $2\mathbf{u} - \frac{1}{2}\mathbf{v} = 2\left\langle -\frac{1}{2}, \frac{7}{2} \right\rangle - \frac{1}{2}\langle 6, -1 \rangle$

$$= \langle -1, 7 \rangle + \left\langle -3, \frac{1}{2} \right\rangle$$

$$= \left\langle -4, \frac{15}{2} \right\rangle$$

17. $\mathbf{u} - 4\mathbf{v} = \langle 6, -1 \rangle - 4\left\langle -\frac{1}{2}, \frac{7}{2} \right\rangle$

$$= \langle 6, -1 \rangle + \langle 2, -14 \rangle$$

$$= \langle 8, -15 \rangle$$

18. $\|\mathbf{v}\|^2 = 500^2 + 8^2 - 2(500)(8)\cos 35°$

$$\|\mathbf{v}\|^2 = 250{,}064 - 8000(0.8192)$$

$$\|\mathbf{v}\|^2 = 243{,}510.4$$

$$\|\mathbf{v}\| = 493$$

$$\frac{\sin\theta}{8} = \frac{\sin 35°}{493}$$

$$\frac{\sin\theta}{8} = \frac{0.5736}{493}$$

$$\sin\theta \approx 0.0093$$

$$\theta \approx 0°32'$$

$$75° - \theta \approx 75° - 0°32' = 74°28'$$

The jet's ground speed is approximately 493 mph, and its true bearing is approximately N 74°28′ E.

19. $\|\mathbf{F}\| = (\sin 25°)(120) = (0.4226)(120) \approx 51$

The required force is approximately 51 pounds.

Copyright © Houghton Mifflin Company. All rights reserved.

Complex Numbers

1. $i^{51} = i^{48} i^3 = (1)(-i) = -i$

2. $(-i)^{35} = (-1)^{35}(i)^{35} = (-1)i^{32} i^3 = (-1)(1)(-i) = -i$

3. $(3i)^4 \cdot i^{44} = 3^4 i^4 i^{44} = 81(1)(1) = 81$

4. $(1-3i) + (-2+i) = (1-2) + (-3+1)i = -1-2i$

5. $(4-3i) - (6-2i) = (4-6) + (-3+2)i = -2-i$

6. $5i(4-2i) = 20i - 10i^2 = 20i - 10(-1) = 10 + 20i$

7. $2i - i(8+3i) = 2i - 8i - 3i^2$
$$= -6i - 3(-1)$$
$$= 3 - 6i$$

8. $(1-4i)(3+2i) = 3 + 2i - 12i - 8i^2$
$$= 3 - 10i - 8(-1)$$
$$= 3 - 10i + 8$$
$$= 11 - 10i$$

9. $(3+5i)(7-4i) = 21 - 12i + 35i - 20i$
$$= 21 + 23i - 20(-1)$$
$$= 21 + 23i + 20$$
$$= 41 + 23i$$

10. $(2-9i)(2+9i) = 4 - 81i^2$
$$= 4 - 81(-1)$$
$$= 4 + 81$$
$$= 85$$

11. $(1+i\sqrt{5})(1-i\sqrt{5}) = 1 - 5i^2 = 1 - 5(-1)$
$$= 1 + 5$$
$$= 6$$

12. $\dfrac{4}{3i} = \dfrac{4}{3i} \cdot \dfrac{i}{i} = \dfrac{4i}{3i^2} = \dfrac{4i}{3(-1)} = \dfrac{4i}{-3}$
$$= -\dfrac{4}{3}i$$

13. $\dfrac{1}{3+11i} = \dfrac{1}{3+11i} \cdot \dfrac{3-11i}{3-11i} = \dfrac{3-11i}{9-121i^2} = \dfrac{3-11i}{9-121(-1)}$
$$= \dfrac{3-11i}{9+121} = \dfrac{3-11i}{130}$$
$$= \dfrac{3}{130} - \dfrac{11}{130}i$$

14. $\dfrac{6-2i}{2+i} = \dfrac{6-2i}{2+i} \cdot \dfrac{2-i}{2-i} = \dfrac{12-6i-4i+2i^2}{4-i^2}$
$$= \dfrac{12-10i+2(-1)}{4-(-1)} = \dfrac{12-10i-2}{4+1}$$
$$= \dfrac{10-10i}{5} = \dfrac{10}{5} - \dfrac{10}{5}i$$
$$= 2 - 2i$$

In Problems 15 and 16, express the complex number in trigonometric form.

15. $3 - i\sqrt{3} \Rightarrow a = 3, \ b = -\sqrt{3}$
$$r = \sqrt{3^2 + (-\sqrt{3})^2} = \sqrt{12} = 2\sqrt{3}$$
$$\tan\theta = -\dfrac{\sqrt{3}}{3}$$
$$\theta = \dfrac{11\pi}{6}$$
$$3 - i\sqrt{3} = 2\sqrt{3}\left(\cos\dfrac{11\pi}{6} + i\sin\dfrac{11\pi}{6}\right)$$
$$= 2\sqrt{3}\ \text{cis}\ \dfrac{11\pi}{6}, \quad \text{or} \quad 2\sqrt{3}\ \text{cis}\ 330°$$

16. $-4 - 4i \Rightarrow a = -4, \ b = -4$
$$r = \sqrt{(-4)^2 + (-4)^2} = \sqrt{32} = 4\sqrt{2}$$
$$\tan\theta = \dfrac{-4}{-4}$$
$$\tan\theta = 1$$
$$\theta = \dfrac{5\pi}{4}$$
$$-4 - 4i = 4\sqrt{2}\left(\cos\dfrac{5\pi}{4} + i\sin\dfrac{5\pi}{4}\right)$$
$$= 4\sqrt{2}\ \text{cis}\ \dfrac{5\pi}{4}, \quad \text{or} \quad 4\sqrt{2}\ \text{cis}\ 225°$$

Copyright © Houghton Mifflin Company. All rights reserved.

17. $z = 4 \text{ cis } 45°$

$z = 4(\cos 45° + i \sin 45°)$

$z = 4\left(\dfrac{\sqrt{2}}{2} + \dfrac{i\sqrt{2}}{2}\right)$

$z = 2\sqrt{2} + 2i\sqrt{2}$

18. $(-2 \text{ cis } 225°)(\sqrt{3} \text{ cis } 75°) = -2\sqrt{3} \text{ cis } (225° + 75°)$

$= -2\sqrt{3} \text{ cis } (300°)$

$= -2\sqrt{3}(\cos 300° + i \sin 300°)$

$= -2\sqrt{3}\left(\dfrac{\sqrt{3}}{2} - \dfrac{i}{2}\right)$

$= -3 + i\sqrt{3}$

19.

$\tan\theta = \dfrac{6}{6\sqrt{3}} = \dfrac{1}{\sqrt{3}} = \dfrac{\sqrt{3}}{3}$

$\theta = 30°$

$r = \sqrt{\left(6\sqrt{3}\right)^2 + (6)^2}$

$= \sqrt{36(3) + 36}$

$= \sqrt{144}$

$= 12$

$\tan\alpha = \dfrac{-\sqrt{3}}{1} = -\sqrt{3}$

$\alpha = -60°$

$\beta = 360° - 60°$

$\beta = 300°$

$r = \sqrt{1^2 + \left(-\sqrt{3}\right)^2}$

$= \sqrt{1 + 3}$

$= \sqrt{4}$

$= 2$

$6\sqrt{3} + 6i = 12 \text{ cis } 30°$ $1 - i\sqrt{3} = 2 \text{ cis } 300°$

$\dfrac{6\sqrt{3} + 6i}{1 - i\sqrt{3}} = \dfrac{12 \text{ cis } 30°}{2 \text{ cis } 300°}$

$= \dfrac{12}{2} \text{ cis } (30° - 300°)$

$= 6 \text{ cis } (-270°)$

$= 6 \text{ cis } 90°$

20. $\left(\dfrac{-\sqrt{3}}{2} + \dfrac{1}{2}i\right)^5 = (\text{cis } 150°)^5$

$= \text{cis } 5(150°)$

$= \text{cis } 750°$

$= (\cos 750° + i \sin 750°)$

$= (\cos 30° + i \sin 30°)$

$= \dfrac{\sqrt{3}}{2} + \dfrac{1}{2}i$

21. $-1 + i\sqrt{3} = 2 \text{ cis } 120°$

$w_k = 2^{1/4} \text{ cis } \dfrac{120° + 360°k}{4}$ $k = 0, 1, 2, 3$

$w_0 \approx 1.19 \text{ cis } \dfrac{120°}{4}$

$w_0 \approx 1.19 \text{ cis } 30°$

$w_1 \approx 1.19 \text{ cis } \dfrac{120° + 360°}{4}$

$w_1 \approx 1.19 \text{ cis } 120°$

$w_2 \approx 1.19 \text{ cis } \dfrac{120° + 720°}{4}$

$w_2 \approx 1.19 \text{ cis } 210°$

$w_3 \approx 1.19 \text{ cis } \dfrac{120° + 1080°}{4}$

$w_3 \approx 1.19 \text{ cis } 300°$

Copyright © Houghton Mifflin Company. All rights reserved.

Topics in Analytic Geometry

1. $x^2 + 4y^2 - 36 = 0$

$x^2 + 4y^2 = 36$

$\dfrac{x^2}{36} + \dfrac{4y^2}{36} = \dfrac{36}{36}$

$\dfrac{x^2}{36} + \dfrac{y^2}{9} = 1$

The graph is an ellipse.

vertices: $(-6, 0)$ and $(6, 0)$
foci: $(-3\sqrt{3}, 0)$ and $(3\sqrt{3}, 0)$

2. $y^2 + 2y - x + 3 = 0$

$y^2 + 2y = x - 3$

$y^2 + 2y + 1 = x - 3 + 1$

$(y + 1)^2 = x - 2$

The graph is a parabola.

$4p = 1$

$p = \dfrac{1}{4}$

vertex: $(2, -1)$

focus: $\left(2 + \dfrac{1}{4}, -1\right) = \left(\dfrac{9}{4}, -1\right)$

3. $x^2 - 4y^2 - 36 = 0$

$x^2 - 4y^2 = 36$

$\dfrac{x^2}{36} - \dfrac{4y^2}{36} = \dfrac{36}{36}$

$\dfrac{x^2}{36} - \dfrac{y^2}{9} = 1$

The graph is a hyperbola.

vertices: $(-6, 0)$ and $(6, 0)$

$c^2 = a^2 + b^2$

$c^2 = 36 + 9$

$c^2 = 45$

$c = \sqrt{45} = 3\sqrt{5}$

foci: $(-3\sqrt{5}, 0)$ and $(3\sqrt{5}, 0)$

4. $y^2 - 6y + x = 0$

$y^2 - 6y = -x$

$y^2 - 6y + 9 = -x + 9$

$(y - 3)^2 = -(x - 9)$

The graph is a parabola

vertex: $(9, 3)$

$4p = -1$

$p = -\dfrac{1}{4}$

focus: $\left(9 - \dfrac{1}{4}, 3\right) = \left(\dfrac{35}{4}, 3\right)$

5. $y^2 - 6y - x + 5 = 0$

$y^2 - 6y = x - 5$

$y^2 - 6y + 9 = x - 5 + 9$

$(y - 3)^2 = x + 4$

$4p = 1$

$p = \dfrac{1}{4}$

vertex: $(-4, 3)$
focus: $\left(-4 + \dfrac{1}{4}, 3\right) = \left(-\dfrac{15}{4}, 3\right)$

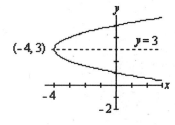

6. $\dfrac{(x - 5)^2}{25} + \dfrac{y^2}{4} = 1$

ellipse with center at $(5, 0)$

$c^2 = a^2 - b^2$

$c^2 = 25 - 4$

$c^2 = 21$

$c = \sqrt{21}$

vertices: $(0, 0)$ and $(10, 0)$
foci: $(5 - \sqrt{21}, 0)$ and $(5 + \sqrt{21}, 0)$

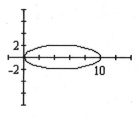

Copyright © Houghton Mifflin Company. All rights reserved.

7. $\dfrac{(y+2)^2}{24} - \dfrac{(x-1)^2}{4} = 1$

hyperbola with vertical transverse axis, center at $(1, -2)$

$a^2 = 24$ $b^2 = 4$ $c^2 = a^2 + b^2$ $\dfrac{a}{b} = \dfrac{2\sqrt{6}}{2}$

$a = 2\sqrt{6}$ $b = 2$ $c^2 = 28$ $= \sqrt{6}$

                          $c = 2\sqrt{7}$

vertices: $(1, -2 - 2\sqrt{6})$ and $(1, -2 + 2\sqrt{6})$

foci: $(1, -2 - 2\sqrt{7})$ and $(1, -2 + 2\sqrt{7})$

asymptotes: $y + 2 = \pm\sqrt{6}(x - 1)$

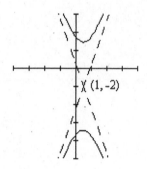

8. $4x^2 - 9y^2 - 8x + 18y = 149$

$4(x^2 - 2x + 1) - 9(y^2 - 2y + 1) = 149 + 4 - 9$

       $4(x-1)^2 - 9(y-1)^2 = 144$

       $\dfrac{4(x-1)^2}{144} - \dfrac{9(y-1)^2}{144} = \dfrac{144}{144}$

       $\dfrac{(x-1)^2}{36} - \dfrac{(y-1)^2}{16} = 1$

hyperbola with horizontal transverse axis, center at $(1, 1)$ a

$a^2 = 36$ $b^2 = 16$ $c^2 = a^2 + b^2$ $\dfrac{b}{a} = \dfrac{4}{6}$

$a = 6$ $b = 4$ $c^2 = 36 + 16$ $= \dfrac{2}{3}$

                        $c^2 = 52$

                        $c = \sqrt{52}$

                         $= 2\sqrt{13}$

vertices: $(-5, 1)$ and $(7, 1)$

foci: $(1 - 2\sqrt{13}, 1)$ and $(1 + 2\sqrt{13}, 1)$

asymptotes: $y - 1 = \pm\dfrac{2}{3}(x - 1)$

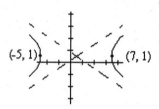

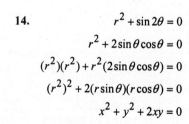

9. $r = \pm\sqrt{(-1)^2 + 1^2} = \pm\sqrt{2}$

$\tan\theta = \dfrac{y}{x} = \dfrac{1}{-1} = -1$

$\theta = \dfrac{3\pi}{4}$

Two polar representations are

$\left(\sqrt{2}, \dfrac{3\pi}{4}\right)$ and $\left(-\sqrt{2}, \dfrac{7\pi}{4}\right)$.

10. $r = \pm\sqrt{(-\sqrt{3})^2 + (-1)^2} = \pm 2$

$\tan\theta = \dfrac{y}{x} = \dfrac{-1}{-\sqrt{3}} = \dfrac{\sqrt{3}}{3}$

$\theta = \dfrac{7\pi}{6}$

Two polar representations are

$\left(2, \dfrac{7\pi}{6}\right)$ and $\left(-2, \dfrac{\pi}{6}\right)$.

11. $r = 3\sin\theta$

       $r^2 = 3r\sin\theta$

    $x^2 + y^2 = 3y$

12. $r = 2\cos\theta$

       $r^2 = 2r\cos\theta$

    $x^2 + y^2 = 2x$

13. $\theta = \pi/6$

    $\tan\theta = \tan\pi/6$

       $\dfrac{y}{x} = \dfrac{\sqrt{3}}{3}$

       $y = \dfrac{\sqrt{3}}{3}x$

14. $r^2 + \sin 2\theta = 0$

       $r^2 + 2\sin\theta\cos\theta = 0$

   $(r^2)(r^2) + r^2(2\sin\theta\cos\theta) = 0$

   $(r^2)^2 + 2(r\sin\theta)(r\cos\theta) = 0$

         $x^2 + y^2 + 2xy = 0$

15.

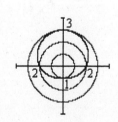

16.

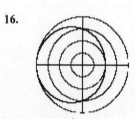

17.

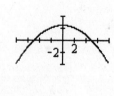

Copyright © Houghton Mifflin Company. All rights reserved.

18.

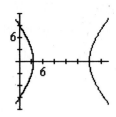

19. $13x^2 + 6\sqrt{3}xy + 7y^2 - 32 = 0$

$A = 13,\ B = 6\sqrt{3},\ C = 7,\ D = 0,\ E = 0,\ F = -32$

$$\cot 2\alpha = \frac{13-7}{6\sqrt{3}} = \frac{6}{6\sqrt{3}} = \frac{1}{\sqrt{3}}$$

$$\csc^2 2\alpha = 1 + \cot^2 2\alpha = 1 + \frac{1}{3} = \frac{4}{3}$$

$\csc 2\alpha = \dfrac{2}{\sqrt{3}}$. Therefore, $\sin 2\alpha = \dfrac{\sqrt{3}}{2}$ and $\cos 2\alpha = \dfrac{1}{2}$.

$$\cos\alpha = \sqrt{\frac{1+\frac{1}{2}}{2}} = \frac{\sqrt{3}}{2} \qquad \sin\alpha = \sqrt{\frac{1-\frac{1}{2}}{2}} = \frac{1}{2}$$

$$A' = 13\left(\frac{\sqrt{3}}{2}\right)^2 + 6\sqrt{3}\left(\frac{\sqrt{3}}{2}\right)\left(\frac{1}{2}\right) + 7\left(\frac{1}{2}\right)^2 = 16$$

$$B' = 0$$

$$C' = 13\left(\frac{1}{2}\right)^2 - 6\sqrt{3}\left(\frac{\sqrt{3}}{2}\right)\left(\frac{1}{2}\right) + 7\left(\frac{\sqrt{3}}{2}\right)^2 = 4$$

$$D' = 0$$
$$E' = 0$$
$$F' = -32$$

$16x'^2 + 4y'^2 - 32 = 0$ or $\dfrac{x'^2}{2} + \dfrac{y'^2}{8} = 1$

The graph is an ellipse.

20. $x = 2t - 1 \qquad\qquad y = 2t^2 + t$

$$t = \frac{x+1}{2} \qquad\qquad y = 2\left(\frac{x+1}{2}\right)^2 + \frac{x+1}{2}$$

$$2y = x^2 + 3x + 2$$

$$y = \tfrac{1}{2}x^2 + \tfrac{3}{2}x + 1$$

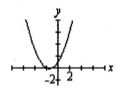

Copyright © Houghton Mifflin Company. All rights reserved.

Exponential and Logarithmic Functions

1.

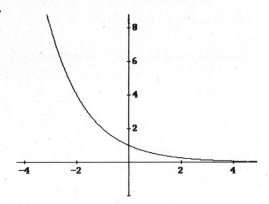

2.

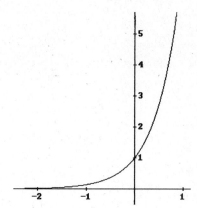

3. $4^d = 3x + 2$

4. $\log_4 y = x + 3$

5. $\frac{1}{2}\log_b y - 3\log_b x - 2\log_b z$

6. $\log_b \dfrac{(x-1)\sqrt{y+2}}{(z+3)^4}$

7. $\log_5 15 = \dfrac{\log 15}{\log 5} = \dfrac{1.176091259}{0.6989700043} \approx 1.6826$

Note that you could also have used natural logarithms.

$\log_5 15 = \dfrac{\ln 15}{\ln 5} = \dfrac{2.708050201}{1.609437912} \approx 1.6826$

8.

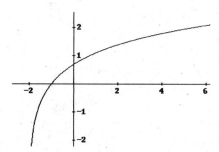

9. $6^x = 60$

$x = \log_6 60$

$x = \dfrac{\log 60}{\log 6} = \dfrac{1.77815125}{0.7781512504} \approx 2.2851$

Note that you could also have used natural logarithms.

$x = \log_6 60 = \dfrac{\ln 60}{\ln 6} = \dfrac{4.094344562}{1.791759469} \approx 2.2851$

10. $\left(\dfrac{10^x}{10^x}\right)\left(\dfrac{10^x - 10^{-x}}{10^x + 10^{-x}}\right) = \dfrac{1}{4}$

$\dfrac{10^{2x} - 1}{10^{2x} + 1} = \dfrac{1}{4}$

$4(10^{2x} - 1) = 10^{2x} + 1$

$4(10^{2x}) - 4 = 10^{2x} + 1$

$4(10^{2x}) - 10^{2x} = 1 + 4$

$10^{2x}(4 - 1) = 1 + 4$

$10^{2x}(3) = 5$

$10^{2x} = \dfrac{5}{3}$

$\log\left(10^{2x}\right) = \log\left(\dfrac{5}{3}\right)$

$2x = \log 5 - \log 3$

$x = \dfrac{1}{2}\log 5 - \dfrac{1}{2}\log 3$

   Copyright © Houghton Mifflin Company. All rights reserved.

11. $\log(2x+1) - \log(3x-1) = 2$

$$\log\frac{2x+1}{3x-1} = 2$$

$$10^2 = \frac{2x+1}{3x-1}$$

$$100(3x-1) = 2x+1$$

$$300x - 100 = 2x + 1$$

$$298x = 101$$

$$x = \frac{101}{298}$$

12. $\ln[(x+1)(x-2)] = \ln(4x-8)$

$$(x+1)(x-2) = 4x-8$$

$$x^2 - x - 2 = 4x - 8$$

$$x^2 - 5x + 6 = 0$$

$$(x-3)(x-2) = 0$$

$$x - 3 = 0 \Rightarrow x = 3$$

$$x - 2 = 0 \Rightarrow x = 2 \quad [\text{No; } 2 \text{ is not in the domain of } \log(x-2)$$
$$\text{and not in the domain of } \log(4x-8).]$$

The solution is 3.

13. a. Use $A = P\left(1+\dfrac{r}{n}\right)^{nt}$ with $P = \$12{,}000$, $r = 0.067$,

$n = 4$, $t = 3$.

$$A = 12{,}000\left(1 + \frac{0.067}{4}\right)^{4(3)}$$

$$= 12{,}000(1 + 0.01675)^{12}$$

$$= 12{,}000(1.01675)^{12}$$

$$= 12{,}000(1.220591027)$$

$$\approx \$14{,}647.09$$

b. Use $A = Pe^{rt}$ with $P = \$12{,}000$, $r = 0.067$,
$t = 3$.

$$A = 12{,}000e^{0.067(3)}$$

$$= 12{,}000e^{0.201}$$

$$= 12{,}000(1.222624772)$$

$$\approx \$14{,}671.50$$

14. a. $M = \log\left(\dfrac{I}{I_0}\right)$

$$= \log\left(\frac{35{,}792{,}000 I_0}{I_0}\right)$$

$$= \log 35{,}792{,}000$$

$$\approx 7.6 \text{ on the Richter scale}$$

b. $\log\left(\dfrac{I_1}{I_0}\right) = 6.4$ and $\log\left(\dfrac{I_2}{I_0}\right) = 5.1$

$$\frac{I_1}{I_0} = 10^{6.4} \qquad\qquad \frac{I_2}{I_0} = 10^{5.1}$$

$$I_1 = 10^{6.4} I_0 \qquad\qquad I_2 = 10^{5.1} I_0$$

To compare the intensities, compute the ratio $\dfrac{I_1}{I_2}$.

$$\frac{I_1}{I_2} = \frac{10^{6.4} I_0}{10^{5.1} I_0} = \frac{10^{6.4}}{10^{5.1}} = 10^{6.4-5.1} = 10^{1.3} \approx 20$$

An earthquake that measures 6.4 on the Richter scale is approximately 20 times as intense as an earthquake that measures 5.1 on the Richter scale.

15. a. $N(t) = N_0 e^{kt}$

$$N_0 = N(0) = 51{,}300$$

$$N(20) = 240{,}500$$

To determine k, substitute $t = 20$ and

$N_0 = 51{,}300$ into $N(t) = N_0 e^{kt}$ to produce

$$N(20) = 51{,}300 e^{k(20)}$$

$$240{,}500 = 51{,}300 e^{20k}$$

$$\frac{240{,}500}{51{,}300} = e^{20k}$$

$$\ln\frac{240{,}500}{51{,}300} = \ln e^{20k}$$

$$\ln\frac{240{,}500}{51{,}300} = 20k$$

$$\frac{1}{20}\ln\frac{240{,}500}{51{,}300} = k$$

$$0.0773 \approx k$$

The exponential growth function is

$$N(t) = 51{,}300 e^{0.0773\,t}.$$

b. Use $N(t) = 51{,}300 e^{0.0773\,t}$ with $t = 30$.

$$N(30) = 51{,}300 e^{0.0773(30)}$$

$$\approx 521{,}500$$

The exponential growth function yields 521,500 as the approximate population of the town in 2010.

Copyright © Houghton Mifflin Company. All rights reserved.

16. The percent of carbon-14 present at time t is
$P(t) = 0.5^{t/5730}$.

$$0.89 = 0.5^{t/5730}$$

$$\ln 0.89 = \ln 0.5^{t/5730}$$

$$\ln 0.89 = \left(\frac{t}{5730}\right)\ln 0.5$$

$$5730\left(\frac{\ln 0.89}{\ln 0.5}\right) = t$$

$$960 \approx t$$

The bone is about 960 years old.

17. a.

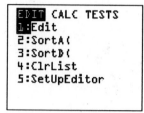

$$y = 1.4473486511(1.79279858)^x$$

b. Use $y = 1.4473486511(1.79279858)^x$ with $x = 8.2$.

$$y = 1.4473486511(1.79279858)^{8.2}$$

$$\approx 174$$

18. a.

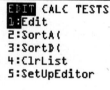 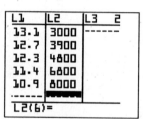

The logarithmic regression function is $h = 72610.5637 - 27040.63079 \ln p$.

When $p = 12.0$, $h = 72610.5637 - 27040.63079 \ln 12.0 \approx 5400$ feet.

b. $6200 = 72610.5637 - 27040.63079 \ln p \Rightarrow \dfrac{6200 - 72610.5637}{-27040.63079} = \ln p$

$\ln p = \dfrac{-66410.5637}{-27040.63079} = 2.455954679 \Rightarrow e^{2.455954679} = e^{\ln p} \Rightarrow e^{2.455954679} = p \approx 11.7$

When the altitude is 6200 feet, the pressure is about 11.7 pounds per square inch.

Copyright © Houghton Mifflin Company. All rights reserved.